# Machine Learning in Farm Animal Behavior using Python

**Natasa Kleanthous**
O&P Electronics & Robotics Ltd
Limassol, Cyprus

**Abir Hussain**
Department of Electrical Engineering,
University of Sharjah, Sharjah, UAE

CRC Press
Taylor & Francis Group
Boca Raton London New York

CRC Press is an imprint of the
Taylor & Francis Group, an **informa** business
A SCIENCE PUBLISHERS BOOK

First edition published 2025
by CRC Press
2385 NW Executive Center Drive, Suite 320, Boca Raton FL 33431

and by CRC Press
4 Park Square, Milton Park, Abingdon, Oxon, OX14 4RN



*CRC Press is an imprint of Taylor & Francis Group, LLC*



*Library of Congress Cataloging-in-Publication Data (applied for)*

ISBN: 978-1-032-62863-9 (hbk)
ISBN: 978-1-032-62871-4 (pbk)
ISBN: 978-1-032-62873-8 (ebk)
DOI: 10.1201/9781032628738

Typeset by Prime Publishing Services

# Dedication

From: Natasa – This book is dedicated to the memory of my loving parents, who are always the source of my inspiration, and to my partner, whose love and support were so essential that they practically deserve co-author status. As for my friends and business partner, no thanks go to you for all those distractions that almost made this book a distant dream!

From: Abir – In the loving memory of my mother, my unwavering source of strength and compassion, and to my father who instilled my curiosity in engineering.

# Acknowledgements

We would like to express our deepest appreciation to everyone who contributed to the realization of this book.

First and foremost, our gratitude goes to our families, who have been the foundation of our strength and patience, supporting us through countless hours of work, providing their love and encouragement.

We would also like to express our thanks to The Douglas Bomford Trust, whose funding and support has been the foundation of our research. Special thanks to Alan Plom and Nick August for their guidance, encouragement, and invaluable assistance.

Additionally, sincere gratitude is due to Dr. Jennifer Sneddon whose expertise in animal behavior has shaped our understanding and approach to animal behaviour research. Her generosity in providing access to her animals for data collection was invaluable, offering practical insights and experiences that have significantly enriched our work on this book.

Our appreciation also goes to Dr. Jacob Kamminga for permitting us to incorporate his dataset in our book. Alongside our own data, his contribution has enriched our analysis and illustration of machine learning applications in farm animal behavior.

Our appreciation extends to the team at Science Publishers and CRC Press ...

We thank our readers as well for joining us on this journey. We hope that our work inspires, informs, and entertains you as much as it has challenged us.

# Preface

In a world where technology and nature increasingly intersect, the potential for machine learning to revolutionize various sectors has become undeniably evident. This book's purpose is to bridge the worlds of farm animal behavior and the vast capabilities of machine learning, using Python as our tool.

Our inspiration for writing this book originated from observing the challenges faced by farmers and researchers in understanding animal behavior. With growing agricultural practices, and the need for more humane livestock management, there is an urgent requirement to combine our knowledge of animals with the predictive and analytic powers of machine learning and deep learning.

This journey will take you through the fundamentals of animal behavior, the core concepts of machine learning, and the usefulness of Python programming. From sensors and data collection to machine learning and deep learning projects using real-world data, this book aims to provide a comprehensive guide for anyone looking to dive into this fascinating combination of both fields.

# Contents

## Who is This Book For?

### *This book is designed for:*

- Students studying animal science, agriculture, or computer science with a curiosity in the convergence of these fields.
- Researchers & academics focused on animal behavior studies wanting to incorporate data-driven methodologies.
- Data scientists & developers wishing to gain new insights for application areas related to animal activity recognition in the agriculture domain.
- Agriculturalists & farmers who need to stay informed about the power of machine learning for insights into animal behavior, health, and productivity.

## Prerequisites and Notes

Before proceeding into the contents of the book, readers are expected to have:

- Python Programming Knowledge: A foundational understanding of Python programming is essential. While the book is accompanied by an online introduction to Python basics, familiarity with its syntax and standard libraries will be advantageous.
- Basic Machine Learning Concepts: An understanding of core machine learning concepts will be beneficial; however, this book provides foundational insights intended for individuals who are not yet familiar with the particulars of machine learning ideas.
- Interest in Animal Behavior and Agriculture: Though not a technical requirement, a genuine interest in animal behavior and the application of technology in agriculture will enhance the learning experience.

Hardware and software requirements:

- Hardware: While many examples are designed for standard personal computers, some deep learning exercises may benefit from a machine equipped with a powerful CPU or a dedicated GPU.
- Python: The book is designed around Python 3.x.
- Jupyter Notebook: All code samples in this book will be demonstrated within the Jupyter Notebook environment. Detailed instructions on how to download and set up Jupyter Notebook are provided in our GitHub repository.
- Python Libraries: Various Python libraries including NumPy, pandas, matplotlib, and scikit-learn will be used throughout the book.
- Deep Learning with PyTorch: PyTorch will be our library for deep learning examples and exercises.

## The Chapters

**Chapter 1: Introduction to Machine Learning for Farm Animal Behavior:** This chapter begins by introducing the world of animal behavior and how machine learning can be utilized to obtain insights from it. This chapter introduces the types of machine learning and their applications in general giving examples in the context of farm animals.

**Chapter 2: Machine Learning Concepts and Challenges:** This chapter provides information related to key machine learning concepts, ensuring a comprehensive understanding of the machine learning workflow.

**Chapter 3: A Practical Example to Building a Simple Machine Learning Model:** In this chapter, a machine learning project is presented from scratch using Python. This project is applied to real-world animal behavior data.

**Chapter 4: Sensors, Data Collection and Annotation:** An overview of the various sensors employed in collecting data on animal behavior, data collection practices, and data annotation is presented in this chapter.

**Chapter 5: Preprocessing and Feature Extraction for Animal Behavior Research:** This chapter highlights the importance of data preprocessing including various techniques to clean, scale, and normalize the data. This chapter also introduces methods to extract meaningful features from raw data providing hands-on python examples.

**Chapter 6: Feature Selection Techniques:** This chapter explores various types of feature selection methods providing theoretical insights coupled with python implementation.

**Chapter 7: Animal Research: Supervised and Unsupervised Learning Algorithms:** This chapter provides theoretical insights of supervised and unsupervised learning methods. Python examples for the implementation of classification, regression, and clustering techniques are presented.

**Chapter 8: Evaluation, Model Selection and Hyperparameter Tuning:** This chapter introduces various evaluation metrics and techniques to enhance model performance accompanied by practical examples using Python.

**Chapter 9: Deep Learning Algorithms for Animal Activity Recognition:** This chapter explores the foundational concepts of deep learning algorithms providing Python implementation and real-world applicability using farm animal data.

## Code examples

All code examples provided in this book are written in Python, utilizing popular libraries such as Scikit-learn and PyTorch. A GitHub repository has been set up allowing readers to access, download, and run the code samples. This ensures a hands-on, interactive learning experience. To access the repository, visit https://github.com/nkcAna/WSDpython.

## Conventions Used in This Book

Throughout this book, we have used a set of conventions to help the reader navigate the content and to understand different types of information. These conventions include:

### *Code Blocks*

Text in Consolas font is used to represent python code scripts.

For example, this code defines a function to read a csv file.

```
import pandas as pd
def read_csv_file(file_path):
    """
    Reads a CSV file and returns a DataFrame.
    Parameters:
        - file_path (str): Path to the CSV file.
    Returns:
        - DataFrame: Data from the CSV file.
    """
    return pd.read_csv(file_path)
# Usage example:
# df = read_csv_file('path_to_your_file.csv')
```

### *Tips and Insights*

Sections that offer additional context.

**Tips and insights appear like this.**

### *Warnings*

Emphasized sections that call attention.

**Warnings appear like this.**

### *Key Terminology*

Key terms or jargons are italicized upon their mention to draw attention and signify their importance.

### *Figures & Diagrams*

Visual aids are occasionally used to illustrate concepts and examples, and they are accompanied by captions or brief explanations.

## Contact Information

For queries, feedback, or further discussions, readers are encouraged to communicate with us through our Book project on GitHub by opening an issue.

CHAPTER 1

# Introduction to Machine Learning for Farm Animal Behavior

Animal behavior is the study of interactions, survival strategies, and communication within the animal kingdom. It encompasses how animals respond to various elements of their environment, including both the surroundings and other organisms, such as their kin, predators, and prey.

Over the years, humans have continually strived to decode this complex web of actions and reactions, primarily for survival and later out of curiosity that forms an integral part of human nature. This understanding, in turn, had a profound impact on diverse domains, from ecology and conservation to psychology and even artificial intelligence.

The study of animal behavior becomes even more vital when we shift our focus to farm animals. These creatures, which include but are not limited to cows, pigs, chickens, hens, and sheep, play a substantial role in human life. They are essential providers of food and raw materials and form the backbone of agricultural economies around the world (Broom, 2010). Understanding their behavior is pivotal in promoting animal welfare and ethics, and crucial for enhancing productivity, disease management, and sustainable farming practices.

In the past, the study of animal behavior was primarily conducted through long hours of observations and manual recordings, a time-consuming process that offered limited scalability and suffered from observer bias. The emergence of machine learning technology, however, has begun to revolutionize this field. With its capability to acquire knowledge and enhance its performance based on previous encounters, machine learning provides a powerful tool to analyze and predict animal behavior on a scale that was previously inconceivable. It can process massive amounts of data quickly and accurately, and also it can discover complex patterns and associations that can go unnoticed by human observers (Neethirajan, 2020).

The integration of machine learning into animal behavior studies stands at the intersection of biology, ethology, computer science, and data science. These interdisciplinary interactions offer unprecedented insights into animal behavior, fundamentally transforming how we approach animal welfare, disease detection, and farm management (Benos et al., 2021; García et al., 2020; Liakos et al., 2018). By researching the world of animal behavior, we begin a journey that enhances our understanding of these creatures and shapes the future of animal farming—making it more efficient, sustainable, and humane.

## Animal Behavior

### *The Essence of Animal Behavior*

The study of animal behavior is a gateway to understanding the interactions, survival strategies, and communication that unfold within the animal kingdom. It is an expedition into interpreting how animals respond to their environment—ranging from the lands they inhabit to the other animals they encounter. This exploration examines patterns of feeding, mating, social dynamics, and even the capacity of animals to learn and remember.

#### Defining Animal Behavior

Animal behavior encompasses the entire domain of an animal's actions and reactions—everything from intricate gestures to profound interactions. This multidisciplinary domain connects biology, psychology, ethology, and now, computer science (Lehner, 1996). It is akin to decoding a secret language—the behavioral cues animals offer provide us an indication into their world, their needs, and their unique ways of life.

> **Imagine animal behavior as a language of gestures, a code animals use to convey their thoughts, emotions, and intentions.**

### *Understanding the Science of Ethology*

#### Ethology: Navigating Natural Behaviors

"Ethology is a branch of zoology that studies the behavior of animals, typically with a scientific focus on behavior under natural conditions and viewing behavior as an evolutionarily adaptive trait" (Lehner, 1996). Throughout history, naturalists have embarked on journeys to interpret the enigmatic world of animal behavior. Ethology traces its line to Charles Darwin and the pioneering work of ornithologists such as Charles O. Whitman, Oskar Heinroth, and Wallace Craig. Yet, it was the emergence of 20th-century Dutch biologist Nikolaas Tinbergen and Austrian biologists Konrad Lorenz and Karl von Frisch that marked the start of modern ethology. Their collective insights earned them the Nobel Prize in Physiology and Medicine in 1973 (*The Nobel Prize in Physiology or Medicine 1973 - NobelPrize.Org*, n.d.).

**Think of ethology as a cultural anthropologist studying animals—observing how they live, interact, and adapt to their environments.**

### The Interplay of Behavior and Ecology

Taking a cue from ethology, *behavioral ecology* extends the narrative. Behavioral ecology is the study of behavioral interactions between individuals within populations and communities, usually in an evolutionary context. It looks at how competition and cooperation between and within species affects evolutionary fitness (Walker & Hill, 2020). It reveals the evolutionary context that forms animal behavior.

## *Farm Animals: An Insightful Glimpse*

### Enhancing Welfare through Behavior

When we shift our focus on farm animals, we uncover a new dimension of significance. Cows, pigs, chickens, and other farm creatures are integral contributors to human societies. Studying their behavior is about understanding their actions and promoting their well-being. Think of farm animal behavior as a guide, helping us create environments where animals can thrive and express their natural behaviors. Farm animal behavior holds the key to detecting health issues early. Changes in behavior often signal underlying problems and by observing how they move, eat, and interact, we gain insights into their well-being (Fogarty et al., 2019).

# Introducing Machine Learning

*Machine learning* (ML) is a subfield of *Artificial intelligence* (AI) that involves teaching machines to identify patterns within data and then use this knowledge to make predictions or decisions without the need for direct programming for specific tasks (Abramson et al., 1963). A machine learning task is a defined problem or objective that a machine learning algorithm attempts to solve. The choice of the *task* will influence many aspects of the *machine learning process*, including the *data* that needs to be collected, the *model* that is selected, and the way *success* is measured.

***Model:* A machine learning model is a mathematical representation of patterns learned from data, designed to make predictions or decisions without direct coding for each specific task.**

> ***Data:*** **This refers to information, often presented in numerical, textual, or visual formats, which is utilized to train, validate, and test machine learning models, ensuring their ability to accurately perform their designated tasks.**
>
> ***Success:*** **This refers to the effectiveness of a model in accurately completing its task, often evaluated using specific metrics.**

Applying machine learning to the study of animal behavior provides an innovative approach to interpret the complexities inherent in the field (Valletta et al., 2017). The reasons for its use are manifold.

- Machine learning enables the handling and processing of vast volumes of data generated in behavioral studies. As technology has advanced, so has our ability to monitor and record animal behavior. Video recordings, GPS tracking, and other sensor-based technologies generate large amounts of data that can be overwhelming to analyze manually. Machine learning algorithms can swiftly examine these extensive data sets, identifying patterns, and trends that might go unnoticed by the human eye.
- Machine learning can help in creating predictive models for animal behavior. By learning from past behavioral data, these algorithms can predict future behavior under different conditions. This capability is immensely useful in farm management. For instance, being able to predict a cow's readiness for milking or an outbreak of disease in a chicken can lead to more efficient and well-timed interventions.
- Machine learning brings objectivity to the study of animal behavior. It minimizes human bias that may influence the interpretation of behavior. This objectivity is particularly important in ensuring that any changes we make to farm environments or practices based on our understanding of animal behavior are grounded in reliable, data-driven insights.

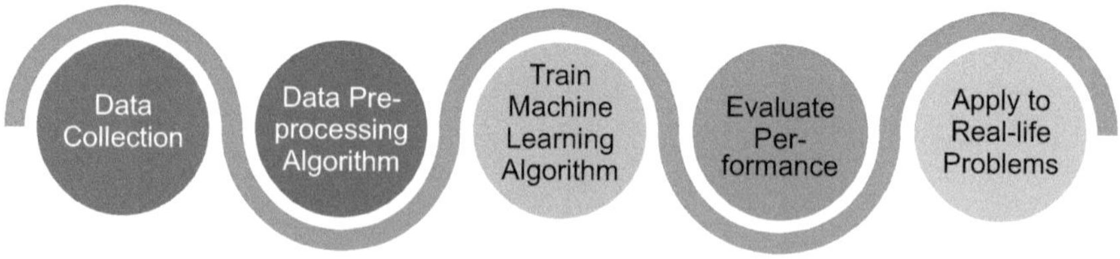


**Figure 1.1: Machine learning approach.**

Figure 1.1 illustrates the principle of machine learning in five steps. Starting with *'Data Collection'* data is gathered from diverse sources. Moving on to *'Data Preprocessing'* the collected data is refined through cleaning, transformation, and normalization processes. In the subsequent *'Train ML Algorithm'* step, a machine learning model is developed and trained using the processed data. The *'Evaluate Performance'* step measures the model's effectiveness using specific metrics. Finally, in the *'Apply to Real-Life Problems'* step, the trained model is put to work, generating valuable insights and predictions to address practical challenges.

## Exploring Types of Machine Learning

Understanding the different types of machine learning techniques is central for their successful application in any domain, including the study of farm animal behavior. The main types are Supervised Learning, Unsupervised Learning, Semi-Supervised Learning, and Reinforcement Learning, each of which is particularly suited to certain kinds of problems and datasets.

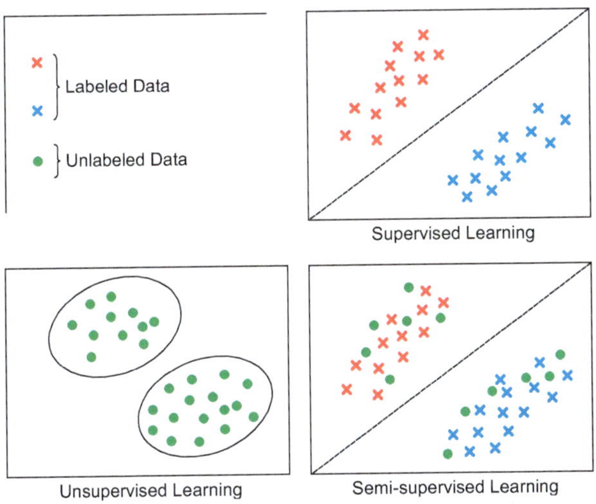


**Figure 1.2: Machine learning types.**

### *Supervised Learning: Learning through Examples*

*Supervised Learning*, as the name suggests, involves training a machine learning model in a supervised manner. In Supervised Learning, the algorithms are *trained* to determine a mapping function from an *input space* $X$ to an *output space* $Y$ using *labeled* datasets. Each entry in a dataset consists of an input vector $x \in X$ and an associated label $y \in Y$. The primary goal is to find a function $f$ such that $y = f(x)$ for any new $x$.

A label $y$, refers to the desired outcome or target for a specific instance (example). It represents the "answer" that the model should ideally produce when presented with a corresponding input vector. For example, in *classification* tasks, where the objective might be determining if an animal exhibits a certain behavior, could be "yes" or "no". On the other hand, in *regression* tasks, where the aim is to predict a continuous value like the weight or height of an animal based on other features, the label could be a numerical value.

The input space $X$, constitutes all input values. Each dimension corresponds to a distinct *feature* of the observed data. Features are distinct measurable properties of the data. For example, for a given feature $x_i$, where $i$ is the $i$th feature, $x_i$ might stand for an animal's age, heart rate, or its daily activity duration. An input vector $x$ is a specific instance within the input space $X$. It is a collection of values representing multiple features of a single instance or observation. Each of its elements, labeled as $x_i$ (where $i$ marks the feature's index), denotes a particular feature value. For instance, if $x_1$ stands for age, $x_2$ for daily activity duration, and $x_3$ for average resting heart rate, a vector could be represented as $x = [x_1, x_2, x_3]$. Thus, the input vector provides a full description of an instance in terms of all its features.

In machine learning, when we feed data into an algorithm, it often comes in the form of many input vectors (i.e., many rows of data, with each row representing one observation across multiple features).

- **A feature is like a column in a spreadsheet or database table.**
- **An input vector is like a row in that spreadsheet or table, which contains values for each of the features.**

Classification and regression are the main categories of supervised learning tasks. Each offers unique methods and applications adapted to different types of data and desired outcomes.

### Classification: Studying Patterns

Classification is an essential supervised learning task. The objective is to appoint a given input vector to one of a fixed set of targets, often referred to as classes or labels.

**Definition:** In classification, given an input space $X$ and a finite set of labels $Y = \{y_1, y_2, \ldots y_k\}$, where $k$ is the number of labels, the goal is to learn a function $f: X \rightarrow Y$ that maps an input vector $x \in X$ to the corresponding label $y \in Y$.

As discussed previously, an input vector $x$ is a collection of values across multiple features for a single instance. In classification, this vector serves as the input that the algorithm uses to determine which class the instance belongs to. For classification, a label $y$ is a discrete value representing one of the possible classes an instance can belong to. For example, in a binary classification task determining whether an animal is carnivorous or herbivorous, $y$ could be {'carnivorous', 'herbivorous'}.

### Decision Boundary

Classification algorithms often work by establishing a decision boundary (or boundaries) in the input space $X$. The decision boundary represents a surface in the feature space that separates instances of different classes. Its complexity and shape depend on the learning algorithm and the type of the data.

> **For many machine learning algorithms, the boundary can be thought as the 'line of decision' that determines how new data is classified.**

In a simple $2D$ space with two features, the decision boundary might literally be a line. In higher dimensions (e.g., $3D$), it might be a plane or even a more complex shape. In general, for a $n$-dimensional space, the decision boundary will be a $n-1$ dimensional hyperplane.

### Applications in Animal Behavior

Consider a dataset capturing various behavioral attributes of sheep, including movement patterns, posture, and gait. Now, consider a specific classification task: differentiating between healthy sheep and those affected by *lameness*—a condition affecting animal mobility. Through the process of classification, the algorithm examines this labeled dataset, learning the *features* in behavior that separate healthy sheep from those with lameness (Figure 1.3).

> **A feature refers to a distinct measurable property extracted from data, acting as input to algorithms for pattern recognition and analysis.**

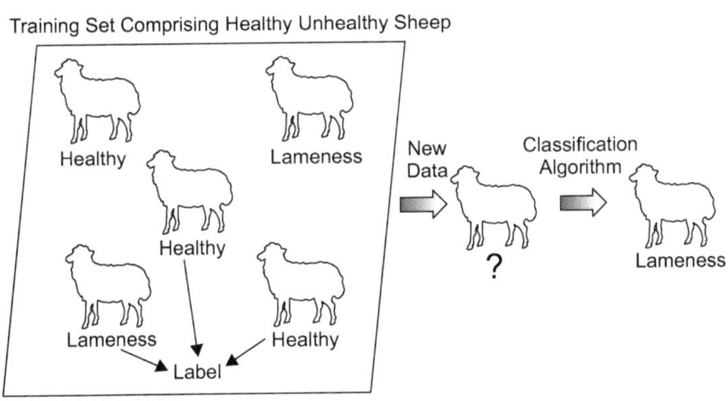


**Figure 1.3: Training set and test instance for healthy sheep classification.**

As the algorithm learns this knowledge, it becomes capable of distinguishing between the two categories based on the observed behavioral cues. When new data about sheep behaviors is *fed* into the system, the trained ML model can make predictions about whether a sheep is healthy or exhibits signs of lameness. This classification task gives farmers and researchers a valuable tool for early detection of health issues, leading to fast and targeted actions related to the well-being of the animals.

### Example: Detecting Threats with Classification

Let's walk through a scenario that demonstrates the ability of classification in protecting farm animals. Picture a pasture housing a group of cows, where their well-being and safety are top priorities of the farmer. In this context, we introduce a classification task aimed at identifying potential threats—thieves or predators—that might intrude upon the habitat of the cows.

To ensure the safety of the cows, we deploy *motion sensors* across the pasture. These sensors *detect* occasions when an animal exhibits fast and irregular movements, suggesting the presence of an intruder.

To make sense of the sensor data, we employ a *classification algorithm*. Our dataset includes *labeled* examples—data points with corresponding labels indicating whether a movement is a threat or not. These labels are based on past events where human observation confirmed the presence of predators or thieves.

As the machine learning algorithm *processes* the labeled data, it *learns* to discriminate between normal movements of cows and irregular ones that may indicate an alarming condition. The algorithm identifies *patterns* in the sensor data that are indicative of potential threats.

With the knowledge extracted from the *training* phase, the algorithm can now take real-time sensor readings and classify the movements as either safe or alarming. When the algorithm detects an anomalous movement pattern, it triggers alerts for immediate action, enabling farmers to instantly respond and ensure the cows' safety (Figure 1.4).

In this example, classification serves as a guard, using machine learning to identify patterns in data and provide appropriate warnings.

### Common Classification Techniques

- **Logistic Regression (LR)** (Hilbe, 2009; Hosmer et al., 1989)**:** Logistic regression is utilized for binary classification. It estimates the probability that an instance belongs to a specific class.
- **Decision Trees (DT)** (Safavian & Landgrebe, 1991; Song & Lu, 2015)**:** These algorithms split the input space into regions and assign a class to each region.

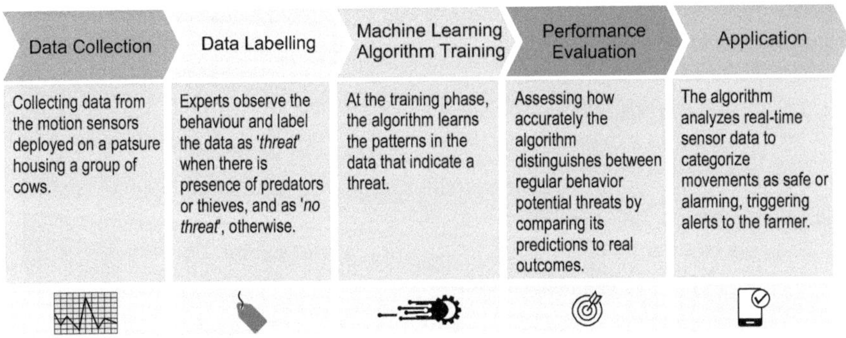


**Figure 1.4: Machine learning process: Detecting threats with classification.**

- **Random Forests** (Breiman, 2001)**:** This is an ensemble method that creates a *forest* of DT and decides based on the majority vote of the trees.
- **Support Vector Machines (SVM)** (Hearst et al., 1998)**:** SVMs aim to find the hyperplane that best divides a dataset into classes.
- **Neural Networks** (Bishop, 1995)**:** Utilize layers of interconnected neurons to classify data.
- **K-nearest Neighbors (KNN)** (Cunningham & Delany, 2021)**:** Classifies an instance based on the majority class among its '*k*' closest instances in the training data.
- **Naive Bayes** (Johnson et al., 2022)**:** Based on Bayes' theorem, this technique assumes independence between features and calculates probabilities to determine class membership.

## Regression: Studying Trends

*Regression*, like classification, is a basis of supervised learning. While classification aims to assign data to distinct categories, regression focuses on predicting continuous numerical values. Understanding regression is crucial, not only for its applications in predicting outcomes, but also for the foundational principles it conveys about the relationship between variables.

**Definition:** In regression, given an input space $X$ and an output space $Y$ of continuous values, the objective is to learn a function $f : X \rightarrow Y$ that maps an input vector $x \in X$ to a continuous value $y \in Y$.

Just as in classification, an input vector $x$ in regression is a collection of values representing different features of a single instance. This vector provides the necessary context based on which the regression algorithm predicts an output value. Unlike the discrete labels in classification, the output in regression, $y$, is a continuous value. For instance, in predicting the age of an animal based on various features, $y$ would represent the age.

Regression algorithms work by adjusting their parameters to minimize the difference between their predictions and actual outcomes. This difference, commonly referred to as the error or loss, is captured by various metrics like *Mean Squared Error (MSE)* or *Mean Absolute Error (MAE)* (Evaluation metrics will be discussed in Chapter 8).

At its core, regression fits a mathematical model to observed data points. Visualize this as sketching a line through a scatter plot of data points. The aim is to draw the line that best captures the general direction of the points. This line becomes a tool that helps predict where new data points might fall on the plot with a certain level of accuracy.

Regression can be used in various ways. Consider a scenario where researchers collect data on the dietary habits of farm animals and their corresponding weight changes. By utilizing regression, the algorithm learns the relationship between dietary intake and weight gain. This knowledge enables researchers to predict weight variations based on dietary adjustments, contributing to informed decisions in farm management practices.

Regression also does well in interpreting patterns over time. Assume we are interested in assessing how an animal's daily activity levels relate with its milk production. Through thorough data collection and regression analysis, we can identify whether increased activity translates to higher milk supply, allowing us to adapt farming practices for maximum productivity.

Moreover, regression plays a pivotal role in anomaly detection. While there are dedicated anomaly detection algorithms, regression models can be employed as part of the process. When using regression for anomaly detection, the core principle revolves around the *residuals*, or errors of the regression model. *Residuals* are the differences between the actual values and the predicted values. If a model is well fitted to most data points, the residuals should, in theory, be randomly distributed and close to zero. Data points with large residuals-values that deviate significantly from the predicted values are considered anomalies.

For instance, if we track an animal's heart rate and notice sudden spikes, regression models can help us identify whether these variations are within the norm or indicative of potential health issues. By studying changes from established tendencies, regression assists in early disease detection, and timely intervention, promoting animal welfare.

**Regression is about understanding how changes in one variable are associated with changes in another.**

## Predicting Weight Gain of Pigs: An Application of Regression

Consider a practical scenario: Predicting the weight gain of pigs. We gather a series of data points that could impact a pig's weight gain. A notable feature in this setting is the daily feed consumption of each pig. This variable plays a significant role as it directly influences the animal's metabolism, which in turn affects its weight gain.

With this data in hand, we construct a dataset where each entry describes a specific timeframe, along with the pig's daily feed consumption and corresponding weight gain. This dataset forms the foundation upon which our regression model will be developed.

The regression process involves identifying the best-fitting line that aligns with the collected data points. The model aims to understand the relationship between daily feed consumption and the animal's weight gain, thereby creating a predictive equation.

Once the regression model is trained on this dataset, it becomes a powerful tool for prediction. Once we acquire a new dataset of the daily feed consumption of each pig for a specific timeframe, by fitting these values into the regression equation, the model estimates the weight gain for each pig during that period (Figure 1.5).

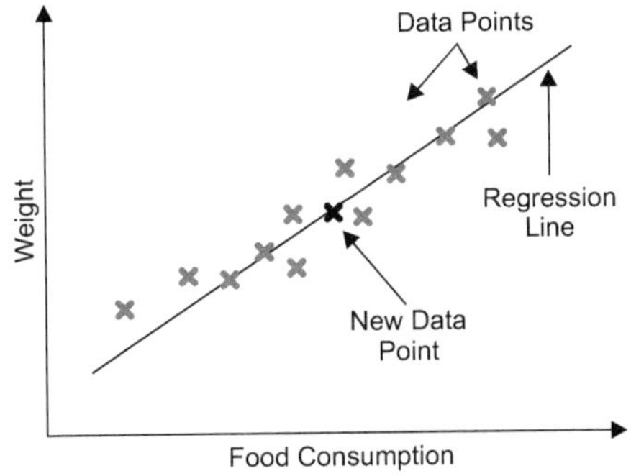


**Figure 1.5: Regression example: Predicting the weight gain of a pig based on food consumption.**

In modern farm management, the integration of advanced sensor technologies significantly enhances the accuracy of such predictions. For example, a smart feeding system could automatically track the daily feed consumption of each pig, eliminating manual data entry errors and providing real-time information for better decision-making.

Through this process, regression enables farmers to predict weight gain trends among pigs based on their daily feed consumption. This knowledge, gained through data analysis and predictive modeling, aids in managing feeding routines, and ensures the overall optimal growth of the pigs' population.

### Common Regression Techniques

- **Linear Regression** (Seber & Lee, 2003)**:** Assumes a linear relationship between input features and the target value, and modifying a line (or, hyperplane in higher dimensions) to fit the data.
- **Polynomial Regression** (Ostertagová, 2012)**:** Extends linear regression to model non-linear relationships by introducing higher-degree terms.
- **Ridge and Lasso Regression:** Linear regression methods with regularization techniques to prevent *overfitting.*
- **Decision Trees for Regression** (Quinlan, 1986)**:** Like classification trees but predict continuous values instead of class labels.
- **Support Vector Regression (SVR)** (Hearst et al., 1998)**:** An adaptation of support vector machines for regression tasks.
- **Random Forest Regression** (Breiman, 2001)**:** Utilizes an ensemble of decision trees to make continuous predictions.
- **Neural Networks** (Bishop, 1995)**:** Neural networks can also be configured for regression tasks.

**In both Classification and Regression, the underlying principle remains the same—utilizing existing data to make informed estimations or categorizations for unseen data.**

## *Unsupervised Learning*

*Unsupervised Learning* (Abramson et al., 1963) is different from supervised learning in its approach and objectives. While supervised learning depends on labeled data to make predictions or classifications, unsupervised learning works with unlabeled data, aiming to discover underlying structures or patterns within the data itself.

**Definition:** In unsupervised learning, given an input space $X$, the aim is to discover patterns, groupings, or structures in the data without any predefined labels. The primary tasks within unsupervised learning are clustering and dimensionality reduction.

Like supervised tasks, an input vector $x$ in unsupervised learning represents the features of a data instance. However, there is no associated label or target value. Unsupervised algorithms work by measuring the similarity or difference between data points, aiming to group similar data together or represent the data in a reduced form while retaining maximum information.

### Common Unsupervised Learning Techniques

- **Clustering:** Clustering aims to partition data into distinct groups or clusters, where data points within the same cluster are more like each other than to those in other clusters. These groupings are based on certain attributes of the data, but the exact definition of *similarity* depends on the specific clustering algorithm used.

#### Clustering Techniques

- **K-Means Clustering** (Hartigan & Wong, 1979)**:** Partitions data into *k* distinct, non-overlapping groups (clusters) according to their similarity.
- **Hierarchical Clustering:** Builds a tree of clusters, useful for understanding hierarchical relationships in the data.
- **DBSCAN** (Density-Based Spatial Clustering of Applications with Noise) (Ester et al., 1996)**:** Groups together closely packed points and marks points in low-density regions as outliers.

#### Example: K-Means Clustering in Animal Behavior

A commonly used unsupervised learning algorithm is K-Means clustering. K-Means attempts to partition a dataset into a set number of clusters, with each data point being assigned to the cluster closest to the mean.

Let's consider a hypothetical study of animal behavior. Assume that we have recorded various behaviors of animals over a period and quantified them along two axes:

- **Activity Level:** Representing the intensity of movement or energy.
- **Interaction Intensity:** Indicating the level of interaction with the environment or other animals.

Our goal is to group these behaviors into distinct clusters to better understand patterns or categorizations within them.

Looking at Figure 1.6, you can see that behaviors are grouped into seven distinct clusters, which might correspond to different types of behaviors, such as:

1. **Resting:** Low activity level and minimal interaction.
2. **Grazing:** Moderate activity level but still relatively low interaction.
3. **Walking:** Moderate activity and interaction levels.
4. **Standing:** Low activity but higher interaction—perhaps the animal is alert.
5. **Running:** High activity and potentially high interaction.
6. **Fighting:** Variable activity but very high interaction.
7. **Scratching:** Moderate to high activity with medium interaction.

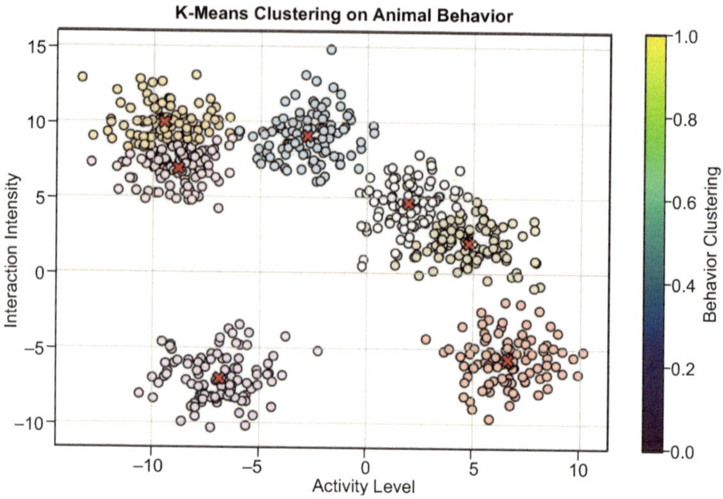


**Figure 1.6: Scatter plot of animal behaviors categorized by Activity Level and Interaction Intensity using K-Means clustering.**

By examining the clusters, we can make informed assumptions about the nature of these behaviors. It is crucial to understand that unsupervised learning, especially clustering, provides an initial exploration of the data. It is up to domain experts, in this case, animal behaviorists, to interpret the significance and implications of these clusters.

> **Note that the labels assigned to clusters are empirical and based on the context provided. In a real-world scenario, domain-specific knowledge would be crucial in labeling and interpreting the clusters.**

- **Dimensionality Reduction:** Dimensionality reduction techniques help to reduce the quantity of features or dimensions within a dataset, retaining a majority of the essential information. By representing the data in fewer dimensions, it becomes easier to visualize and analyze. This leads to an improved performance with certain algorithms.

### Dimensionality Reduction Techniques

- **Principal Component Analysis (PCA)** (Reddy et al., 2020)**:** Reduces the dimensionality of data by projecting it onto principal components that acquire the greatest amount of variance.
- **t-Distributed Stochastic Neighbor Embedding (t-SNE)** (Belkina et al., n.d.)**:** A nonlinear dimensionality reduction method especially effective for the visualization of high-dimensional data.

- **Autoencoders** (Y. Wang et al., 2015)**:** These are neural network architectures designed to compress data into a lower-dimensional form and then reconstruct it.

## Example: PCA in Animal Behavior

In an animal behavior study, imagine researchers have collected numerous metrics on animals such as heart rate, body temperature, speed of movement, frequency of vocalizations, duration of sleep, and many others, resulting in a dataset with tens or even hundreds of dimensions. While each of these metrics provides valuable insights, visualizing such high-dimensional data is challenging.

PCA transforms the original high-dimensional data into a new coordinate system. The first coordinate (or the first principal component) will acquire the highest amount of variance in the data, the second coordinate (or the second principal component) acquires the second highest, and so on.

By using PCA, the many dimensions of our animal behavior dataset might be reduced to just two principal components, which can be easily visualized on a scatter plot.

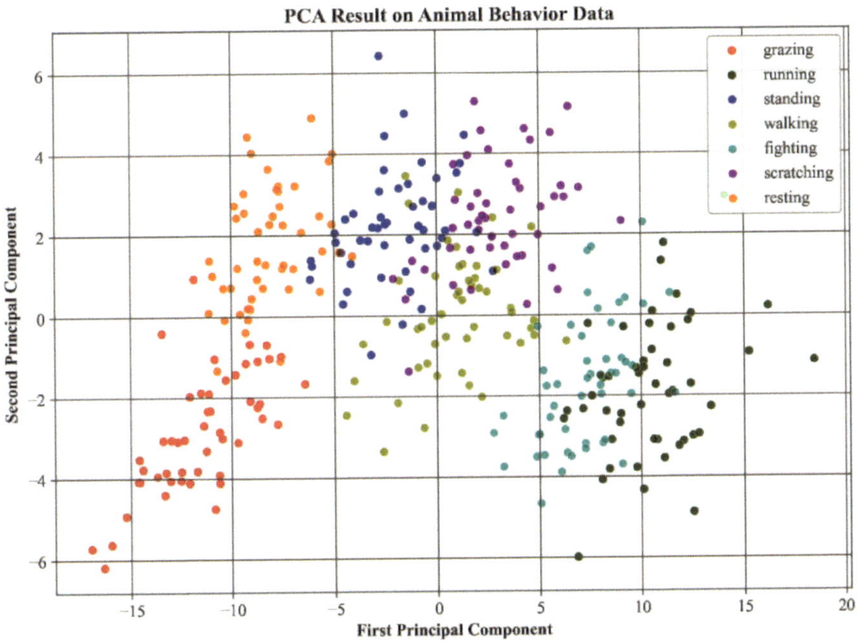


**Figure 1.7: Scatter plot showing animal behaviors based on the first two principal components derived from PCA. Each point represents an individual animal's behavior pattern, plotted according to the two most significant patterns of variance in the data.**

Observing Figure 1.7, the $x$-axis might represent a pattern related to the overall activity level of the animals (combining metrics like movement speed, heart rate, and vocalizations), while the $y$-axis might capture another significant pattern, perhaps related to the animal's social behaviors (like duration and frequency of interactions with other animals).

What is great about dimensionality reduction is that while the axes do not correspond directly to any original measurement, they capture the primary modes of variation within the dataset. So, animals that are close to each other on the scatter plot have similar behaviors across all the metrics collected, while those far apart have different behaviors.

- **Association Rule Learning:** *Association rule learning* (Kaur & Madan, 2015; Telikani et al., 2020) is a technique primarily used to discover relationships and patterns between items in large datasets. In the setting of animal behavior, this can be applied to find patterns in sequences or combinations of behaviors exhibited by animals under certain conditions or environments.

There are several key types of association rule learning techniques, including:

- **Apriori Algorithm** (Santosh Kumar & Balakrishnan, 2019)**:** This is a widely recognized method for mining frequent itemsets for Boolean association rules. It progressively extends the size of the frequent itemsets by one item at a time, checks a candidate set of itemsets for frequency, and then prunes those candidate sets which are found infrequent.
- **Eclat Algorithm** (Hahsler et al., 2005)**:** This is an alternative depth-first technique which uses set intersection to identify frequent itemsets.
- **FP-Growth** (Borgelt, 2005)**:** Differing from Apriori, this approach avoids the step of candidate generation. It adopts a divide-and-conquer strategy facilitated by a prefix tree representation of the database.

Imagine observing a group of animals over, a period of time. Each specific behavior they exhibit can be seen as an 'item'. Whenever an animal exhibits a sequence of behaviors, that sequence can be treated as a transaction. By analyzing numerous sequences or transactions, association rule mining can help deduce rules such as, "When an animal displays behavior A, it is likely to show behavior B shortly after".

### Association Rule Mining: An Example Using the Apriori Algorithm

Consider a hypothetical scenario in a dairy farm. Let's say that we have been observing certain behaviors in our cows and recording them over a period. We have identified some patterns that occur frequently together. For instance:

- Whenever a cow has increased vocalization (perhaps mooing more frequently), it tends to show restlessness.
- When a cow is restless, it often kicks when being milked.

- The kicking during milking is strongly associated with a decrease in milk production.
- It is also observed that an increase in vocalization can lead directly to decreased milk production.

Based on these observations, we can form the following association rules:

1. If [Increased Vocalization] then [Restlessness].
2. If [Restlessness] then [Kicks when Milked].
3. If [Kicks when Milked] then [Decreased Milk Production].
4. If [Increased Vocalization] then [Decreased Milk Production].

These rules are indicative and can provide insights to farm managers about potential stressors or health issues the cows might be experiencing. Such insights can lead to early interventions, ensuring the well-being of the cows and maintaining production levels.

Using the Apriori algorithm, we can extract such association rules from large datasets. It works by identifying the commonly occurring individual items within the database and extending them into bigger item sets if they appear frequently in the database. The graphical representation of these association rules is shown in Figure 1.8.

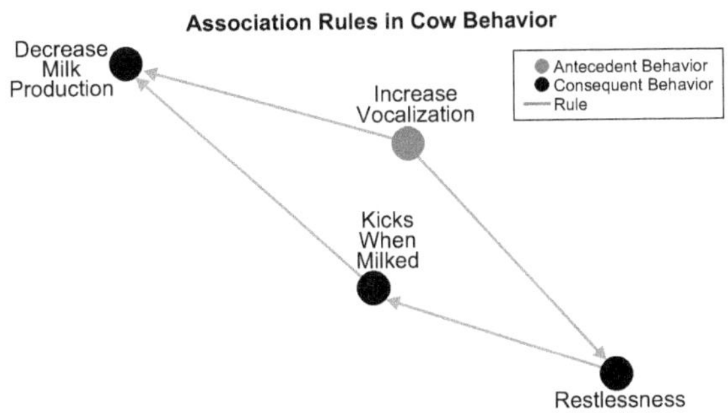


**Figure 1.8: Graphical representation of association rules in cow behavior.**

This directed graph (Figure 1.8) visualizes the association rules based on cow behaviors. Nodes represent different behaviors, and directed edges signify an association rule between them.

- **Nodes:** Represent behaviors. Antecedent behaviors (starting points of rules) are shown in light grey, while consequent behaviors (end points of rules) are in black.
- **Edges:** The grey arrows represent the directionality of the association rules, from antecedent to consequent behavior.

The arrows show the direction of the association rule, for instance, there is an arrow from "Increased Vocalization" to "Restlessness", indicating the rule "If

[Increased Vocalization] then [Restlessness]". The color coding and directional arrows offer a clear visual guide to understanding the rules and their implications.

### Anomaly Detection in Unsupervised Learning

Anomaly detection (Chandola et al., 2009) is another critical technique that aims to identify data points that differ significantly from the norm—these are known as anomalies or outliers. Anomalies could be due to variability in the data or could indicate errors, extreme events, or a particular issue that requires attention.

Mathematically, given a dataset $D$ consisting of $n$ samples, each datapoint can be represented in a multidimensional feature space. The objective is to find whether $x_i$ is an outlier based on a certain metric of similarity or distance. If the measure exceeds a predetermined threshold, the data point is considered an anomaly.

$$f(x_i) = \begin{cases} 1, \text{ } if\ d(x_i, D) > threshold \\ 0, \text{ } otherwise. \end{cases}$$

Here $f(x_i)$ is a function demonstrating if a data point $x_i$ is an outlier, where $d(x_i, D)$ might be a distance or difference measure between $x_i$ and some dataset or reference point $D$. The output is binary (either 1 or 0) based on whether this measure exceeds a particular threshold.

This is a high-level, basic rule, and in practice, the measure of *difference* and the choice of threshold would be critical and may need to be determined through data analysis, domain knowledge, and validation.

Anomaly detection has numerous applications across various domains, including credit card fraud detection, network security, and health monitoring. In the context of animal behavior, it can be helpful in detecting unusual behaviors which might indicate distress, sickness, or other potential concerns.

### Types of Anomaly Detection

- **Statistical Anomaly Detection** (Paschalidis & Chen, 2010)**:** This approach assumes that normal data instances occur in high probability regions of a stochastic model, while anomalies occur in the low probability regions of the stochastic model.
- **Clustering-based Anomaly Detection** (Pu et al., 2021)**:** Data is organized into clusters where instances of similar data are grouped together. Points which do not belong to any cluster or belong to small clusters can be considered anomalies.
- **Density-based Anomaly Detection** (Samariya & Thakkar, 2023)**:** It computes the density of data points around a particular instance. Regions of low density are typically indicative of anomalies.
- **Isolation Forest** (Samariya & Thakkar, 2023)**:** Targets anomalies by isolating them due to their rarity and distinctness. It iteratively selects random features

and split values, with anomalies often isolated quicker, indicating fewer steps signify abnormality.

### Example: Anomaly Detection in Animal Activity Monitoring

Horses have a consistent speed range for their gaits, whether they are trotting, cantering, or galloping. Suppose we are monitoring the activity levels of a group of horses using sensor technology. Over time, we have gathered a stable pattern of their activity levels.

To illustrate, consider the running speed of a horse over a specific duration as shown in Figure 1.9.

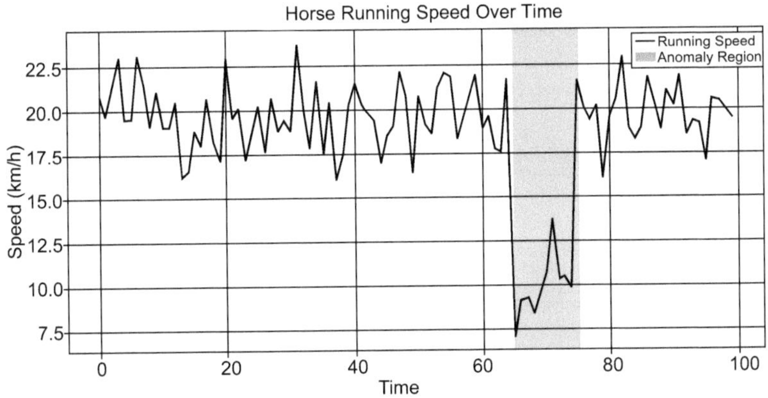


**Figure 1.9: Consistent running speed of a horse with a specific segment highlighting a potential anomaly in the speed.**

Based on the graph, a possible anomaly can be detected in the horse's running speed between the time intervals 65 to 75. Such examples, identified using anomaly detection techniques, could suggest issues like health problems, environmental disturbances, or other concerns that might impact the horse's usual behavior.

## *Semi-supervised Learning*

*Semi-supervised learning* (van Engelen & Hoos, 2020) incorporates both labeled and unlabeled data, particularly in scenarios where obtaining labeled data is challenging or insufficient. By leveraging both types of data, semi-supervised learning endeavors to enhance learning performance, achieving improved accuracy and generalization compared to using either data type independently.

**Definition:** In semi-supervised learning, given an input space $X$, where some data is labeled and some is not, the objective is to improve the performance of a model by utilizing the information contained within the unlabeled data. This learning paradigm often focuses on classification, regression, and sometimes clustering tasks.

An input vector $x$ in semi-supervised learning, represents the features of a data instance, like supervised and unsupervised learning. While a portion of these input vectors come with associated labels, a significant amount of data remains unlabeled. The algorithms in this paradigm aim to use the structure and distribution of the unlabeled data to enhance the learning process derived from the labeled data.

### Common Semi-supervised Learning Techniques (Goldberg, 2009)

- **Self-training:** In self-training, an initial model is trained using the labeled data. This approach is subsequently applied to assign labels to the unlabeled data, where predictions made with the highest confidence are incorporated into the training set along with their predicted labels. The model is retrained on the combined dataset, and this process can be iteratively repeated.
- **Multi-view Learning (Co-training):** This assumes that the data can be seen from multiple views or feature sets. Each view can train a separate classifier. Views are then combined to collaboratively decide on the label.
- **Consistency Regularization:** This technique involves training a model to produce consistent predictions even when small perturbations or noise are introduced to the input data. By leveraging unlabeled data as a source of regularization, this method promotes stability and enhances generalization.
- **Label Propagation and Label Spreading:** Label Propagation works by constructing a similarity graph over all data points. Labels are then propagated through this graph. Label spreading is similar, but it uses a modified algorithm that normalizes the relationships between points.
- **Pseudo-labeling:** This technique is somewhat like the self-training approach. In deep learning (Deep learning is addressed in Chapter 9), the idea is to label the unlabeled data using the current model and then use these pseudo-labels for further training. However, due care must be taken to avoid amplifying the model's mistakes.

For additional insights into semi-supervised learning techniques, readers may find information in (Goldberg, 2009).

### Hen Behavior Classification: A Semi-supervised Learning Example

Consider a scenario where we are trying to categorize hens based on their behaviors. For simplicity, let's assume we have two categories—'Hen' and 'Not Hen'. While we have clear labels for some instances (either 'Hen' or 'Not Hen'), the majority remain unlabeled. This is where semi-supervised learning can be utilized to predict the behavior of the unlabeled instances.

Figure 1.10 illustrates a scatter plot showcasing labelled and unlabeled data points for hens' behaviors. Labelled points for 'Hen' and 'Not Hen' are marked in blue and orange squares, respectively. Unlabeled points are shown in gray dots. After

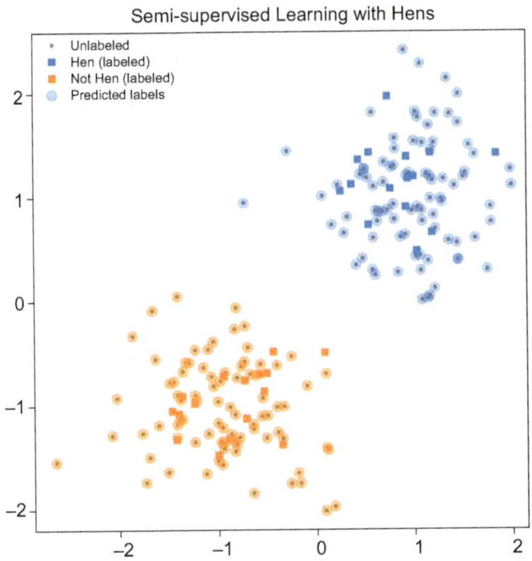


**Figure 1.10: Visualization of Semi-supervised Learning with Hens.**

using the semi-supervised learning algorithm, predicted labels are overlaid on the initial data. Initially, only a few data points are labelled, but with the use of the *Label Spreading* algorithm, we can effectively predict the behavior of the unlabeled instances. This demonstrates the power of semi-supervised learning in scenarios with limited labeled data.

It is important to note that while we offer a brief overview of this technique, the scope of semi-supervised learning is vast and comprises many methods and approaches. As the subsequent chapters investigate deeper into specific aspects of machine learning for farm animal behavior, we will not explore semi-supervised learning in further detail. Nevertheless, it stands as a significant field in machine learning, and we recommend additional exploration for those interested in a more comprehensive understanding.

### Reinforcement Learning

Reinforcement learning (RL) (Kaelbling et al., 1996) is a specialized domain within ML but is outside the scope of this book. However, we will briefly introduce it in this section. Unlike supervised and unsupervised learning, which rely on labeled data or characteristic patterns, RL revolves around agents making sequential decisions to maximize some notion of cumulative reward in uncertain environments. RL represents a machine learning paradigm where an agent learns how to behave in an environment by performing certain actions and receiving rewards or penalties in exchange. It's akin to teaching a dog a new trick: the dog represents the agent, the environment serves as the setting where the dog can execute tricks, and the rewards (or penalties) are the treats (or lack thereof).

**Definition:** Reinforcement learning seeks to train agents to make optimal decisions by interacting with an environment. These decisions are not based on explicitly provided correct answers, but rather feedback in the form of rewards or penalties based on the actions the agent takes (refer to Figure 1.11).

**Figure 1.11: Reinforcement learning process.**

### Core Concepts

- **Agent:** The decision maker or the learner.
- **Environment**: Everything that the agent interacts with.
- **Action (A):** What the agent can do. For example, in a game, actions might be moving left or right, jumping, or some other activity.
- **State (S):** Current situation returned by the environment.
- **Reward (R):** Feedback from the environment. Can be immediate or delayed.

In RL, an agent takes an action according to its state. The action influences the environment, which returns to the next state and a reward for the taken action. The aim of the agent is to learn a policy that optimizes the estimated cumulative reward over time.

### Key Components

- **Policy ($\pi$):** Strategy that defines the agent's way of taking actions. It can be deterministic or stochastic.
- **Value Function:** Estimation of future rewards. There are two types:
  - **State Value (V):** Expected reward initiated from state $s$ and then following policy $\pi$.
  - **Action Value (Q):** Expected reward initiated from state $s$, then action is taken, and then following policy $\pi$.
- **Model of the Environment:** This is optional. If the agent knows the model, it can predict the next state and reward for a given action.

**Deterministic:**

- **Environment: If for a given state $s$ and an action $a$, the resultant next state $s'$, and reward $r$ are always the same, the environment is said to be deterministic. There is no uncertainty about the outcome of an action.**

- **Policy:** A deterministic policy provides a specific action for a given state. For state *s*, there is always one action a that the policy will output. This means, every time the agent is in state *s*, it will take action *a*. Mathematically, a deterministic policy is represented as $a = \pi(s)$.

Stochastic:

- **Environment:** If for a given state *s* and an action *a*, the resultant next state *s'*, and reward *r* might vary with different probabilities, then the environment is stochastic. In such an environment, the same action in the same state might lead to different results at different times.
- **Policy:** A stochastic policy produces a probability distribution across possible actions for a given state. Instead of taking the same action for a state *s*, the agent will take an action a based on some probability. This allows for exploration and can handle uncertainties in the environment better. A stochastic policy is typically represented as $P(a|s)$, where *P* is the probability of taking action *a* when in state *s*.

### Types of Reinforcement Learning

- **Model-free vs Model-based:** In model-free reinforcement learning, the agent does not have a model of the environment and learns by trial and error. In model-based reinforcement learning, the agent knows the model of the environment.
- **Value-based vs Policy-based vs Actor-critic:** In value-based reinforcement learning, the focus is on finding the optimal value function, and not the policy. In policy-based reinforcement learning, the focus is on finding the optimal policy. Actor-critic reinforcement learning combines both.
- **Exploration and Exploitation:** In this category of reinforcement learning the agent needs to explore the environment initially. Once it knows enough, it can start exploiting its knowledge to get the maximum reward.

### Key Algorithms

- **Q-learning** (Sayed, 2023)**:** Model-free, value-based method.
- **Deep Q Networks (DQN)** (Varga et al., 2023)**:** Combines Q-learning with deep neural networks.
- **Policy Gradients** (R. S. Sutton et al., 2000)**:** Policy-based method.
- **Proximal Policy Optimization (PPO):** An optimization for policy gradient methods.
- **A3C (Asynchronous Advantage Actor-critic)** (Liu et al., 2018)**:** Combines value-based and policy-based learning.

### Q-learning

Q-learning is a value-based RL algorithm. In this method, the agent learns a value function, or 'Q-value', which represents the expected future reward for each possible action in each possible state. The agent is assigned with the task of learning the policy that will maximize this expected future reward.

The learning process commences with arbitrary Q-values, as the agent explores the environment, taking actions and receiving rewards, it uses these experiences to update the Q-values. With time, these updates should converge on the true Q-values under the optimal policy.

An application of Q-learning in a farm setting might be to optimize the feeding schedule of farm animals. The agent's state could be the current time, the hunger level of the animals, and the amount of available feed. The actions could be to feed a certain amount of food or to wait. The reward could be a measure of the animals' health and productivity, with penalties for overfeeding or underfeeding. The Q-learning algorithm would iteratively learn the best action to take in each state to maximize the animals' well-being and productivity, resulting in an optimized feeding schedule.

### Policy Gradients

Policy Gradients, on the other hand, are policy-based RL where the agent directly learns the policy, i.e., the mapping of states to actions. The policy is typically represented as a probability distribution over actions, and the learning process involves adjusting the parameters of this distribution to increase the expected future reward.

### Example: Policy Gradients in Environment Control

Consider the task of controlling the environment in a chicken coop to maximize egg production. The state could be variables like temperature, humidity, and light levels. The actions could be to adjust these variables, and the reward could be the number of eggs produced.

A Policy Gradient algorithm could learn a policy that maps the current environmental conditions to the optimal adjustments to maximize egg production. The algorithm would gradually refine this policy, exploring different actions and observing their effects on egg production.

While the use of reinforcement learning in the study of farm animal behavior might be less straightforward compared to supervised and unsupervised learning, its potential is significant. By simulating an environment and allowing an agent to interact with it, reinforcement learning algorithms can find innovative solutions to complex problems and help optimize decision-making processes to enhance farm efficiency and animal welfare.

## Summary

In the first chapter, the concepts of Animal Behavior were explored, and how machine learning can help understand farm animal behavior. We learned about animal behavior basics and how technology, especially machine learning, can give us new insights. We also broke down the main types of machine learning techniques: supervised, unsupervised, semi-supervised, and reinforcement learning, and discussed how they can be used in studying animals.

CHAPTER 2

# Foundational Concepts and Challenges in Machine Learning

In this chapter, we delve into key ML concepts, ensuring a comprehensive understanding of the workflow, from data collection to model deployment. We will explore the challenges faced in machine learning, and address common traps, ranging from overfitting and underfitting to the importance of bias and variance. In the last section, we outline the typical phases of a machine learning project.

## Machine Learning Concepts and Challenges

### *The Central Challenge in Machine Learning: Generalization*

In principle, machine learning is about making predictions. Whether we are classifying animals as active or inactive, predicting the animals' weight, or identifying patterns in data, the aim is for our model to perform well not just on the data it has seen (training set), but on new, unseen data (test set or real-world data). This ability to perform well on unseen data is known as *generalization* (Bishop, 2006).

Generalization refers to the model's ability to apply what it has learned from the training set to unseen data. A good machine learning model should be able to generalize well from the training dataset to any data from the problem domain. This ensures that our predictions are not just fitting in with the training data but are obtaining underlying patterns that can be applied to new data.

*Training error, also known as training loss,* is the error that we get from our machine learning algorithm on the data that it was trained on. It gives us a measure of how well our model fits the training data. A low training error indicates that our model has learned the patterns in the training data well. However, a very low training error might also indicate that the model has become too complex and has fit to the noise in the data, a phenomenon known as *overfitting.*

*The generalization error, known as the out-of-sample error or test error,* is the measure of how accurately an algorithm can predict the outcome values for previously unseen data. It gives us a measure of the difference between the model's predictions on new data and the true values for that data. Ideally, we want our model's generalization error to be low, indicating that it performs well on new, unseen data. A model with a low test error can be considered to be well generalized, whereas a model with a high test error is likely either overfitting or underfitting the training data.

While training error gives us insight into how well our model has learned the training data, it is the generalization (or test) error that truly matters, as it indicates how our model will perform in real-world scenarios. We need to achieve a balance, ensuring our models are neither too simple (leading to underfitting and high bias) nor too complex (leading to overfitting and high variance), to achieve good generalization.

### Model Complexity

In machine learning, a model's *complexity* often refers to its ability to fit a wide variety of functions. A more complex model generally has a greater number of parameters and, consequently, a higher *capacity* to adapt its shape to fit the training data.

**Capacity refers to a model's ability to fit a variety of functions.**

Advantages of High Model Complexity:

- **Flexibility:** A complex model can fit the training data more closely, capturing complex patterns that simpler models might miss.
- **Reduced Bias:** Since a complex model has a high capacity to fit to the training data, it is less likely to make assumptions that deviate from the actual patterns in the data.

Disadvantages of High Model Complexity:

- **Overfitting:** This is a common problem with complex models. While they might fit the training data very closely, they might do so at the expense of generalizing poorly to unseen data. For example, they may capture the noise in the training data, mistaking it for a true pattern.
- **Computational Cost:** Complex models often require more computational resources, both in terms of memory and processing time. Training and inference can be slower compared to simpler models.
- **Difficult Interpretability:** Highly complex models can become black boxes, making it challenging to understand how decisions or predictions are made.

To further understand the relationship between model complexity, overfitting, and underfitting, consider fitting a curve to a set of data points.

Using polynomial regression as an example, for a single input feature $x$, a polynomial regression of degree $n$ is given by:

$$y = \beta_0 + \beta_1 x + \beta_2 x^2 + \cdots + \beta_n x^n + \varepsilon$$

where,

- $y$ is the predicted output.
- $x$ is the input feature.
- $\beta_0, \beta_1, \ldots, \beta_n$ are the model's parameters.
- $\varepsilon$ is the error term.

Now, the degree of the polynomial determines the capacity of the model:

- A low-degree polynomial (e.g., linear or 2nd-degree) might be too simple to capture the structure of the data, leading to *underfitting*.
- A very high-degree polynomial might fit almost every data point perfectly, including its noise. While this results in low training error, it can lead to high error on unseen data, or *overfitting*.
- An optimal model would reach a balance, fitting the general trend of the data without being influenced by noise.

We can visualize the effect of different polynomial degrees on a dataset:

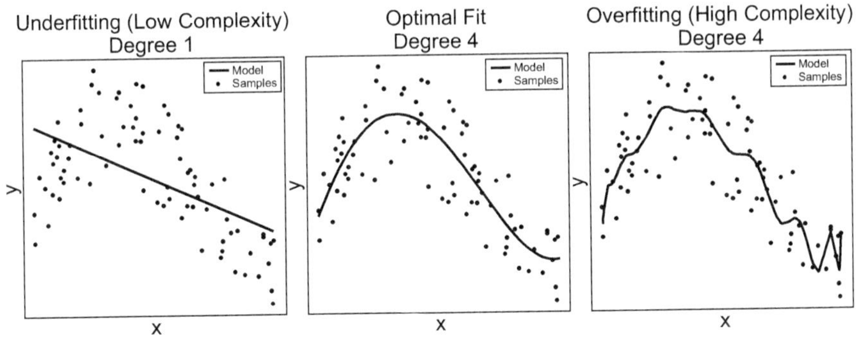


**Figure 2.1: Demonstration of model complexity using polynomial regression. The progression from underfitting with a low-degree polynomial, through an optimal fit with a moderate degree, to overfitting with a high-degree polynomial is clearly illustrated.**

Figure 2.1 shows three scatter plots to illustrate the concepts of underfitting, optimal fit, and overfitting in the context of model complexity and data fitting. In the first plot, labeled "Underfitting (Low Complexity) Degree 1", a straight line attempts to model the data points but fails to capture the underlying trend, indicating that the model is too simple. The middle plot, "Optimal Fit Degree 4", shows a curve that fits the data points well, suggesting that the model's complexity is just right to capture the essential patterns without being overly simplistic or too complex. The third plot, "Overfitting (High Complexity) Degree 15", portrays a highly wavy line that passes through large number of points, that may indicate a complex model that fits the training data too closely. This could be an overfitting model that may perform poorly on unseen data due to its excessive sensitivity to noise or minor fluctuations in the training dataset.

## Bias and Variance in Machine Learning

### *Bias*

While constructing a ML model, a problem that we have to overcome is the problem of *bias*. At a fundamental level, bias can be considered as the model's simplified assumptions aimed at making the dependent variable easier to estimate. It is the difference between the predictions our model makes, and the actual values (true values). A high bias can demonstrate itself in the form of oversimplifying the model or making assumptions that a linear behavior exists when it is not appropriate. Such models usually fail to capture the actual trends that exist in the data, leading to errors in predictions.

Bias is represented as:

$$Bias(\hat{f}(x)) = \mathrm{E}[\hat{f}(x) - f(x)]$$

where,

- $f(x)$ is the true function we aim to approximate.
- $\hat{f}(x)$ is our model's prediction.
- E is the expected value.

### *Variance*

While bias is related to errors resulting from incorrect assumptions, variance is associated with errors introduced by the model's sensitivity to minor variations in the training set. A high variance suggests that the algorithm shapes itself too closely to the training data, obtaining the noise along with the underlying pattern. When this happens, while our model might perform very well on the training data, it will have a harder time generalizing to new, unseen data.

In a more technical sense, variance captures how much the predictions $\hat{f}(x)$ for a given point $x$ would vary between different training sets.

Variance can be defined as:

$$Var(\hat{f}(x)) = \mathrm{E}\,[(\hat{f}(x) - \mathrm{E}[\hat{f}(x)])^2].$$

While examining bias and variance individually, the relationship between these two becomes a foundation in machine learning. Achieving a balance between them is influential in creating models that reflect both the training data and new, unseen data.

### *Bias and Variance Trade-off*

As we mentioned previously, *capacity* refers to the model's capability to fit a wide variety of functions. It can be thought of as the flexibility of a model. This is a crucial concept in machine learning that is related to capacity. High-capacity models can have low bias but high variance, meaning they fit the training data very closely but can vary significantly on different training sets. Low-capacity models might have high bias but low variance, meaning they might consistently miss the mark but are more stable across different training sets.

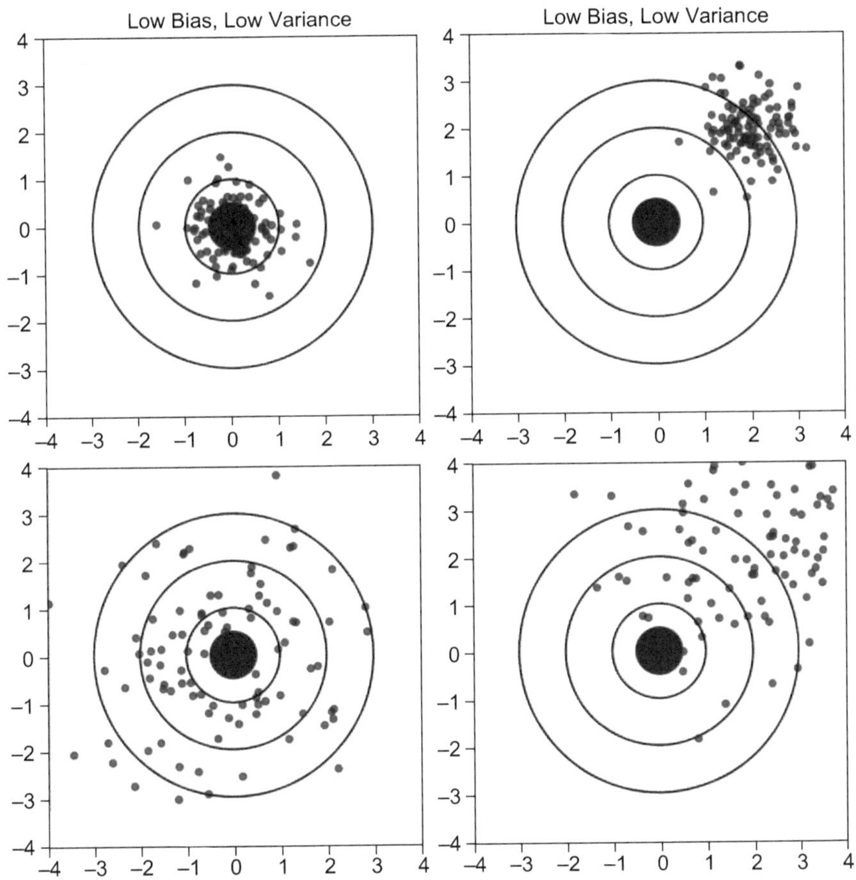


**Figure 2.2: Illustration of the bias-variance trade-off using a dartboard analogy. The position and spread of the darts in each plot represent different combinations of bias and variance.**

Figure 2.2 is a classic way to represent the bias-variance trade-off in ML.

- **Low Bias, Low Variance:** The dots are closely clustered around the center.
  - The model's predictions are largely accurate and consistent across different datasets. It means the model is well-calibrated to predict the correct outcomes and does so consistently.
  - This is an ideal scenario for a model, indicating that it generalizes well to new data and captures the underlying patterns correctly without being overly sensitive to small variations.
- **High Bias, Low Variance:** The dots are closely clustered but miss the center.
  - The model's predictions consistently miss the mark, but they do so in a predictable manner across different datasets. It implies the model may be too simple, missing important patterns in the data.
  - Such a model is underfitting. While it is consistently off-target, its predictions are stable. However, its simplicity means it is not capturing the true patterns in the data, leading to systematic errors.
- **Low Bias, High Variance:** The dots are spread out but are aligned around the center.
  - While the model gets predictions right on average, those predictions can vary widely depending on the specific dataset it is trained on. It is sensitive to small changes or noise in the data.
  - Such a model is overfitting to its training data. It captures the underlying patterns and fits noise or random fluctuations in the training set. As a result, its performance can fluctuate significantly on different datasets.
- **High Bias, High Variance:** The dots are spread out and miss the center.
  - The model's predictions are off-target on average and can be wildly different depending on the specific dataset it is trained on.
  - This is the least desirable scenario. The model neither captures the underlying patterns of the data well (due to high bias) nor produces stable predictions across different datasets (due to high variance). The model's poor calibration and inconsistency make it unreliable.

Figure 2.3 explains the fundamental trade-off between bias and variance as we adjust the complexity of a machine learning model. As the model's complexity increases:

- Bias (shown in —) decreases, indicating that the model becomes more adaptable and starts fitting the training data more accurately.
- Variance (shown in – –) increases, suggesting that the model becomes more sensitive to minor changes and noise in the training data, potentially capturing patterns that don't generalize well to new, unseen data.

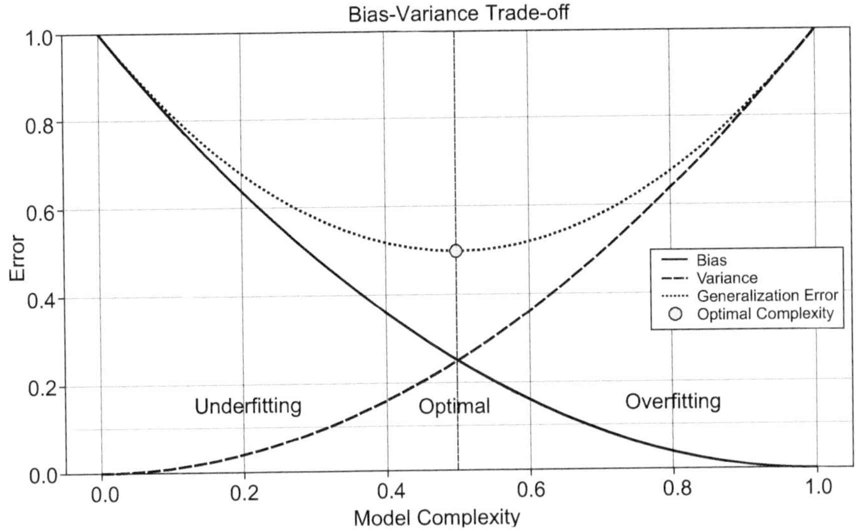


**Figure 2.3: The Bias-variance trade-off.**

The point of "Optimal Complexity" is where the generalization error (highlighted in ...) is the minima, finding a balance between bias and variance. To the left of this point, the model tends to underfit. To the right, the model overfits, implying it is too complex, that even fits the noise in the training data.

## Regularization: Keeping Complexity in Check

In machine learning, the challenge of overfitting appears frequently (Bishop, 2006). A complex model might fit the training data almost perfectly but perform badly on new data. This is where regularization comes into the picture.

### *What is Regularization?*

Regularization is a method employed to deter overfitting by incorporating a penalty into the model's loss function during training. The objective is to restrict too complex models which have the tendency to overfit the dataset by introducing a term to the loss function that increases with complexity. By doing so, regularization attempts to ensure that the model is as simple as possible, while still fitting the data reasonably well.

**A loss function measures how well our model's predictions are in line with the true data. In machine learning, we often aim to minimize this loss.**

Types of regularization techniques used in linear regression:

- **L1 Regularization (Lasso Regression) (Friedman et al., 2010):** L1 regularization includes a penalty equivalent to the absolute value of coefficients' magnitude. This can lead to some coefficients becoming exactly zero, effectively selecting a simpler model that does not include those features.
- **L2 Regularization (Ridge Regression) (Friedman et al., 2010):** L2 regularization introduces a penalty proportional to the square of the coefficients' magnitude. It tends to drive the coefficients to small, non-zero values.
- **Elastic Net (Zou & Hastie, 2005):** Combines both L1 and L2 regularization.

A critical point to remember is that regularization can be highly effective, however, it is not a universal solution. The choice of the regularization strength, type, and even whether to use it at all, often requires careful *tuning* and *validation.*

While we have looked at regularization in the context of traditional machine learning, it is worth noting that this concept is also fundamental in deep learning. Techniques such as *weight decay*, *dropout*, and *batch normalization*, to name a few, act as regularization mechanisms for neural networks. We about discuss that in Chapter 9 on deep learning. For now, it is essential to understand the foundational concept and its role in creating stable and generalized models, reducing the risks of overfitting.

**The terms *error* and *loss* in ML have distinct meanings.**
**Error:**

- **Refers to the difference between the predicted output of the model and the actual output for a specific data point.**
- **For example, in regression problems, the error for a single data point can be defined as: *error* $= y_i - \hat{y}_i$, where $y_i$ is the actual output, and $\hat{y}_i$ is the predicted output of the $i$th example.**
- **It is a measurement of 'how off' our prediction is for a single instance.**

**Loss:**

- **Refers to a scalar quantity used to evaluate the extent to which the model's predictions align with the true labels, typically averaged over all instances.**
- **The loss function calculates the error to adjust the model's weights during training. The goal during training is to minimize this loss.**
- **The loss gives a general view of the model's performance across all data points.**

## *Other Machine Learning Challenges*

So far, in our introduction to machine learning, we have looked at several fundamental challenges, such as *generalization*, *model complexity*, the *bias-variance trade-off*, and issues of *overfitting* and *underfitting*. As we continue, it becomes evident that the subject of machine learning is extensive and contains more aspects. For clarity, we will further divide these challenges into two primary categories: those occurring from machine learning algorithms, and those originating from the data.

### Challenges Related to the Machine Learning Algorithm

The algorithms, regardless of their ability, come with their set of challenges that practitioners need to be aware of.

#### *Model Assumptions*

- Every algorithm comes with its own set of assumptions about data. If the data does not adhere to these assumptions, the model performance can be compromised.
- **Solution:** Understand the assumptions of each algorithm and preprocess the data to fit these assumptions, or choose an algorithm better aligned with the data characteristics.
- **Note:** For instance, linear regression assumes a linear relationship between the input and output variables. If this is not met, its predictions can be off mark.

Imagine a scenario where we are trying to predict the growth of plants based on the amount of water they receive. Intuitively, one might think that the more water a plant gets, the more it grows. But after a certain threshold, too much water can drown the roots, causing the growth to decline. This relationship is not linear but rather *curvilinear*[1].

If we try to fit a linear regression model to this data, it might perform poorly because it will attempt to fit a straight line to a curve. The assumptions of linearity are violated in this scenario. Figure 2.4 presents a plot where the true data follows a curvilinear trend, and the linear regression model is trying to fit a straight line to it. The linear model clearly is not capturing the degrees of the true relationship between the amount of water and plant growth.

#### *Hyperparameter Tuning*

- Algorithms have hyperparameters that are not learned from the training process but influence performance.

[1] Curvilinear: consisting of a curved line

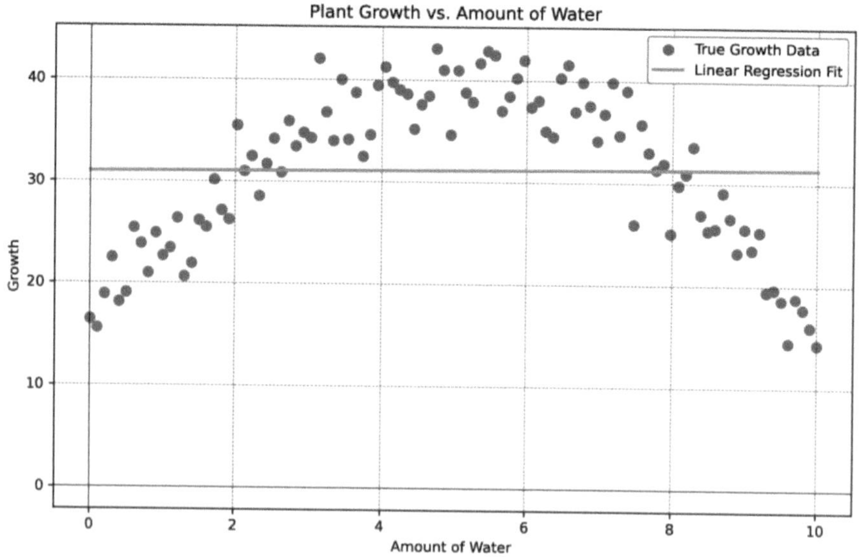


**Figure 2.4 Example of wrong model assumptions.**

- **Solution:** Techniques like grid search, random search, and Bayesian optimization can aid in tuning hyperparameters.
- **Note:** Hyperparameters could be as simple as learning rates or more complex structures in algorithms.

### *Scalability*

- As data grows, algorithms may face difficulties in processing them within reasonable time frames.
- **Solution:** Distributed computing, data sampling, dimensionality reduction, online learning, and cloud-based ML platforms can help tackle the scalability challenges.
- **Note:** Choosing the optimal method or a combination of methods will depend upon the specific objectives of the study and the nature of the data collected.

Tracking animal behavior on a large scale can provide invaluable insights in wildlife research and conservation efforts (Rast et al., 2020). With the introduction of Internet of Things (IoT), devices and camera traps, it has become feasible to collect large amounts of data on animal movements, behaviors, and interactions (Marjani et al., 2017; Tran et al., 2021).

Imagine a nature reserve that has deployed thousands of camera traps and tracking devices across extensive environments, aiming to monitor the behavior of a particular endangered species. Over a year, these devices capture millions of hours of footage and track data, generating terabytes of information. The reserve intends to use machine learning to analyze these patterns, understand migration habits, detect anomalies, and predict potential threats. However, given the volume of data, traditional machine learning models would take excessively long to process and analyze the footage, making timely interventions nearly impossible.

This scenario highlights the scalability challenge in the scope of animal behavior monitoring via machine learning. As the data expands, the computational power, memory, and time needed to train models can increase rapidly, posing practical challenges in achieving timely insights.

### *Interpretability*

- Complex models, especially deep learning ones, can be hard to interpret, leading to a lack of trust or understanding.

As the algorithms become more complex, they often become analogous to a *black box*, where the inner workings and decision-making processes are not directly transparent to the users. While a highly complex model may offer impressive predictive accuracy, understanding why it makes certain decisions can be challenging. This lack of transparency, termed as the issue of *interpretability*, is significant in many applications.

Interpretability is important for the user or stakeholder to trust the model especially in critical applications such as healthcare or finance. They must understand how the model makes decisions in order to have confidence in its predictions. Additionally, if the model is making inaccurate decisions, understanding the decision-making process can help in diagnosing the underlying issues.

In certain sectors, there may be legal requirements for decisions made by algorithms to be explainable. Beyond the legalities, ensuring that the algorithms are not maintaining biases or making unreasonable decisions is an ethical constraint.

Consider a machine learning model designed to predict the health status of cattle based on various behavioral and physiological indicators. If the model determines a particular cow is likely to be unhealthy, the farmer or vet will want to understand the reasons behind this prediction before taking action. Simply knowing the model's accuracy rate is not enough; understanding which specific behaviors or indicators led to the prediction can guide more effective interventions and treatments.

### *Potential Solutions*

- **Model-specific Tools:** For models like decision trees or random forests, *feature importance* scores can show which features most heavily influence predictions.
- **Model-agnostic Tools:** Techniques like LIME (Local Interpretable Model-agnostic Explanations) (Ribeiro et al., 2016) or SHAP (SHapley Additive

exPlanations) (Lundberg & Lee, 2017) can be used to understand predictions from models by approximating them locally with simpler, interpretable models.

- **Simpler Models for Critical Decisions:** In situations where understanding decisions is important, choosing simpler models might be practical.
- **Visualization Techniques:** Visual tools can help illustrate how individual features impact model predictions.

In conclusion, while the attraction of highly accurate predictive models is undeniable, ensuring they remain interpretable is crucial for trust, utility, and ethical considerations. As machine learning continues to integrate into various sectors, the demand for transparent and understandable models will grow.

### *Sensitivity to Initialization*

- Some algorithms, especially iterative ones, can be sensitive to initial parameter values, affecting their convergence or final model quality.

This phenomenon is observed mostly in iterative optimization algorithms, particularly in the context of training deep neural networks. It concerns the influence of the initial values (starting points) of parameters on the outcome of the optimization process.

In many machine learning algorithms, especially in neural networks, the choice of initial weights can significantly influence the training process. The *initial weights* are the starting values of these parameters before training begins. When the optimization landscape consists of multiple local *minima*, *saddles*, or other complex structures, different initializations can lead the optimization process towards different *local optima*. The *optimization landscape* refers to a conceptual visualization of how the possible solutions to an optimization problem are distributed with respect to their performance or error. Imagine a surface where each point represents a possible set of parameters (weights) of the model, and the height represents the error or cost associated with those parameters. The landscape might contain valleys, peaks, and plateaus, corresponding to low-error (optimal) and high-error regions.

> ***Optimization Landscape:*** **This term refers to a conceptual visualization of how the possible solutions to an optimization problem are distributed with respect to their performance or error.**
>
> ***Local Minima:*** **In the optimization landscape, a local minimum is a point where the model's parameters yield a lower error than the neighboring points, but there might be other points with even lower error values elsewhere in the landscape. The presence of multiple local minima can make it challenging for optimization algorithms to find the** ***global minimum,*** **which represents the best possible solution.**

> *Local Optima:* **Similar to local minima, local optima refers to points in the optimization landscape where the solution is optimal relative to nearby points but not necessarily the best overall solution. Local optima can be *minima* or *maxima*, depending on whether the goal is to minimize or maximize the objective function.**
>
> *Saddles:* **A saddle point in the context of an optimization landscape is a point that acts as a minimum along one dimension but a maximum along another. These points can complicate the optimization process, as gradient-based methods might slow down or get stuck when encountering them.**

Why initialization is important?

- **Convergence to Poor Minima:** Poor initialization can lead the optimization process to converge to a suboptimal local minimum, which could lead to a poorly performing model.
- **Slow Convergence:** Starting far away from any minimum (good or bad) might result in a very slow convergence[2], wasting computational resources and time.
- **Vanishing and Exploding Gradients:** Especially in deep networks, certain initializations can worsen the issues of vanishing or exploding gradients, which makes training extremely challenging or even impossible.

Some strategies to address the sensitivity to initialization include heuristic initializations, pre-training, regularization, and batch normalization. While sensitivity to initialization might seem like a minor detail, it plays an important role in the effective training of machine learning models. Proper initialization is often the difference between a model converging quickly and effectively and one that fails to learn at all.

## Data-Related Challenges in Machine Learning

Having talked about challenges associated with machine learning algorithms, we now shift to an equally significant part: *the data*. Even the most sophisticated algorithm will fail if the data at hand is filled with issues. While we previously highlighted challenges that arise from the workings of algorithms, it is equally essential to address challenges that originate from data, as these can block the ability of our models. Data presents its own unique set of challenges. These challenges, if not addressed, can compromise the model's performance, regardless of the quality of the algorithm.

2 convergence: A machine learning model converges when its training loss stabilizes, indicating that additional training will not improve its performance.

## Some of the Most Common Data-related Challenges

### *Insufficient Data*

For many real-world problems, gathering large enough datasets for training can be a challenge and sometimes is one of the most time-consuming steps of a ML project. The scarcity of data can limit the potential of the model, making it underperform.

The issue of insufficient data is manifold and brings forth several complications:

- **Poor Generalization:** One of the primary concerns with limited data is that the trained model may not generalize well to new instances. With fewer data points, the model may fail to capture the broader patterns in the data.
- **Overfitting/Underfitting:** With limited data, models, especially complex ones, are prone to overfitting. As a result, the model becomes less flexible in accommodating new data patterns. On the other hand, if the model fails to identify the relationships within the dataset, it can lead to underfitting.
- **Limited Model Choices:** Not all machine learning algorithms are well-suited to scenarios with sparse data. Complex models like deep neural networks require substantial amounts of data to train effectively without overfitting. On the other hand, a simpler model like decision trees might do better with smaller datasets, but they may not capture complex relationships as effectively as others.
- **Challenges in Model Evaluation:** With fewer data points, splitting the data into training, validation, and test sets becomes problematic. A small test set might not be representative of the broader data distribution, leading to inaccurate estimations of a model's performance.

However, there are several strategies to tackle the challenges related to insufficient data:

- **Data Augmentation:** Techniques such as random rotations, flips, or even synthetic data generation using tools like Generative Adversarial Networks (GANs) (Goodfellow et al., 2014) can help in artificially expanding the dataset.
- **Transfer Learning** (Pan & Yang, 2010)**:** Instead of training a model from scratch, one can use a pre-trained model (typically on a larger dataset) and fine-tune it for the specific task at hand. This approach allows the model to leverage the knowledge gained from the larger dataset.
- **Ensemble Methods:** Techniques like bootstrapping (C.D. Sutton, 2005) can be used to create multiple datasets from the available data, and then different models can be trained on each of these datasets. The final output can be a combination (e.g., average or majority voting) of outputs from these models, leading to more robust predictions.

While insufficient data causes genuine challenges in machine learning, with the right techniques and considerations, one can still develop models that are effective and reliable.

### *Noisy Data*

In the real-world scenario of machine learning, especially when focusing on specific domains like animal behavior, the data is often far from perfect. This imperfection frequently manifests as noisy data, a type of data corrupted by irrelevant or false information. Data that seems innocuous can significantly harm the performance and reliability of ML models.

Various factors can lead to the generation of noisy data. Common causes include errors during the data acquisition phase, such as malfunctioning sensors in animal tracking devices. Outliers, which are data points that exhibit considerable variation from the rest of the data, often add another layer of noise.

The presence of noisy data in a dataset can misdirect a machine learning model in several ways:

- **Distorted Patterns:** A model trained on noisy data might misinterpret the noise as valid patterns, leading it away from understanding the actual underlying relationships in the data.
- **Compromised Model Performance:** As models try to fit noisy data, they might end up aligning too closely to the dataset's imperfections, making them less effective on new data.
- **Increased Model Complexity:** To accommodate noise, models might unnecessarily become more complex, making them harder to interpret.

Addressing the challenges inquired by noisy data requires a mix of strategies and methodologies:

- **Data Cleaning:** This primary step involves inspecting the dataset for errors and correcting them. It might involve the removal of duplicates, addressing outliers, or relabeling mislabeled data points.
- **Noise-Resilient Algorithms:** Certain algorithms, like Random Forests, have inherent mechanism s that allow them to handle noise better than others.
- **Domain Expertise:** Particularly in niche domains, expertise can differentiate between valid data points and noise. For instance, in animal behavior studies, an understanding of the conditions under which data was collected can filter out anomalies.

Fundamentally, while noisy data remains a persistent challenge in the world of machine learning, a combination of thorough data preprocessing and strategic modeling can reduce its bad effects.

### *Imbalanced Datasets*

In machine learning, we frequently encounter issues with imbalanced datasets, where certain classes of data are significantly more common than others. For example, in a dataset comprising animal behaviors, the examples of common behaviors like grazing or walking might outnumber rare behaviors like fighting or mating.

The challenges presented by imbalanced datasets can influence the effectiveness of a ML model in the following manner:

- **Biased Model Predictions:** A model trained on imbalanced data may exhibit a strong bias towards the majority class. As a result, the model might frequently misclassify instances from the minority class, simply because it has not encountered them often enough during training.
- **Compromised Model Performance Metrics:** Traditional metrics, such as accuracy, can be misleading for imbalanced datasets. A model could achieve high accuracy by predicting the majority class, even if it consistently misclassifies the minority class.
- **Overlooking Significant Insights:** In many situations, especially in animal behavior studies, the minority class (like certain rare behaviors) might carry significant importance. Imbalanced datasets can lead models to neglect these critical insights.

Here are some methods and strategies that can be employed to tackle those challenges:

- **Resampling Techniques:** This involves either oversampling the minority class, undersampling the majority class, or a combination of both, to balance out the class distribution. Techniques such as the Synthetic Minority Over-sampling Technique (SMOTE) (Chawla et al., 2002) can be used to generate artificial instances of the minority class.
- **Cost-sensitive Training:** By assigning different misclassification costs to different classes, a model can be guided to treat each misclassification instance based on its associated cost.
- **Ensemble Methods:** Techniques like bagging and boosting (C.D. Sutton, 2005), can often lead to better performance on imbalanced datasets.

It is important to note that while addressing dataset imbalance can improve model performance, it is not always necessary to achieve a perfect balance. In some real-world scenarios, certain classes are naturally less frequent, and achieving a perfect balance might not be reflective of the true data distribution.

### *Missing Values*

An issue that consistently emerges across various domains, including animal behavior studies, is the presence of missing values in datasets. Whether due to sensor malfunctions, human errors during data entry, or any other reasons, gaps in data are a common occurrence.

Missing values in the dataset presents distinct challenges, especially when it comes to ensuring the reliability and accuracy of machine learning models:

- **Compromised Model Integrity:** A model trained on data with significant missing values might not capture the underlying relationships effectively, leading to inaccurate performance.
- **Reduction in Dataset Size:** Simply removing rows or columns with missing values can significantly reduce the available dataset size, limiting the amount of information available for training the model.

As with other challenges, there are numerous techniques available to effectively handle missing values, ensuring the resultant models remain robust:

- **Imputation Techniques** (Donders et al., 2006)**:** Depending on the nature and structure of the data, various imputation methods, such as mean, median, mode imputation, or techniques like K-nearest Neighbors imputation, can be applied.
- **Multiple Imputations** (Li et al., 2015; Schafer & Olsen, 1998)**:** Instead of filling missing values once, multiple imputations involve creating several datasets with different imputations and averaging the results, offering a more robust solution to the missing data issue.
- **Utilizing Model Algorithms that Handle Missing Values:** Some algorithms, such as *decision trees* or *random forests*, can handle missing values inherently, making them a good choice for datasets with such gaps.
- **Using Domain Knowledge:** Leveraging domain-specific knowledge can assist in making educated guesses about missing values.

Handling missing values is crucial, as the chosen method can influence the machine learning model's outcomes and interpretations. As we progress through this book, we will investigate some of these strategies, offering hands-on techniques and considerations for dealing with missing data effectively.

### *Non-representative Data*

For the purpose of developing reliable and efficient ML models, ensuring that the training set represents the broader population is of paramount importance. Non-representative data arises when the dataset used to train a model does not accurately reflect the reality or does not capture the full spectrum of variations present in the real-world scenario.

Relying on non-representative data can cause several issues:

- **Skewed Predictions:** If a model is trained on data that is not representative of the larger population or context, its predictions can be biased towards the data it has seen, resulting in inaccurate and unreliable outcomes.

- **Loss of Model Credibility:** Stakeholders or end-users might lose confidence in a model if it consistently provides outputs that are not reflected with real-world observations or expectations.
- **Inaccurate Decision Making:** In applications where machine learning models drive decision-making processes, non-representative data can lead to decisions that might not be optimal or could even be damaging.
- **Challenges in Model Validation:** Validating the performance of a model becomes complex if the data is not complete. A model might obtain good performance on the training data but is unsuccessful when exposed to new, real-world data.

Despite these challenges, there are systematic approaches to handle the issues presented by non-representative data:

- **Stratified Sampling** (Iliyasu & Etikan, 2021)**:** This method involves dividing the broader population into homogenous subgroups and ensuring that samples from each subgroup are included proportionally in the training data. By doing so, one can create a more representative dataset.
- **Domain Expertise:** Engaging with domain experts, especially in specialized fields like animal behavior, can provide insights into whether the data in hand is reflective of the broader scenario.
- **Continuous Model Updating:** Regularly updating the model as new data becomes available can help ensure that it remains aligned with the evolving real-world scenario.
- **Diversity in Data Sources:** Combining data from multiple sources or diverse environments can enhance the robustness and representativeness of the dataset.

These are just a few of the many challenges that data can present. While addressing these challenges, it is important to remember that there is no one-size-fits-all solution in machine learning. Each problem requires its own approach, based on the characteristics of the data and the specific task involved.

## *Understanding the ML Workflow*

Having explored the various challenges associated with machine learning, both related to algorithms and data, we turn our attention to the structured approach that guides practitioners through machine learning projects. This sequence of steps, termed the machine learning workflow, is essential in ensuring the successful execution of a project.

Understanding the machine learning workflow is crucial for several reasons:

- **Systematic Approach:** It ensures that practitioners do not miss important steps and provides a road map that can be revisited and refined as needed.

- **Efficiency:** A well-defined workflow allows for resource optimization, ensuring that time and computational resources are used effectively.
- **Reproducibility:** Following a structured approach aids in the documentation process, allowing others to replicate and understand the methodology used.

To accomplish this objective, a systematic workflow is employed, facilitating the transformation of raw data into meaningful predictions.

The machine learning workflow is composed of several stages:

- **Problem Definition:** Start by clearly identifying the problem. This means understanding the context, setting clear objectives, and defining any constraints.
- **Data Collection:** This is the phase where relevant data is gathered. The quality and quantity of this data can directly influence the performance of the model. Depending on the problem domain, this could range from collecting images for a computer vision task to gathering sensor-based data from animals for signal processing. Ethical considerations and approval are requirements for collecting any data.
- **Data Preprocessing:** Once the data is collected, it often requires cleaning and transformation. This could involve handling missing values, removing outliers, or encoding categorical values, ensuring that the subsequent model training phase is effective and efficient.
- **Exploratory Data Analysis:** Before moving into modeling, it is vital to explore and understand the nature of your data. Exploratory data analysis involves visualizing data, understanding its distribution, and identifying potential trends, patterns, and relationships.
- **Feature Engineering:** Here, we either modify existing features or create new ones to better represent the underlying patterns in the data. Well-crafted features can substantially improve model performance.
- **Model Selection and Training:** With preprocessed data ready, we iteratively train various ML models and select the most suitable algorithm. This phase involves feeding our data to the algorithm and optimizing its parameters (*hyperparameter tuning*) to achieve the best results.
- **Model Evaluation:** After training, it is vital to evaluate how well the model is performing on the new data using various quality measures.
- **Deployment:** If a model performs well, it can be deployed in a real-world environment. This means integrating it into a production system where it can take in new data and make predictions. The goal is often to implement the trained model in a real-world scenario, whether it is a web application, a mobile app, or any other platform.

- **Monitoring and Maintenance:** Post-deployment, models need regular monitoring to ensure their performance does not degrade over time. If a model's accuracy drops or if it starts making incorrect predictions, it might require retraining.

## Summary

In Chapter 2, we have covered the essentials of machine learning, focusing on key concepts like how models generalize from data, the balance between complexity and accuracy, and the common issues of overfitting and underfitting. We have also highlighted the importance of evaluating model performance to ensure it matches expectations closely. The challenges discussed fall into two categories: those related to model functionality and those related to data quality and availability. We examined the impact of the size of datasets, and problems like noisy, incomplete, or biased data. Additionally, we walked through the steps involved in ML workflow, from defining the problem to deploying the solution.

CHAPTER 3

# A Practical Example to Building a Simple Machine Learning Model

In this chapter, we will build a machine learning model from scratch. We will guide you through each step of the process, providing practical examples and hands-on experience. Upon completing this chapter, you will have a comprehensive understanding of creating and evaluating machine learning models utilizing Python.

### *Note to the Reader*

Introduction to Python, how to install Jupyter notebook and Python on your machine, and Python basics is not present in this chapter. To ensure that both novice and experienced readers can follow along without disruption, we have compiled a comprehensive guide covering the Python basics, including installation instructions for Python, Anaconda, and Jupyter Notebook.

This guide is designed to support readers new to Python, providing a solid foundation in the programming language that underpins the machine learning techniques discussed throughout the book.

We have made this Introduction to Python and Python basics guide available as a free downloadable PDF hosted on GitHub. You will also find Python examples and code snippets that will be used throughout the book. This resource allows us to maintain the clarity and coherence of the machine learning pipeline discussion in the book, while still offering detailed support for those who are new to Python.

To access the Python basics guide and examples, please visit the following GitHub repository: https://github.com/nkcAna/WSDpython. We encourage readers to download this material and refer to it as needed to optimize their understanding and application of the machine learning concepts covered in this book.

## Data Collection and Preprocessing

Before we start reading and investigating the dataset, in this section we introduce the dataset utilized for our farm animal behavior recognition project. It is essential to acknowledge that the datasets employed in this work are made available with the explicit requirement of citing the following academic paper:

**Title:** Generic Online Animal Activity Recognition on Collar Tags

**Authors:** Jacob W. Kamminga, Helena C. Bisby, Duv V. Le, Nirvana Meratnia, and Paul J.M. Havinga

**Publication:** Proceedings of the 2017 ACM International Joint Conference on Pervasive and Ubiquitous Computing (UbiComp/ISWC'17)

**Publication Date:** September 2017

**Publisher:** ACM

The data collection procedure and characteristics of this dataset are detailed in the paper (Kamminga et al., 2017). Our work builds upon the insights and contributions of the authors.

### *Short Description of the Dataset*

The dataset is focused on the behavior of four goats and two sheep, using multi-sensor data streams. Sensors, including accelerometers, gyroscopes, and others, were attached to collars around the animals' necks to monitor activities such as lying, standing, grazing, and running, among others (refer to Table 3.1 for all included activities). Data collection involved synchronized sensors sampling at 200 samples per second. The data was labeled manually using a MATLAB GUI application, ensuring accurate representation of various animal behaviors.

**Table 3.1: Activities of the animals.**

| | |
|---|---|
| Lying | The animal is resting on the ground. |
| Standing | The animal is stationary, occasionally moving its head or taking slow steps. |
| Grazing | The animal is feeding on fresh grass, hay, or twigs. |
| Fighting | The animal is engaging in aggressive interactions with another animal, often involving head or body contact. |
| Shaking | The animal performs rapid, whole-body shaking, sometimes accompanied by head shaking. |
| Scratch-biting | The animal nibbles on its own skin using its teeth, occasionally using its hoofs. |
| Walking | The animal is moving at various paces, from slow walking to nearly trotting. |
| Trotting | A phase between walking and running, characterized by energetic walking. |
| Running | The animal is galloping/running. |

### *Data File Formats*

The dataset is provided in multiple file formats for each animal:

- **Matlab Variable Files (.mat):** These files contain sensor data for each animal from the neck position in random orientations. Each animal is represented by a unique .mat file.
  Example: S1.mat, S2.mat, G1.mat, G2.mat, G3.mat, G4.mat
- **CSV Files (.csv):** The sensor data is also available in CSV file format for each animal, capturing the same neck position with random orientations.
  Example: S1.csv, S2.csv, G1.csv, G2.csv, G3.csv, G4.csv

### *Raw Data Columns Within Files*

Each data file contains the following columns:

- **Label:** Describes the behavior associated with each row's data.
- **Animal ID:** Identifies the animal corresponding to the data. For instance, 'S1' represents Sheep 1, and 'G1' represents Goat 1, and so on.
- **Segment ID:** Activities have been sorted into segments, and data within one segment is continuous. Segments are numbered incrementally for each animal and are not consecutive.
- **Timestamp (ms):** The timestamp column indicates the time of data capture.
- **Accelerometer Raw Data (ax, ay, az):** These columns contain raw data from the accelerometer along the $x$-axis, $y$-axis, and $z$-axis, respectively. The data is sampled at 200 Hz, with a range of ±8 g.
- **High G Accelerometer Raw Data (axhg, ayhg, azhg):** These columns contain raw data from the high-intensity accelerometer along the $x$-axis, $y$-axis, and $z$-axis, respectively. The data is sampled at 200 Hz, with a range of ±100 g.
- **Compass (Magnetometer) Raw Data (cx, cy, cz):** These columns contain raw data from the compass (magnetometer) along the $x$-axis, $y$-axis, and $z$-axis, respectively. The data is sampled at 100 Hz.
- **Gyroscope Raw Data (gx, gy, gz):** These columns contain raw data from the gyroscope along the $x$-axis, $y$-axis, and $z$-axis, respectively. The data is sampled at 200 Hz, with a range of ±2000°/s.
- **Barometer Raw Data (pressure):** This column contains raw data from the barometer, sampled at 25 Hz, with a range of 260 to 1260 hPa.
- **Temperature Raw Data (temp):** This column contains raw temperature data, sampled at 200 Hz, within a range of –40°C to 85°C.

The data files are available in the book's repository in folder *GSData*, ensuring accessibility for further analysis and experimentation. This folder contains only the following files: S1.csv, S2.csv, G1.csv, G2.csv, G3.csv, G4.csv for the purpose of our experiments.

Note that, the readers can download the data from its original source:

(https://lifesciences.datastations.nl/dataset.xhtml?persistentId=doi:10.17026/dans-zp6-fmna).

### *Reading and Preparing the Data Using Python*

The first crucial step in any machine learning project is obtaining access to the dataset. In our case, we have the data readily available in the book's repository. We will begin by reading this data to set the stage for our farm animal behavior recognition project. To work with data files and perform data preprocessing, we need to import essential Python libraries. These libraries will provide the tools necessary for data manipulation and analysis.

Below is a Python function designed to read multiple CSV files from a specified folder and merge them into a single Pandas DataFrame. It also includes error handling to handle potential issues.

```
import pandas as pd
import os
import glob

def read_csv_files_from_folder(folder_path):
    """

    Reads all CSV files from a specified folder and merges them into a
    single DataFrame.

    Parameters:
    folder_path (str): The path to the folder containing CSV files.

    Returns:
    pd.DataFrame: A Pandas DataFrame containing the merged data or
      None
    if no data is found.
    """

    try:
        # Create an empty list to store DataFrames
        dfs = []

        # Use glob to get a list of all CSV files in the folder
        csv_files = glob.glob(os.path.join(folder_path, '*.csv'))

        if not csv_files:
            print("No CSV files found in the specified folder.")
            return None
```

```
        # Read and append each CSV file to the list
        for csv_file in csv_files:
            df = pd.read_csv(csv_file)
            if not df.empty:
                dfs.append(df)

        if not dfs:
            print("No valid data found in the CSV files.")
            return None

        # Concatenate all DataFrames vertically
        merged_df = pd.concat(dfs, ignore_index=True)

        return merged_df

    except FileNotFoundError:
        print("The specified folder or CSV files were not found.")
        return None

    except Exception as e:
        print(f"An error occurred: {str(e)}")
        return None
```

Here is a breakdown of what the code does:

- Importing Libraries: The code begins by importing the necessary Python libraries: pandas for data manipulation, os for working with the file system, and glob for file pattern matching.
- Function Definition: The read_csv_files_from_folder function is defined with a docstring that explains its purpose, parameters, and return value.
- Parameter: The function takes one parameter, folder_path, which is a string representing the path to the folder containing the CSV files that you want to read and merge.
- Try-Except Block: The code is enclosed in a try block with exception handling to manage potential errors gracefully.
- Empty DataFrames List: An empty list named dfs is created to store individual DataFrames from each CSV file.
- Glob for CSV Files: The glob.glob function is used to obtain a list of all CSV files in the specified folder. It searches for files with the .csv extension within the folder.
- Checking for CSV Files: It checks if any CSV files were found in the specified folder. If no CSV files are found, it prints a message and returns None.
- Reading CSV Files: It iterates through the list of CSV files obtained from glob.glob. For each file, it reads the CSV data using pd.read_csv and appends it to the dfs list if the DataFrame is not empty.

- Checking for Valid Data: After reading all CSV files, it checks if any valid data was found. If no valid data is found (all DataFrames are empty), it prints a message and returns None.
- Concatenating DataFrames: If valid data is found, it uses pd.concat to concatenate all the individual DataFrames stored in the dfs list vertically, creating a single merged DataFrame named merged_df. The ignore_index = True argument ensures that the index is reset for the merged DataFrame.
- Returning the Merged DataFrame: Finally, the function returns the merged DataFrame (merged_df) if data is successfully read and merged. If any errors occur during the process, it prints appropriate error messages and returns None.
- This function provides a convenient and robust way to read and merge multiple CSV files from a specified folder into a single DataFrame.

## *Data Inspection and Exploration*

At this stage, we are going to inspect the dataset that we have just read from the CSV files. The purpose is to provide an overview of the data, check for missing values, summarize numerical features, list column names, examine data types, and count the classes (behaviors of the animals).

```
df = read_csv_files_from_folder("GSdata/Data")
df.head()
```

In this code snippet, we are using the previously defined read_csv_files_from_folder function to read and merge the CSV files located in the GSdata directory. The function reads the CSV files, merges them into a Pandas DataFrame, and assigns the resulting DataFrame to the variable df. To look at the top five columns of our dataset, we use the .head() method on df (Figure 3.1). This DataFrame will be used for data inspection and exploration in the following steps.

| | label | animal_ID | segment_ID | timestamp_ms | ax | ay | az | axhg | ayhg | azhg | cx | cy | |
|---|---|---|---|---|---|---|---|---|---|---|---|---|---|
| 0 | walking | G1 | 1 | 1 | 1.57538 | 4.34787 | -9.27514 | 5.61615 | 8.19897 | 10.21120 | -0.5685 | -0.7875 | 2.52 |
| 1 | walking | G1 | 1 | 6 | 1.47962 | 4.30477 | -9.31105 | 4.98546 | 8.64947 | 4.47490 | NaN | NaN | N |
| 2 | walking | G1 | 1 | 11 | 1.36469 | 4.24492 | -9.42118 | 5.19569 | 9.25012 | 5.79634 | -0.5850 | -0.7830 | 2.52 |
| 3 | walking | G1 | 1 | 16 | 1.21386 | 4.22816 | -9.59835 | 4.23463 | 8.82966 | 9.28016 | NaN | NaN | N |
| 4 | walking | G1 | 1 | 21 | 1.07021 | 4.29520 | -9.67257 | 5.13562 | 7.89864 | 5.88644 | -0.5835 | -0.7815 | 2.52 |

**Figure 3.1: Top 5 rows of the dataset.**

There are 18 features (12 are visible in the image Figure 3.1): ‘label’, ‘animal_ID’, ‘segment_ID’, ‘timestamp_ms’, ‘ax’, ‘ay’, ‘az’, ‘axhg’, ‘ayhg’, ‘azhg’, ‘cx’, ‘cy’, ‘cz’, ‘gx’, ‘gy’, ‘gz’, ‘pressure’, ‘temp’.

Below a function is created to inspect the dataset:

```
# Creating a function to inspect the dataset for missing values,
checking the features, data types, and counting the classes

import pandas as pd

def inspect_dataset(dataset):
    """
    Inspects a dataset and prints various statistics and
    information.

    Parameters:
    dataset (pd.DataFrame): The dataset to inspect.

    Returns:
    None
    """
    # Basic information about the dataset
    print("Dataset Information:")
    print(dataset.info())

    # Summary statistics of numerical features
    print("\nSummary Statistics:")
    print(dataset.describe())

    # Check for missing values
    missing_values = dataset.isnull().sum()
    print("\nMissing Values:")
    print(missing_values[missing_values > 0])

    # List of features (columns)
    print("\nFeatures:")
    print(dataset.columns.tolist())

    # Data types of features
    print("\nData Types:")
    print(dataset.dtypes)

    # Count the classes
    if 'label' in dataset.columns:
        class_counts = dataset['label'].value_counts()
        print("\nClass Counts:")
        print(class_counts)

    else:
        print("\nNo target column found for class counting.")
```

Here, we define a Python function named inspect_dataset. This function is designed to inspect and explore a given dataset. It takes one parameter, dataset, which should be a Pandas DataFrame that you want to inspect.

Breakdown of the key aspects of the inspect_dataset function:

- Dataset Information: It prints basic information about the dataset using dataset.info(). This includes the number of non-null entries in each column, data types, and memory usage.
- Summary Statistics: It provides summary statistics of numerical features using dataset.describe(). This includes count, mean, standard deviation, minimum, and maximum values.
- Missing Values: It checks for missing values in the dataset using dataset.isnull().sum(). If any missing values are found, it prints the columns with missing values and their respective counts.
- Features (Columns): It lists all the features (columns) in the dataset using dataset.columns.tolist().
- Data Types: It prints the data types of each feature using dataset.dtypes.
- Class Counts (Optional): Using the column named 'label', it calculates how many times each class occurs and then prints the class counts.

The inspect_dataset function provides a comprehensive overview of the dataset, allowing us to understand its structure and characteristics.

Below in the next step, we execute the inspect_dataset function with our df DataFrame to obtain insights into the farm animal behavior dataset.

```
# Executing the inspect_dataset function with our df DataFrame
# Obtain insights into the farm animal behavior dataset
inspect_dataset(df)
```

Once we call the function inspect_dataset(df), we get the following information on the screen:

```
Dataset Information:
<class 'pandas.core.frame.DataFrame'>
RangeIndex: 13778153 entries, 0 to 13778152
```

**Figure 3.2: Overview of the dataframe structure.**

The information displayed here provides an overview of the dataset.

- **<class 'pandas.core.frame.DataFrame'>:** This line indicates that we are working with a Pandas DataFrame.

- **RangeIndex: 13778153 entries, 0 to 13778152:** This is the index of the DataFrame, essentially the row numbers.
- **13778153 entries:** This is the total number of entries or rows in the DataFrame. It tells us that there are 13,778,153 rows in this dataset.
- **0 to 13778152:** This specifies the range of the index. It starts from 0 and goes up to 13,778,152, indicating that the index spans this range.

This information gives us a basic knowledge of the dataset's dimensions and structure. It tells us that we are dealing with a dataset comprising more than 13 million rows, and each row is assigned a numeric index from 0 to 13,778,152.

```
Data columns (total 18 columns):
 #   Column        Dtype
---  ------        -----
 0   label         object
 1   animal_ID     object
 2   segment_ID    int64
 3   timestamp_ms  int64
 4   ax            float64
 5   ay            float64
 6   az            float64
 7   axhg          float64
 8   ayhg          float64
 9   azhg          float64
 10  cx            float64
 11  cy            float64
 12  cz            float64
 13  gx            float64
 14  gy            float64
 15  gz            float64
 16  pressure      float64
 17  temp          float64
dtypes: float64(14), int64(2), object(2)
memory usage: 1.8+ GB
None
```

**Figure 3.3: Details of the DataFrame columns and data types.**

This section provides information about the columns (features) in the dataset, their names, and their respective data types.

Data columns (total 18 columns): This line indicates that the following information refers to the dataset's columns, and there are a total of 18 columns in the dataset.

This part of the output is a table that describes each column. It consists of three columns:

- #: This is the column index, starting from 0 for the first column and incrementing by 1 for each subsequent column.
- Column: This is the name of the column.
- Dtype: This is the data type of the column.

dtypes: This line provides a summary of the data types present in the dataset. It indicates that there are: 14 columns with the data type float64, 2 columns with the data type int64, 2 columns with the data type object.

Memory usage: This line tells us about the memory usage of the dataset. In this case, it indicates that the dataset consumes approximately 1.8+ gigabytes (GB) of memory.

None: This is the return value of the info() function. It is displayed because the info() function does not return a value; it prints the information to the console.

```
Summary Statistics:
          segment_ID  timestamp_ms            ax            ay            az
count   1.377815e+07  1.377815e+07  1.377815e+07  1.377815e+07  1.377815e+07
mean    3.902753e+02  5.775226e+06  5.384297e-01 -2.309943e-01 -5.648953e+00
std     1.579438e+02  3.373652e+06  5.610789e+00  4.477043e+00  6.209313e+00
min     1.000000e+00  1.000000e+00 -7.845320e+01 -7.845320e+01 -7.845320e+01
25%     2.980000e+02  2.870446e+06 -4.271260e+00 -3.287240e+00 -9.248800e+00
50%     4.020000e+02  5.740896e+06  2.123660e+00 -1.699880e-01 -7.548920e+00
75%     4.780000e+02  8.611346e+06  4.355050e+00  2.928110e+00 -4.244920e+00
max     9.290000e+02  1.288802e+07  7.845080e+01  7.845080e+01  7.845080e+01

                axhg          ayhg          azhg            cx            cy
count   1.377815e+07  1.377815e+07  1.377815e+07  6.889073e+06  6.889073e+06
mean    2.763919e+00  3.512998e+00  7.725999e+00  6.491124e-01 -1.459219e-01
std     6.407745e+00  6.141199e+00  8.072271e+00  1.340528e+00  3.931574e-01
min    -2.029020e+02 -4.889650e+02 -1.459600e+02 -1.737000e+00 -1.032000e+00
25%    -4.084470e+00 -9.610520e-01  1.982170e+00 -5.610000e-01 -5.220000e-01
50%     5.405920e+00  4.264670e+00  5.616150e+00  6.240000e-01 -5.100000e-02
75%     7.448150e+00  7.688410e+00  1.480620e+01  1.980000e+00  1.545000e-01
max     2.096590e+02  7.089860e+02  2.071370e+02  2.352000e+00  6.630000e-01

                  cz            gx            gy            gz      pressure
count   6.889073e+06  1.377815e+07  1.377815e+07  1.377815e+07  1.722231e+06
mean    6.682943e-01 -9.485811e-02  1.061554e-01 -1.455922e-01  1.008268e+03
std     9.386851e-01  6.232186e+01  4.281269e+01  4.229533e+01  1.119303e+00
min    -7.755000e-01 -1.998050e+03 -1.948660e+03 -1.962930e+03  1.005070e+03
25%    -1.560000e-01 -4.451220e+00 -3.719510e+00 -3.902440e+00  1.008030e+03
50%     5.145000e-01  0.000000e+00  0.000000e+00  0.000000e+00  1.008690e+03
75%     1.093500e+00  4.207320e+00  3.780490e+00  4.146340e+00  1.009020e+03
max     2.778000e+00  1.997990e+03  1.778290e+03  1.993050e+03  1.009730e+03

                temp
count   1.377815e+07
mean    3.558117e+01
std     5.898740e+00
min     2.547000e+01
25%     3.098000e+01
50%     3.400000e+01
75%     4.075000e+01
max     4.962000e+01
```

**Figure 3.4: Summary statistics output.**

The above output provides summary statistics for the numerical columns in the dataset.

1. Summary Statistics: This line indicates that the following information pertains to the summary statistics of the dataset.

2. Count: The count row under each column shows how many non-null values exist for each numerical feature. This tells us how many data points are available for each feature.
3. Mean: The mean row displays the mean value for each feature.
4. Std: The std row represents the standard deviation, which measures the spread or variability of the data points around the mean.
5. Min: The "min" row shows the minimum value observed for each feature.
6. 25%: This row corresponds to the 25th percentile, indicating the value below which 25% of the data falls.
7. 50%: The 50th percentile, also known as the median, represents the middle value of the data.
8. 75%: The 75th percentile indicates the value below which 75% of the data falls.
9. Max: The "max" row shows the maximum value observed for each feature.

The last section of the function's output is the following:

```
Missing Values:
cx             6889080
cy             6889080
cz             6889080
pressure      12055922
dtype: int64

Features:
['label', 'animal_ID', 'segment_ID', 'timestamp_ms', 'ax', 'ay', 'az', 'axhg', 'ayhg', 'azhg', 'cx',
'cy', 'cz', 'gx', 'gy', 'gz', 'pressure', 'temp']

Data Types:
label            object
animal_ID        object
segment_ID        int64
timestamp_ms      int64
ax              float64
ay              float64
az              float64
axhg            float64
ayhg            float64
azhg            float64
cx              float64
cy              float64
cz              float64
gx              float64
gy              float64
gz              float64
pressure        float64
temp            float64
dtype: object

Class Counts:
standing          6031069
grazing           3735835
walking           2109794
lying              984396
trotting           408051
running            324775
scratch_biting      92715
fighting            75209
shaking             16309
Name: label, dtype: int64
```

**Figure 3.5: Missing values, features, data types, and class counts information.**

The above output provides information related to the missing values of the dataset, feature names, data types for each feature, and the count of each class in the dataset.

- Missing Values: It provides the names of the columns that have missing values. We have identified missing values in several columns, including cx, cy, cz, and pressure.
- Class Counts: It shows how many instances belong to each behavior class.

## Visualizing the Dataset with Histograms

### *Interpretation of Histograms*

- Feature Distribution: Histograms provide a visual representation of the distribution of numerical features. The *x*-axis represents the range of values, and the *y*-axis represents the frequency or count of data points falling within each range.
- Normality: The shape of the histogram can give insights into the normality of the feature's distribution. A bell-shaped curve indicates a roughly normal distribution, while skewed distributions may show a tail on one side (refer to Figure 3.6).
- Spread and Range: The width of the histogram (spread) indicates the range of values the feature can take. A wider spread means a broader range of values.
- Peaks and Modes: Multiple peaks in a histogram may suggest the presence of multiple modes (clusters) in the data. Each mode corresponds to a group of data points with similar values.
- Outliers: Histograms can help identify outliers, which are data points significantly distant from the bulk of the data. Outliers appear as isolated bars far from the main distribution.
- Data Concentration: The height of the bars indicates the concentration of data points within specific value ranges. Higher bars represent higher data density in those ranges.

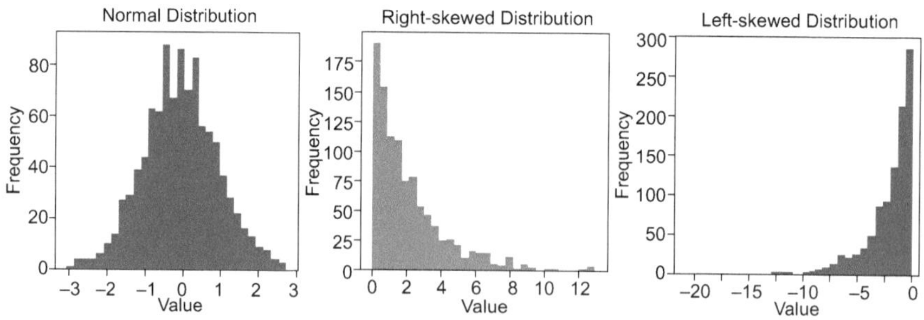


**Figure 3.6: Histograms of different distributions.**

We will now visualize the dataset using histograms.

Below is the code for visualization:

```
import matplotlib.pyplot as plt
%matplotlib inline

# Select columns to create histograms
selected_columns = ['ax', 'ay', 'az', 'axhg', 'ayhg', 'azhg', 'cx',
'cy', 'cz', 'gx', 'gy', 'gz', 'pressure', 'temp']

df[selected_columns].hist(bins=50, figsize=(30, 20))
plt.show()
```

***Code Breakdown***

- import matplotlib.pyplot as plt: This line imports the pyplot module from the matplotlib library.
- selected_columns: This is a list holding the column names from the dataset that you want to create histograms for. These columns represent numerical features from the dataset, such as sensor readings and measurements.
- df[selected_columns].hist(bins = 50, figsize = (30, 20)):
  - df[selected_columns]: This part selects the subset of the DataFrame df that includes only the columns specified in selected_columns.
  - .hist(bins = 50, figsize = (30, 20)): This part applies the .hist() method to the selected subset of the DataFrame.
  - bins = 50: This parameter specifies the number of bins or intervals into which the data will be divided for creating the histogram. In this case, it is set to 50, meaning that the histogram will have 50 bars.
  - figsize = (30, 20): This parameter sets the size of the figure (the plot) in inches. It determines the dimensions of the histogram plot.
- plt.show(): This line is used to display the histogram plot.

### *The Output of df.hist()*

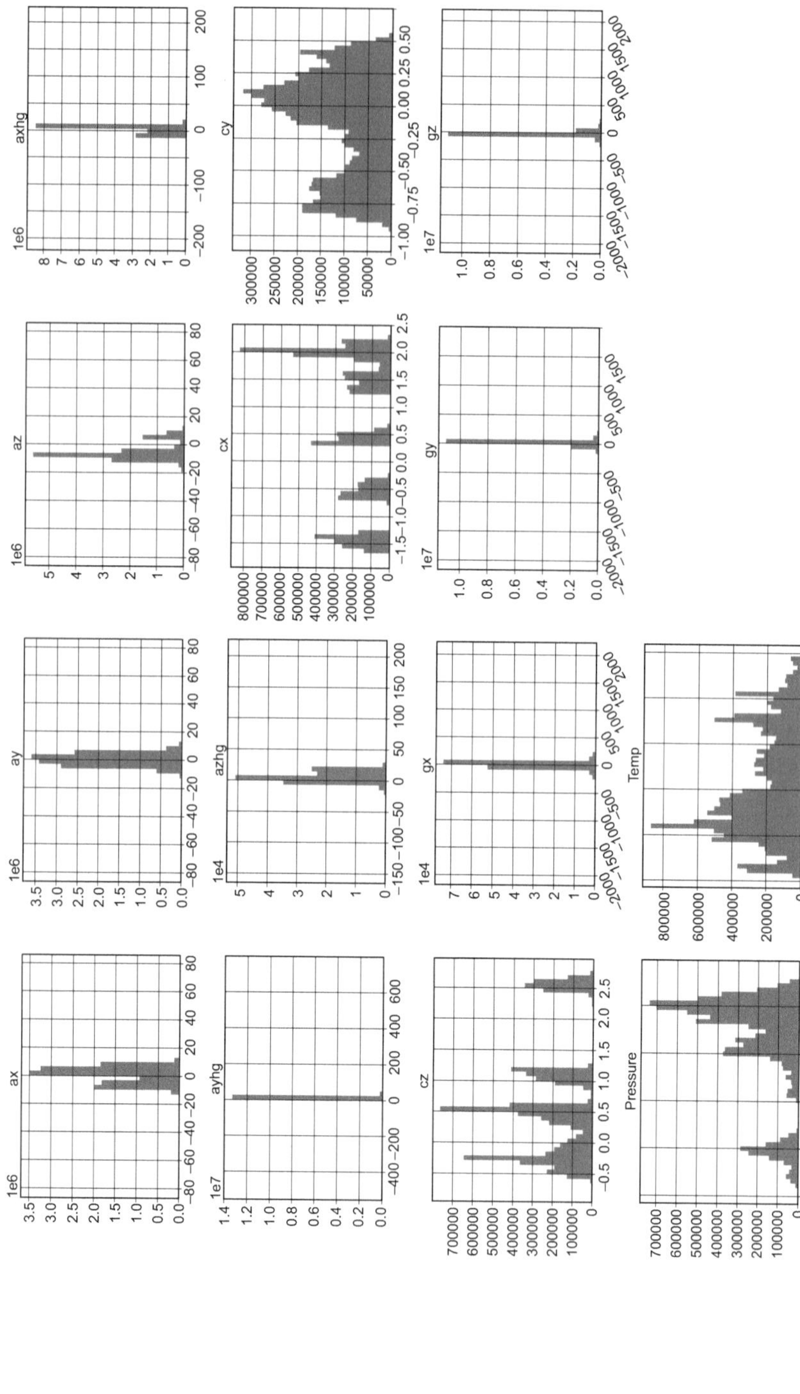


**Figure 3.7: Histograms of selected sensor data features.**

### Handling Missing Values

Our dataset contains some missing values in 'cx', 'cy', 'cz', and 'pressure' and we will remove them using the following python code:

```
# Defining the function to identify and remove all columns that
contain missing values
def remove_columns_with_missing_values(df):
    """
    Removes columns with missing values from a DataFrame.
    Parameters:
    df (pd.DataFrame): The input DataFrame.
    Returns:
    pd.DataFrame: The DataFrame with missing value columns removed.
    """

    # Identify columns with missing values
    columns_with_missing_values = df.columns[df.isnull().any()]

    # Remove columns with missing values
    df_cleaned = df.drop(columns = columns_with_missing_values)

    return df_cleaned

# Call the function to remove the columns that contain missing values
df_cleaned = remove_columns_with_missing_values(df)

# have a look at the new cleaned dataset
df_cleaned.head()
```

#### *Function Definition*

- def remove_columns_with_missing_values(df): This line begins the definition of a function named remove_columns_with_missing_values. The function takes one argument, df, which is expected to be a pandas DataFrame
- Docstring
  - " " " : The function definition contains a docstring (enclosed within triple quotes) that describes what the function does, its parameters, and what it returns.

#### *Identifying Columns with Missing Values*

- columns_with_missing_values = df.columns[df.isnull().any()]: This line of code identifies which columns in the DataFrame have any missing values. df.isnull() creates a boolean mask where True indicates missing values. The .any() method is then applied to this mask to check each column; if any value in a column is True (indicating a missing value), .any() will return True for that column. Finally, df.columns[...] selects the names of the columns that have missing values.

#### *Removing Columns with Missing Values*

- df_cleaned = df.drop(columns = columns_with_missing_values): This line removes the identified columns from the DataFrame. The drop method is used with the columns argument specifying which columns to remove. The result is stored in a new variable df_cleaned, which represents the DataFrame after the columns with missing values have been removed.

#### *Returning the Cleaned DataFrame*

- return df_cleaned: The function returns df_cleaned, the DataFrame that no longer contains the columns with missing values.

#### *Function Usage*

After defining the function, it is called with the following line:

- df_cleaned = remove_columns_with_missing_values(df): This line applies the function to a DataFrame df. The result is a cleaned version of df that has had all columns with any missing values removed.

The cleaned dataframe is displayed as below:

- df_cleaned.head(): The .head() shows the first 5 rows of the df_cleaned by default. If we want to see the first 20 rows of the dataset we can call the .head(20).

## Exploratory Data Analysis (EDA)

Here we will conduct EDA of the farm animal behavior dataset by utilizing Pandas, Matplotlib, and Seaborn libraries.

### *Class Distribution*

We will begin by inspecting the distribution of animal behaviors in our dataset. Understanding the balance or imbalance between behavior classes is essential for model training and evaluation.

To examine the class distribution of behaviors in our dataset, we can use the following Python code:

```
import matplotlib.pyplot as plt
import seaborn as sns

# Count the occurrences of each behavior class
class_counts = df_cleaned['label'].value_counts()
```

```
# Plot the class distribution
plt.figure(figsize=(10, 6))
sns.barplot(x=class_counts.index, y=class_counts.values)
plt.xlabel('Behavior')
plt.ylabel('Count')
plt.title('Class Distribution')
plt.xticks(rotation=45)
plt.show()
```

The code visualizes the distribution of classes in a dataset using matplotlib and seaborn libraries.

Key steps include:

1. Importing libraries: matplotlib.pyplot as plt for plotting and seaborn as sns for easier plot creation.
2. Calculating class frequencies: Using value_counts() on the label column of df_cleaned to get frequencies.
3. Plotting: Setting the figure size, using sns.barplot for the bar plot, labelling axes as 'Behavior' and 'Count', titling the plot 'Class Distribution', and rotating $x$-axis labels for better readability.
4. Saving and displaying the plot: The plot is saved with a high resolution (300 dpi) and then displayed.

Output of the Code:

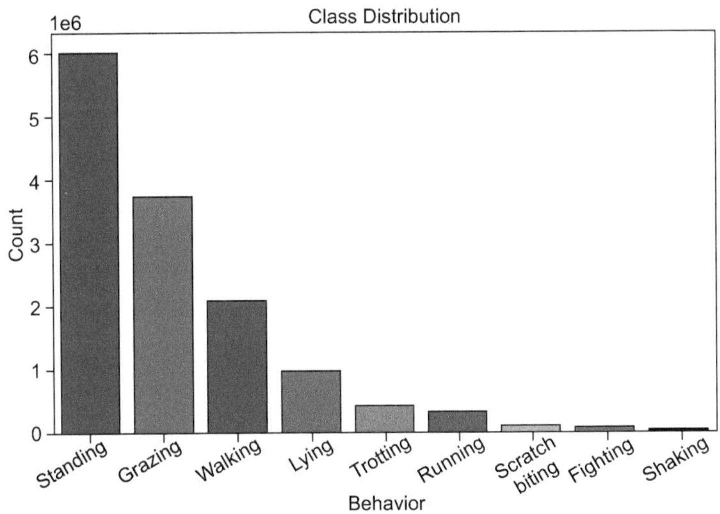


**Figure 3.8: Barplot showing the distribution of animal behaviors (classes) in the dataset. Some classes are more prevalent than others, indicating a class imbalance.**

In our dataset, we observe (Figure 3.8) varying class distribution among the different animal behaviors.

## *Correlation Analysis*

The correlation coefficient, ranging from –1 to 1, is a statistical metric that expresses the extent of a linear association between two quantitative variables.

### Interpreting Correlation

- Positive Correlation: When the correlation coefficient is positive (closer to +1), it indicates a positive linear relationship. This indicates a tendency for one variable to increase in conjunction with an increase in the other variable.
- Negative Correlation: A negative correlation coefficient (closer to –1) indicates a negative linear relationship. As one feature increases, the other tends to decrease.
- No Correlation: A correlation coefficient near 0 suggests no linear relationship between the features. Changes in one feature do not significantly impact the other.

### Limitations of Correlation

While correlation is a valuable tool for identifying linear relationships, it has some limitations:

- Linearity Assumption: Correlation measures only linear relationships. It may not capture non-linear associations between variables. For instance, two variables might have a strong quadratic relationship that correlation cannot detect.
- Outliers: Outliers influence correlation, where even a single outlier can significantly impact the correlation coefficient, potentially leading to misleading results.
- Other Factors: Correlation does not account for other factors that might influence the relationship between variables. It cannot establish causation, and false correlations can occur when a third variable influences both.
- Limited to Numerical Data: Correlation works only for numerical data. It cannot quantify relationships between categorical variables.

The following code calculates the correlation coefficients between selected numerical columns in a pandas DataFrame and plots the correlation matrix.

```
import pandas as pd
import numpy as np
```

```
import seaborn as sns
import matplotlib.pyplot as plt

numerical_features = ['ax', 'ay', 'az', 'gx', 'gy', 'gz']
# Calculate the correlation coefficients
correlation_matrix = df_cleaned[numerical_features].corr()
correlation_matrix

# Visualize the correlation matrix using a heatmap
plt.figure(figsize=(10, 8))
sns.heatmap(correlation_matrix,  annot=True,  fmt=".2f",
cmap='coolwarm',  cbar=True)
plt.title("Correlation Matrix Heatmap")
plt.show()

Output:
          ax         ay         az         gx         gy         gz
ax  1.000000   0.113400   0.243339   0.038723  -0.012482  -0.027463
ay  0.113400   1.000000  -0.132366   0.117453  -0.043101  -0.018925
az  0.243339  -0.132366   1.000000   0.025681  -0.024909  -0.033320
gx  0.038723   0.117453   0.025681   1.000000  -0.065109  -0.035213
gy -0.012482  -0.043101  -0.024909  -0.065109   1.000000   0.076284
gz -0.027463  -0.018925  -0.033320  -0.035213   0.076284   1.000000
```

- import numpy as np: This line imports the numpy library and aliases it as np. numpy
- numerical_features = ['ax', 'ay', 'az', 'gx', 'gy', 'gz']: This line creates a list of strings where each string represents the name of a numerical feature in the dataset.
- correlation_matrix = df_cleaned[numerical_features].corr():
  - df_cleaned[numerical_features] is indexing into a pandas DataFrame called df_cleaned, selecting only the columns listed in numerical_features.
  - .corr() is a pandas DataFrame method that calculates the correlation coefficients between the columns in the DataFrame.
- correlation_matrix: This line displays the resulting correlation matrix (pandas DataFrame).
- sns.heatmap(correlation_matrix, annot = True, fmt = ".2f", cmap = 'coolwarm', cbar = True): This code creates a heatmap using Seaborn to visualize the correlation matrix correlation_matrix, with annotations displaying the correlation values formatted to two decimal places, using a 'coolwarm' color map, and including a color bar for reference (Figure 3.9).
- The output is a correlation matrix displaying the linear relationship between pairs of variables (ax, ay, az, gx, gy, gz). Each cell shows the correlation coefficient for a pair, ranging from –1 to 1. Diagonal cells, which compare each variable to themselves, are all 1, indicating a perfect positive correlation.

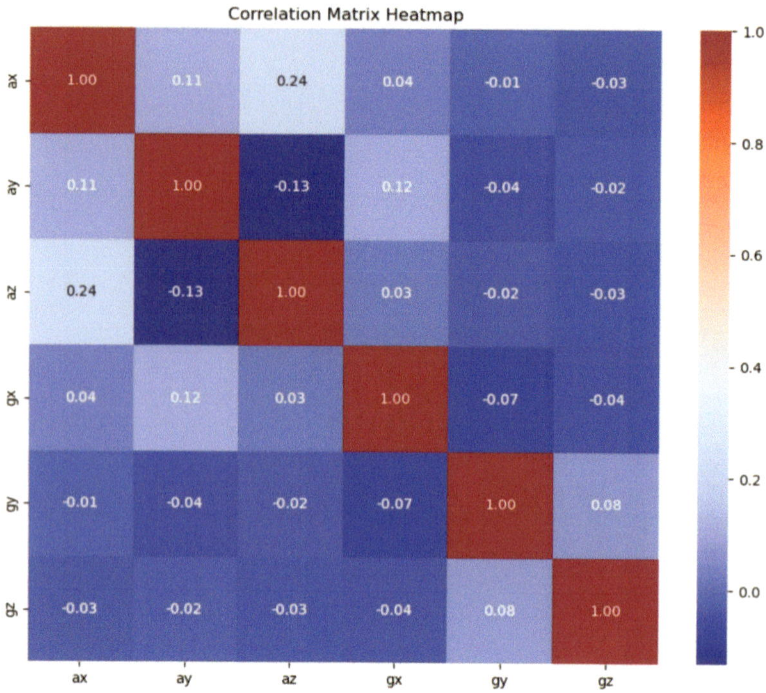


**Figure 3.9: Correlation matrix heatmap.**

## *Outlier Detection Using Boxplots*

Outlier detection is an essential step in data preprocessing, especially in statistical analyses where outliers can significantly skew the results. A common graphical method for identifying outliers is the use of boxplots, which utilize a five-number summary (minimum, first quartile Q1, median, third quartile Q3, and maximum). They offer a common graphical technique to detect outliers and display data distribution. They reveal outlier values, data symmetry, grouping tightness, and skewness. In a boxplot, outliers are typically indicated by points that are plotted outside the whiskers, extensions of the box that represent the variability outside the upper and lower quartiles.

The following code snippet creates boxplots for a set of numerical features in our dataset.

```
# Create box plots for numerical features to detect outliers
import seaborn as sns
plt.figure(figsize=(12, 8))
for feature in numerical_features:
```

```
    plt.subplot(3, 2, numerical_features.index(feature) + 1)
    sns.boxplot(x=df[feature])
    plt.xlabel(feature)
plt.tight_layout()
plt.show()
```

The code uses Seaborn's boxplot function. This function calculates the quartiles and plots the box and whiskers, accordingly, showing any outliers as individual points. Figure 3.10 is the result from the code and illustrates the distribution of six numerical features (ax, ay, az, gx, gy, and gz) using boxplots. The boxplots for 'ax', 'ay', and 'az' suggest a central clustering of data points with a few outliers, whereas the 'gx', 'gy', and 'gz' boxplots reveal a more dispersed distribution with numerous outliers.

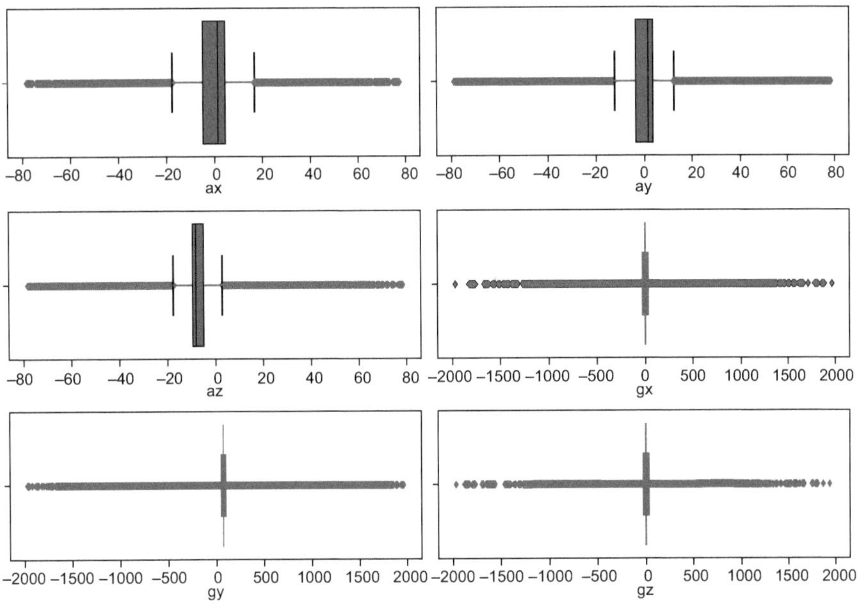


**Figure 3.10: Boxplot illustration of six variables highlighting symmetry and outlier distribution.**

## *Feature Extraction Using Sliding Windows*

Time-series data, particularly accelerometer measurements, plays a pivotal role in feature extraction. These measurements are often collected at various time intervals, spanning milliseconds to seconds, forming a rich time-series dataset. To effectively analyze such data, researchers commonly employ a technique known as *windowing*. Windowing involves dividing the time-series data into overlapping or disjoint segments of fixed size.

In this project, we focus on accelerometer readings, and we have a fixed sampling rate of 200Hz. Our windowing approach involves selecting a window size of 5 seconds as a compromise between capturing detailed activity patterns and maintaining computational efficiency.

With our window size selected, we can then move into the process of feature extraction. In this section, we extract a specific set of features from our sensor data, which include the following: mean, crest factor, root mean square velocity, skewness, kurtosis, madogram, zero crossing rate, squared integrals, and signal entropy. These features have been chosen for their proven effectiveness in previous research within the field of activity recognition. It is important to note that while these features provide valuable insights, the upcoming chapter on feature extraction (Chapter 5) will explore a more comprehensive range of features and their significance in greater detail.

Before proceeding with feature extraction, it is essential to calculate the magnitude of the accelerometer readings. The magnitude provides valuable information about the overall intensity of motion, which can be useful for activity recognition.

The magnitude of acceleration, often denoted as $|A|$, is computed as the Euclidean norm of the three accelerometer axes ($x$, $y$, and $z$):

$$|A| = \sqrt{x^2 + y^2 + z^2}.$$

Now, we proceed with calculating the magnitude and extracting the desired features for each window of data. We will create a function extract_features_from_window to perform these operations. Before that, we will create two functions; one to drop the columns that we do not need for the project, and the second function that calculates the magnitude.

```
# Select the columns you need from df
# We will select only accelerometer readings for this project

def drop_unneeded_columns(df):
    """
    Drop columns that are not needed for feature extraction.

    Parameters:
    df (pd.DataFrame): DataFrame containing accelerometer data.

    Returns:
    pd.DataFrame: DataFrame with unnecessary columns dropped.
    """
    # List of columns to keep (accelerometer columns)
    # Include 'label' column to keep classes

    columns_to_keep = ['label', 'ax', 'ay', 'az']
```

```
    # Drop columns not in the list
    df = df[columns_to_keep]

    return df

def calculate_magnitude(df):
    """
    Calculate the magnitude of accelerometer readings and add it
    as a new column.

    Parameters:
    df (pd.DataFrame): DataFrame containing accelerometer data.

    Returns:
    pd.DataFrame: DataFrame with magnitude column added.
    """

    # Calculate magnitude
    df['magnitude'] = np.sqrt(df['ax']**2 + df['ay']**2 +
       df['az']**2)

    return df

# Select the columns and define the new dataset
df_new = drop_unneeded_columns(df_cleaned)

# Calculate the magnitude and add it to the new dataset
df_mag = calculate_magnitude(df_new)
```

In the above code:

- The drop_unneeded_columns takes as input a pandas DataFrame and returns a new DataFrame with only the specified columns (ax, ay, az).
- Next, the calculate_magnitude function calculates the magnitude of our accelerometer readings resulting in a new DataFrame df_mag that includes the computed magnitude of the accelerometer data, the accelerometer readings and the label column.

Then we define the sample rate, the window size, and the overlapping step. The following code calculates the window size and the overlap size.

```
# Defining the sample rate, window size and overlap
sample_rate = 200  # Sample rate in Hz
window_duration = 5  # Desired window duration in seconds
overlap_percent = 50  # Desired overlap percentage

# Calculate the window size in samples
window_size = int(sample_rate * window_duration)
```

```
# Calculate the step size for overlapping windows
overlap_size = int(window_size * (overlap_percent / 100))

print(f"Window size: {window_size} samples")
print(f"Overlap size: {overlap_size} samples")

# Output
Window size: 1000 samples
Overlap size: 500 samples
```

Now, let's create the function to extract the desired features using the sliding window.

```
from scipy.stats import skew, kurtosis
from scipy.integrate import simps

def extract_features_with_windows(dataset, window_size,
overlap_percent):
    """

    Extracts features from a dataset using sliding windows.

    Parameters:
    dataset (pd.DataFrame): The dataset containing sensor data
    with 'label'column.
    window_size (int): The size of the sliding window in samples.
    overlap_percent (int): The overlap percentage between
    consecutive windows.

    Returns:
    pd.DataFrame: A DataFrame containing extracted features for
    each window with corresponding labels.
    """

    features = []
    labels = []

    # Calculate the step size for overlapping windows
    step_size = int(window_size * (overlap_percent / 100))

    for i in range(0, len(dataset) - window_size + 1, step_size):
       # Create a copy of the window_data
       window_data = dataset.iloc[i:i + window_size].copy()

       # Extract features from the window_data here
       window_features = []

       # Calculate mean for accelerometer axes and magnitude
       for axis in ['ax', 'ay', 'az', 'magnitude']:
          window_features.append(window_data[axis].mean())
```

```
# Calculate crest factor for accelerometer axes
for axis in ['ax', 'ay', 'az']:
    crest_factor = window_data[axis].max() / window_data[axis].
        std()
    window_features.append(crest_factor)

# Calculate root mean square velocity
rms_velocity = np.sqrt(np.mean(window_data['ax']**2 +
    window_data['ay']**2 + window_data['az']**2))
window_features.append(rms_velocity)

# Calculate skewness for accelerometer axes
for axis in ['ax', 'ay', 'az']:
    window_features.append(skew(window_data[axis]))

# Calculate kurtosis for accelerometer axes
for axis in ['ax', 'ay', 'az']:
    window_features.append(kurtosis(window_data[axis]))

# Calculate the madogram
madogram = simps(window_data['ax'], dx=1/sample_rate)
window_features.append(madogram)

# Calculate zero crossing rate
zero_crossings = np.sum(np.diff(np.sign(window_
    data['ax'])) != 0)
window_features.append(zero_crossings)

# Calculate squared integrals
for axis in ['ax', 'ay', 'az']:
    squared_integral = simps(window_data[axis]**2, dx=1/
        sample_rate)
    window_features.append(squared_integral)

# Calculate signal entropy, checking for invalid values
  and divide by zero
entropy_values = window_data['ax'] / np.sum(window_data['ax'])
entropy_values = entropy_values.replace([np.inf, -np.inf],
    np.nan)
entropy_values = entropy_values.dropna()

signal_entropy = -np.sum(entropy_values * np.log(entropy_
    values))
window_features.append(signal_entropy)

# Determine the majority label within the window
majority_label = window_data['label'].mode().iloc[0]

features.append(window_features)
labels.append(majority_label)
```

```
    # Create a DataFrame with the extracted features and labels
    feature_names = ['mean_ax', 'mean_ay', 'mean_az', 'mean_
        magnitude',
            'crest_factor_ax', 'crest_factor_ay', 'crest_factor_az',
            'rms_velocity', 'skewness_ax', 'skewness_ay',
            'skewness_az', 'kurtosis_ax', 'kurtosis_ay',
            'kurtosis_az', 'madogram', 'zero_crossing_rate',
            'squared_integral_ax', 'squared_integral_ay',
            'squared_integral_az', 'signal_entropy']

    feature_df = pd.DataFrame(features, columns=feature_names)
    label_df = pd.DataFrame(labels, columns=['label'])

    return pd.concat([feature_df, label_df], axis=1)

    extracted_features = extract_features_with_windows(df_mag,
        window_size, overlap_percent)
```

## Code Key Points

- We define the extract_features_with_windows function, which takes three parameters: dataset, window_size, and overlap_percent. This function extracts features from the sensor data using sliding windows.
- We initialize two empty lists, features, and labels, to store the extracted features and their corresponding labels.
- We calculate the step size for overlapping windows based on the specified window_size and overlap_percent. This step size determines how much the windows will overlap.
- We iterate through the dataset using a loop. For each iteration, we create a window of data. The loop ensures that we cover the entire dataset with sliding windows.
- We create a copy of the window data to avoid modifying the original dataset.
- Inside the loop, we initialize an empty list, window_features, to store the features extracted from the current window.
- We calculate the mean for the accelerometer axes ('ax', 'ay', 'az') and for the magnitude ('magnitude') of acceleration. These values represent the average acceleration in each direction within the window.
- We calculate various features for the accelerometer axes. These features include crest factor, root mean square velocity, skewness, kurtosis, madogram, zero crossing rate, squared integrals, and signal entropy.
- We determine the majority label within the current window. If there is more than one activity in the window, we select the label that appears most frequently. This ensures that each window is linked to one label.

- We store the extracted features and the majority label in the features and labels lists, respectively, for each window.
- We then create pandas DataFrames for the extracted features and labels.
- Finally, the function returns a pandas DataFrame containing the extracted features for each window, along with their corresponding majority labels.
- extract_features_with_windows function, processes the DataFrame df_mag using specified window size and overlap percentage parameters, to the variable extracted_features.

The following code snippet will help us to understand the balance or imbalance of the different activities within our dataset. By using the .value_counts() method on the 'label' column of the extracted_features DataFrame the code evaluates how many instances of each unique class label are present.

```
# Checking the distribution of the classes
extracted_features['label'].value_counts()

# Output
label
standing          12063
grazing            7472
walking            4220
lying              1968
trotting            814
running             649
scratch_biting      186
fighting            150
shaking              33
```

From the result, it is clear that we have a highly imbalanced dataset, with 'standing' and 'grazing' behaviors significantly more represented than activities like 'shaking' or 'fighting'. This imbalance could lead to a model that is overly proficient at recognizing the majority classes but performs poorly on the minority classes, which might be equally or more important for the predictive task at hand. To address this, one might consider implementing techniques such as resampling the underrepresented classes, applying class weights during model training, or choosing evaluation metrics that are sensitive to class imbalance, ensuring the model's predictive performance is balanced across all classes.

For the scope of our project, while acknowledging the presence of class imbalance in our dataset, we will not apply any specific techniques to address this issue. Instead, we will proceed with the analysis as is, which will allow us to maintain the natural distribution of behaviors as they occur.

## *Splitting the Dataset*

Dividing the dataset into training, validation, and test subsets is essential in the development of machine learning models. It ensures that our model is trained on one set of data, validated on another, and finally tested on a completely independent set. This separation is essential for the following reasons:

- **Model Generalization:** The goal is to develop a machine learning model capable of effectively generalizing to new, unseen data. By testing our model on a separate test set, we can evaluate its performance on data it has never encountered during training.
- **Avoiding Data Snooping Bias:** Data snooping bias occurs when we peek at the test set or use information from the test set to make decisions during model development. This can lead to overly optimistic performance estimates and models that do not perform well on new data.

### Dataset Splitting Strategies

There are different strategies for splitting a dataset and it is important to follow best practices to ensure unbiased model evaluation:

- **Random Splitting:** Randomly dividing the dataset into training, validation, and test sets is a common approach. However, it is crucial to avoid repeatedly generating different random splits, as this can lead to testing the entire dataset over time, which is not ideal.
- **Fixed Splitting:** A better practice is to create a fixed split at the beginning of the project and stick with it throughout model development. This way, the test set remains unseen, and we can assess model performance accurately.
- **Stratified Sampling:** Stratified sampling is beneficial in scenarios involving imbalanced datasets. It ensures that each class is proportionally represented in each subset. This helps prevent a situation where a particular class is underrepresented in the test set, leading to biased evaluation.
- **Cross-Validation:** Cross-validation involves partitioning the data into multiple subsets (folds) and systematically rotating which fold is used for validation while the rest are used for training. This method provides a more robust estimate of model performance and is effective in situations where the dataset is small.
- **Time-Based Splitting:** In time-series data, it is essential to split the data chronologically. This ensures that the model is trained on past data and validated/tested on future data, simulating real-world scenarios.

In our project, we will employ stratified sampling to divide the dataset, allocating 70% for training and splitting the remainder equally between validation and testing, with a 15% each.

Here is the Python code for stratified sampling:

```
from sklearn.model_selection import train_test_split

# Define the features (X) and labels (y)
X = extracted_features.drop(columns=['label'])
y = extracted_features['label']

# Perform the initial split into training (70%) and temp (30%)
X_train, X_temp, y_train, y_temp = train_test_split(X, y, test_
    size=0.30, random_state=42, stratify=y)

# Further split the temp set into validation (50%) and test (50%)
X_validation, X_test, y_validation, y_test = train_test_split(X_
    temp, y_temp, test_size=0.50, random_state=42, stratify=y_temp)
```

### *Code Breakdown*

- We import the train_test_split function from sklearn.model_selection.
- We define our feature matrix X as all columns in the dataset except for the 'label' column, and our target vector y as the 'label' column.
- We perform the initial split using train_test_split:
  - X_train and y_train contain 70% of the data for training.
  - X_temp and y_temp contain the remaining 30% temporarily.
- We further split the X_temp and y_temp sets into validation and test sets using another train_test_split:
  - X_validation and y_validation contain 50% of the data for validation.
  - X_test and y_test contain the remaining 50% for testing.
- We use the stratify parameter to ensure that the class proportions are maintained in both the training/validation and validation/test splits.

## *Feature Scaling*

Feature scaling guarantees that each feature equally influences the model training process and can prevent certain features from dominating others simply because of their scale. In this section, we will explore the importance of feature scaling and demonstrate how to perform it using Python for our machine learning project.

Why feature scaling is essential:

- Equal Contribution: Scaling ensures that all features have a similar influence on the learning algorithm. Without scaling, features with larger scales can dominate those with smaller scales.

- Convergence: Gradient-based optimization algorithms converge faster and more reliably when features are scaled. This leads to quicker model training.
- Distance Metrics: Algorithms that rely on distance metrics, like KNN or clustering algorithms, are sensitive to feature scales. Scaling helps them work effectively.

### Common Scaling Techniques

There are two primary techniques for feature scaling:

- **Min-Max Scaling (Normalization):** Min-Max scaling, also known as normalization, scales the features to a specific range, typically between 0 and 1.
- **Standardization (Z-score Scaling):** Standardization, also known as Z-score scaling, transforms the features to have a mean and standard deviation of 0 and 1, respectively.

The choice between StandardScaler and MinMaxScaler depends on the characteristics of your dataset and the ML algorithms you plan to use. Feature scaling and normalization will be discussed in more detail in Chapter 5.

For our project, we will apply both standardization and normalization to our dataset and see which one leads to better model performance. We will then train various machine learning models to determine the most suitable feature scaling method based on their performance.

```
# Importing the libraries
from sklearn.preprocessing import MinMaxScaler, StandardScaler

# Initialize scalers for Min-Max scaling and standardization
min_max_scaler = MinMaxScaler()
standard_scaler = StandardScaler()

# Perform Min-Max scaling on the training, validation and test sets
X_train_min_max_scaled = min_max_scaler.fit_transform(X_train)
X_validation_min_max_scaled = min_max_scaler.transform(X_
    validation)
X_test_min_max_scaled = min_max_scaler.transform(X_test)

# Perform standardization on the training, validation and test sets
X_train_standardized = standard_scaler.fit_transform(X_train)
X_validation_standardized = standard_scaler.fit_transform(X_
    validation)
X_test_standardized = standard_scaler.transform(X_test)
```

```
# Creating the new dataframes with the standardized and normalized
features
X_train_s = pd.DataFrame(X_train_standardized, columns = X_train.
    columns)
X_validation_s = pd.DataFrame(X_validation_standardized,
    columns = X_validation.columns)
X_test_s = pd.DataFrame(X_test_standardized, columns = X_test.
    columns)
X_train_n = pd.DataFrame(X_train_min_max_scaled, columns = X_train.
    columns)
X_validation_n = pd.DataFrame(X_validation_min_max_scaled,
    columns = X_validation.columns)
X_test_n = pd.DataFrame(X_test_min_max_scaled, columns = X_test.
    columns)
```

This code snippet shows that we import two preprocessing classes from scikit-learn, MinMaxScaler and StandardScaler. We then initialize Min-Max scaling and standardization processes, perform scaling on the datasets, and then we create new data frames with the scaled features.

### Model Training and Evaluation

Now, that we have our scaled datasets ready, we proceed with the subsequent python code. The code is designed to compare the models' performance across two different feature scaling methods: StandardScaler and MinMaxScaler.

```
# Importing the libraries
from sklearn.preprocessing import LabelEncoder
from sklearn.ensemble import RandomForestClassifier,
GradientBoostingClassifier
from sklearn.neighbors import KNeighborsClassifier
from sklearn.svm import SVC
from sklearn.metrics import classification_report, accuracy_score
from sklearn.metrics import confusion_matrix
from sklearn.neural_network import MLPClassifier

# Define results dictionaries for standardized and normalized
datasets
results_std = {}
results_minmax = {}

# Choose some ML models
models = {
    "Random Forest": RandomForestClassifier(n_estimators=100,
        random_state=42),
```

```
    "Gradient Boosting": GradientBoostingClassifier(n_
       estimators=100, random_state=42),
    "K-Nearest Neighbors": KNeighborsClassifier(n_neighbors=5),
    "Support Vector Machine (Radial)": SVC(kernel="rbf", C=1),
    "Multilayer perceptron (MLP)": MLPClassifier(hidden_layer_
       sizes=(128,),
                                      activation='relu',
                                      solver='adam',
                                      max_iter=100,
                                      random_state=42)
}

# Function to train and evaluate models with LabelEncoder
def evaluate_models(X_train, X_validation, y_train, y_validation,
scaler_name, results_dict):
    label_encoder = LabelEncoder()
    y_train_encoded = label_encoder.fit_transform(y_train)
    y_validation_encoded = label_encoder.transform(y_validation)

    for model_name, model in models.items():
        model.fit(X_train, y_train_encoded)
        y_pred_encoded = model.predict(X_validation)
        accuracy = accuracy_score(y_validation_encoded,
           y_pred_encoded)
        classification_rep = classification_report(y_validation_
           encoded, y_pred_encoded)

        confusion_matrix_encoded = confusion_matrix(y_validation_
           encoded, y_pred_encoded)
        results_dict[model_name] = {"Confusion Matrix": confusion_
           matrix_encoded,
              "Accuracy": accuracy,
              "Classification Report": classification_rep,
              "Scaler": scaler_name}

# Evaluate models on the standardized dataset
evaluate_models(X_train_s, X_validation_s, y_train, y_
validation,"StandardScaler", results_std)

# Evaluate models on the normalized dataset
evaluate_models(X_train_n, X_validation_n, y_train, y_validation,
"MinMaxScaler", results_minmax)

# Print the results for both datasets
print("Results for Standardized Dataset:")
for model_name, metrics in results_std.items():
    print(f"Model: {model_name} (Scaler: {metrics['Scaler']})")
    print(f"Accuracy: {metrics['Accuracy']:.2f}")
```

```
    print(f"Classification Report:\n{metrics['Classification
       Report']}\n")

print("Results for Normalized Dataset:")
for model_name, metrics in results_minmax.items():
    print(f"Model: {model_name} (Scaler: {metrics['Scaler']})")
    print(f"Accuracy: {metrics['Accuracy']:.2f}")
    print(f"Classification Report:\n{metrics['Classification
       Report']}\n")

# Output
Results for Standardized Dataset:
Model: Random Forest (Scaler: StandardScaler)
Accuracy: 0.92
Classification Report:

              precision    recall  f1-score   support
           0       0.95      0.87      0.91        23
           1       0.92      0.93      0.93      1121
           2       0.94      0.58      0.72       295
           3       0.97      0.98      0.97        97
           4       0.86      0.21      0.34        28
           5       1.00      0.80      0.89         5
           6       0.90      0.96      0.93      1809
           7       0.95      0.94      0.95       122
           8       0.95      0.95      0.95       633

    accuracy                           0.92      4133
   macro avg       0.94      0.80      0.84      4133
weighted avg       0.92      0.92      0.91      4133

Model: Gradient Boosting (Scaler: StandardScaler)
Accuracy: 0.88
Classification Report:

              precision    recall  f1-score   support
           0       0.59      0.57      0.58        23
           1       0.90      0.91      0.90      1121
           2       0.87      0.63      0.73       295
           3       0.96      0.96      0.96        97
           4       0.04      0.14      0.06        28
           5       0.25      0.20      0.22         5
           6       0.91      0.91      0.91      1809
           7       0.93      0.91      0.92       122
           8       0.92      0.92      0.92       633

    accuracy                           0.88      4133
   macro avg       0.71      0.68      0.69      4133
weighted avg       0.90      0.88      0.89      4133
```

```
Model: K-Nearest Neighbors (Scaler: StandardScaler)
Accuracy: 0.88
Classification Report:
              precision    recall  f1-score   support
           0       0.79      0.65      0.71        23
           1       0.86      0.90      0.88      1121
           2       0.76      0.74      0.75       295
           3       0.97      0.98      0.97        97
           4       0.60      0.11      0.18        28
           5       1.00      0.60      0.75         5
           6       0.90      0.89      0.89      1809
           7       0.95      0.94      0.95       122
           8       0.93      0.94      0.93       633

    accuracy                           0.88      4133
   macro avg       0.86      0.75      0.78      4133
weighted avg       0.88      0.88      0.88      4133

Model: Support Vector Machine (Radial) (Scaler: StandardScaler)
Accuracy: 0.89
Classification Report:
              precision    recall  f1-score   support
           0       0.92      0.52      0.67        23
           1       0.89      0.91      0.90      1121
           2       0.78      0.65      0.71       295
           3       0.98      0.90      0.94        97
           4       1.00      0.11      0.19        28
           5       1.00      0.80      0.89         5
           6       0.87      0.93      0.90      1809
           7       0.95      0.86      0.91       122
           8       0.95      0.91      0.93       633

    accuracy                           0.89      4133
   macro avg       0.93      0.73      0.78      4133
weighted avg       0.89      0.89      0.89      4133

Model: Multilayer perceptron (MLP) (Scaler: StandardScaler)
Accuracy: 0.91
Classification Report:
              precision    recall  f1-score   support
           0       0.89      0.74      0.81        23
           1       0.92      0.91      0.92      1121
           2       0.79      0.75      0.77       295
           3       1.00      0.89      0.94        97
           4       0.71      0.43      0.53        28
           5       1.00      1.00      1.00         5
           6       0.92      0.94      0.93      1809
           7       0.89      0.94      0.92       122
           8       0.94      0.94      0.94       633
```

```
    accuracy                           0.91      4133
   macro avg       0.90      0.84      0.86      4133
weighted avg       0.91      0.91      0.91      4133

Results for Normalized Dataset:
Model: Random Forest (Scaler: MinMaxScaler)
Accuracy: 0.94
Classification Report:
              precision    recall  f1-score   support
           0       0.91      0.87      0.89        23
           1       0.92      0.94      0.93      1121
           2       0.93      0.85      0.89       295
           3       0.98      0.98      0.98        97
           4       0.89      0.29      0.43        28
           5       1.00      1.00      1.00         5
           6       0.94      0.96      0.95      1809
           7       0.97      0.93      0.95       122
           8       0.95      0.94      0.95       633

    accuracy                           0.94      4133
   macro avg       0.94      0.86      0.88      4133
weighted avg       0.94      0.94      0.94      4133

Model: Gradient Boosting (Scaler: MinMaxScaler)
Accuracy: 0.92
Classification Report:
              precision    recall  f1-score   support
           0       0.84      0.70      0.76        23
           1       0.90      0.93      0.92      1121
           2       0.85      0.81      0.83       295
           3       0.97      0.96      0.96        97
           4       0.60      0.32      0.42        28
           5       0.75      0.60      0.67         5
           6       0.94      0.94      0.94      1809
           7       0.93      0.93      0.93       122
           8       0.93      0.93      0.93       633

    accuracy                           0.92      4133
   macro avg       0.86      0.79      0.82      4133
weighted avg       0.92      0.92      0.92      4133

Model: K-Nearest Neighbors (Scaler: MinMaxScaler)
Accuracy: 0.89
Classification Report:
              precision    recall  f1-score   support
           0       0.88      0.65      0.75        23
           1       0.87      0.91      0.89      1121
           2       0.82      0.75      0.78       295
```

```
           3       0.98      0.99      0.98        97
           4       1.00      0.21      0.35        28
           5       0.75      0.60      0.67         5
           6       0.91      0.90      0.90      1809
           7       0.97      0.95      0.96       122
           8       0.91      0.92      0.91       633

    accuracy                           0.89      4133
   macro avg       0.90      0.77      0.80      4133
weighted avg       0.89      0.89      0.89      4133

Model: Support Vector Machine (Radial) (Scaler: MinMaxScaler)
Accuracy: 0.84
Classification Report:
              precision    recall  f1-score   support
           0       1.00      0.48      0.65        23
           1       0.81      0.90      0.85      1121
           2       0.83      0.28      0.42       295
           3       0.95      0.97      0.96        97
           4       0.00      0.00      0.00        28
           5       0.00      0.00      0.00         5
           6       0.82      0.92      0.87      1809
           7       0.92      0.88      0.90       122
           8       0.91      0.79      0.84       633

    accuracy                           0.84      4133
   macro avg       0.69      0.58      0.61      4133
weighted avg       0.83      0.84      0.82      4133

Model: Multilayer perceptron (MLP) (Scaler: MinMaxScaler)
Accuracy: 0.87
Classification Report:
              precision    recall  f1-score   support
           0       1.00      0.74      0.85        23
           1       0.87      0.89      0.88      1121
           2       0.79      0.50      0.61       295
           3       0.96      0.97      0.96        97
           4       0.00      0.00      0.00        28
           5       1.00      0.40      0.57         5
           6       0.85      0.94      0.89      1809
           7       0.94      0.93      0.93       122
           8       0.94      0.86      0.90       633

    accuracy                           0.87      4133
   macro avg       0.82      0.69      0.73      4133
weighted avg       0.87      0.87      0.87      4133
```

Below is a detailed explanation of each segment of the code:

1. The code imports the libraries and modules to handle data preprocessing, machine learning models, and evaluation metrics.
2. Two dictionaries (results_std and results_minmax) are defined to store the performance metrics of the models for the standardized and normalized datasets, respectively.
3. Choose Appropriate Models: The code defines a dictionary (models) that maps model names to their corresponding scikit-learn classifiers. Various hyperparameters for each model are set at this stage.
4. A function named evaluate_models is defined, which accepts training and validation datasets, the name of the scaler used, and a results dictionary. This function does the following:
    - Uses LabelEncoder to transform the categorical target labels into numerical form.
    - Iterates over each model defined in the models dictionary to:
        - Train the classification models with the training dataset.
        - Predict outcomes using the validation dataset.
        - Evaluate the predictions using accuracy_score, classification_report, and confusion_matrix.
5. Stores these evaluation metrics in the results dictionary.
6. The evaluate_models function is then called twice: once with the standardized dataset and once with the normalized dataset.
7. Finally, the code prints the performance metrics for each model, grouped by the scaling method used.

Once we run the above code, we get the results for each model based on each scaling method. From the output of the code, we can see that some models trained on the normalized dataset perform slightly better. For instance:

1. Random Forest's accuracy changes from 0.92 to 0.94. This suggests that MinMax normalization may be more suitable for this dataset than Z-score standardization. The model is especially good at classifying majority classes but seems to struggle a bit on one of the minority classes which has a recall rate of only 29%. This paints a picture of a model that is generally robust but has specific limitations in recognizing certain minority classes.
2. Gradient Boosting performs reasonably well, especially on the normalized dataset with an accuracy of 0.92. It is slightly less effective than Random Forest, but its macro-average scores are still decent.

3. KNN performs similarly on both types of datasets, but it is generally less accurate than Random Forest. It shows a particular struggle with minority classes, which might be due to the 'curse of dimensionality' or the limitations of distance-based metrics in higher dimensions.
4. SVM, with radial basis function (RBF) kernel, performs reasonably well. However, its performance is less satisfactory on the normalized data. It seems to struggle mainly with classes that have fewer instances, like KNN.
5. Finally, MLP offers good performance, but with a downgrade in accuracy on the normalized data. This suggests that standardization might be more suitable for neural network-based models in this case.

Following the insights gained from our prior analysis, we move forward with our project by utilizing the datasets that have been normalized and selecting Random Forest as our preferred model.

The code below is designed to train Random Forest using the training set, and then evaluate its performance on the validation dataset:

```
# Random forest training

def train_random_forest(X_train, y_train, X_val, y_val, n_
    estimators=100, max_depth=None, random_state=42):

    """
    Train Random Forest on the provided training set and evaluate
    it on validation set.

    Parameters:
    - X_train: Training data
    - y_train: Training data labels
    - X_val: Validation data
    - y_val: Validation data labels
    - n_estimators: Number of trees in the Random Forest
    - max_depth: Maximum depth of the trees
    - random_state: Seed for reproducibility

    Returns:
    - rf_model: Trained Random Forest model
    - performance_metrics: Classification report for the model on
      validation data
    """

    # Initialize the Random Forest Classifier
    rf_model = RandomForestClassifier(n_estimators=n_estimators,
        max_depth=max_depth, random_state=random_state)
```

```
    # Train the model on the training data
    rf_model.fit(X_train, y_train)

    # Predict the labels on validation data
    y_val_pred = rf_model.predict(X_val)

    # Generate performance metrics
    performance_metrics = classification_report(y_val, y_val_pred)
    print("Random Forest Model Performance on Validation Data:")
    print(performance_metrics)
    print(f"Accuracy: {accuracy_score(y_val, y_val_pred) *
       100:.2f}%")

    return rf_model, performance_metrics

# Use the function train_random_forest
rf_model, metrics = train_random_forest(X_train_n, y_train,
   X_validation_n, y_validation)

# Output
Random Forest Model Performance on Validation Data:

              precision    recall  f1-score   support
      fighting      0.91      0.87      0.89        23
       grazing      0.92      0.94      0.93      1121
         lying      0.93      0.85      0.89       295
       running      0.98      0.98      0.98        97
 scratch_biting      0.89      0.29      0.43        28
       shaking      1.00      1.00      1.00         5
      standing      0.94      0.96      0.95      1809
      trotting      0.97      0.93      0.95       122
       walking      0.95      0.94      0.95       633

      accuracy                          0.94      4133
     macro avg      0.94      0.86      0.88      4133
  weighted avg      0.94      0.94      0.94      4133

Accuracy: 93.90%
```

This code defines a function, train_random_forest, designed to train Random Forest on a specified training data, and then evaluate its performance on a separate validation dataset. The function is parameterized to allow customization of the Random Forest's key attributes, such as n_estimators, max_depth, and random_state for reproducible results. Upon training the model on X_train, and y_train, it predicts outcomes on the X_val, and y_val and calculates its performance metrics. The printed output provides a detailed look at the model's effectiveness on the validation data, highlighted by an accuracy percentage.

The output provides the performance evaluation of the Random Forest on validation data, based on various metrics The model displays high precision and recall across most classes, indicating strong predictive capabilities, with particularly perfect performance in shaking. However, scratch_biting shows a significant drop in performance, suggesting difficulty in accurately classifying this behavior. Overall, the model achieves a good accuracy of 93.90%.

## Feature Selection

At this point of our project, we want to examine the features that influence our model's predictions. To achieve this, we employ code to extract and rank the features according to their importance as determined by the Random Forest classifier.

The reasoning for performing feature selection is twofold. Firstly, it enhances model interpretability, allowing us to understand which factors are pivotal in distinguishing between behaviors like grazing, running, or lying. Secondly, by minimizing the quantity of features, we might improve the model's performance and reduce possible overfitting, as the classifier concentrates on the most relevant information.

While the current approach directly utilizes the built-in feature importance of the Random Forest model, it is essential to note that this is just one way to perform feature selection. There are numerous other techniques, each with its advantages and applications, ranging from univariate statistical tests to model-based methods and iterative selectors. These methods will be comprehensively explored in Chapter 6, which is dedicated to the subject of feature selection.

The following code is designed to explain the significance of each feature used in training our Random Forest classifier:

```
# Extract feature importances from rf_model
feature_importances = rf_model.feature_importances_

# Get sorted indices based on importance scores
sorted_indices = feature_importances.argsort()[::-1]

# Display the feature rankings
print("Feature rankings based on importance:")

for i, index in enumerate(sorted_indices, 1):
    print(f"{i}. Feature {index} - Importance: {feature_
    importances[index]:.4f}")

# We select the top 10 features (you can choose any number)
top_features = [X_train_n.columns[i] for i in sorted_indices[:10]]
```

```
# Rebuild the model with top-ranked features
X_train_top_features = X_train_n[top_features]
rf_model.fit(X_train_top_features, y_train)

# Output
Feature rankings based on importance:
1.     Feature  7 - Importance:   0.1588
2.     Feature  6 - Importance:   0.1404
3.     Feature  3 - Importance:   0.0923
4.     Feature  4 - Importance:   0.0821
5.     Feature  5 - Importance:   0.0632
6.     Feature 16 - Importance:   0.0519
7.     Feature 18 - Importance:   0.0499
8.     Feature  0 - Importance:   0.0479
9.     Feature  1 - Importance:   0.0461
10.    Feature 14 - Importance:   0.0436
11.    Feature  2 - Importance:   0.0367
12.    Feature 17 - Importance:   0.0362
13.    Feature 15 - Importance:   0.0303
14.    Feature 12 - Importance:   0.0264
15.    Feature 13 - Importance:   0.0232
16.    Feature 10 - Importance:   0.0178
17.    Feature 11 - Importance:   0.0158
18.    Feature  8 - Importance:   0.0151
19.    Feature  9 - Importance:   0.0147
20.    Feature 19 - Importance:   0.0079
```

Here is a breakdown:

1. We begin by extracting the importance of each feature as determined by the Random Forest model (rf_model.feature_importances_).
2. Then, we sort these importances in descending order (argsort()[::–1]) to identify which features are most influential.
3. We then iterate over the sorted indices, printing out the rank, the feature index, and its corresponding importance score (refer to the output).
4. For demonstration purposes, we choose the top 10 features according to their importance scores. This selection is arbitrary and can be adjusted based on the desired model complexity or specific domain knowledge. The selected features are then used to create a subset of the original training data (X_train_n[top_features]).
5. Finally, we retrain the Random Forest with only the selected top features. By doing so, we can see how well the model performs when it is focused solely on the most significant features, potentially leading to improved efficiency and, possibly, performance due to reduced noise and complexity.

Now that we have trained our model on the top 10 important features, we evaluate its performance on the validation set using the following code:

```
# Ensure that your validation data has the same features as X_
    train_smote
# Select only the top-ranked features from the validation dataset
    X_val_top_features = X_validation_n[top_features]

# Predict using the new model
y_val_pred = rf_model.predict(X_val_top_features)

# Display model performance
print("Random Forest Model Performance on Validation Data
    (Top Features):")
print("Accuracy:", accuracy_score(y_validation, y_val_pred))
print("Classification Report:")
print(classification_report(y_validation, y_val_pred))

# Output
Random Forest Model Performance on Validation Data (Top Features):
Accuracy: 0.9363658359545125
Classification Report:
                precision    recall  f1-score   support
     fighting        0.95      0.78      0.86        23
      grazing        0.93      0.93      0.93      1121
        lying        0.93      0.84      0.88       295
      running        0.97      0.97      0.97        97
scratch_biting       0.75      0.32      0.45        28
      shaking        1.00      0.60      0.75         5
     standing        0.94      0.96      0.95      1809
     trotting        0.92      0.92      0.92       122
      walking        0.94      0.94      0.94       633

     accuracy                            0.94      4133
    macro avg        0.93      0.81      0.85      4133
 weighted avg        0.94      0.94      0.94      4133
```

Here is an explanation of the code:

1. Preparing Validation Data: The validation dataset, X_validation_n, is prepared by selecting only the top-ranked features (top_features) that were used to train the model. This step ensures that the model is evaluated on the same feature set it was trained on, maintaining consistency in the feature space.
2. Model Prediction: With the validation dataset prepared, the model (rf_model) predicts the outcomes (y_val_pred) for the validation data (X_val_top_features).

3. Displaying Model Performance: Finally, the performance of the model on the validation data is displayed. The accuracy score provides a single metric representing the overall correctness of the predictions.

The results obtained using the top 10 features in our Random Forest model show the effectiveness of feature selection. With an overall accuracy of approximately 93.64%, the model demonstrates good performance, closely mirroring the accuracy achieved when all features were considered. This indicates that the selected features retain most of the predictive power of the full feature set, showing the effectiveness of focusing on the most informative attributes.

Breaking down the output:

1. For most behaviors, the model's ability to correctly identify and classify observations is worthy.
2. The model shows strong performance in identifying 'running' with precision and recall both at 0.97, indicating that the features critical for distinguishing this behavior are well-represented among the top 10.
3. 'Scratch_biting' and 'shaking', however, exhibit lower performance metrics compared to others, highlighting potential areas where the reduced feature set may lack some information necessary for optimal classification. The drop in recall for 'shaking' (down to 0.60) suggests that while the model can accurately identify this class when it predicts it, it often misses it in other instances.

Comparing these results to those achieved using all features, it is evident that feature selection has streamlined the model without substantially compromising its performance. This makes the model simpler and potentially faster but also emphasizes the importance of the selected features in making predictions. The slight variances in performance for certain classes could direct further investigation into which features are most critical for those specific behaviors and whether any additional features should be considered.

### *Dimensionality Reduction Using PCA*

After evaluating the performance of the Random Forest model with feature selection, it is valuable to explore how dimensionality reduction techniques, such as PCA, impact the model's performance. PCA will transform the original features into a new set of features, which are linear combinations of the original ones. By selecting a subset of the new features, which capture most of the data's variance, we can potentially reduce the dimensionality of the dataset while retaining most of its information.

To apply PCA to our dataset, we take the following steps:

- Standardize the original dataset.
- Apply PCA.
- Transform the data into the first few principal components.

The following apply_pca custom function implements and evaluates a Random Forest model with PCA.

```
from sklearn.decomposition import PCA

def apply_pca(X_train, y_train, X_validation, y_validation, n_
components=0.95):
    """
    Apply PCA on the training dataset and validate using a Random
    Forest model.

    Parameters:
    - X_train: Training features.
    - y_train: Training labels.
    - X_validation: Validation features.
    - y_validation: Validation labels.
    - n_components: Number of PCA components or explained variance.

    Returns:
    - RandomForest classifier trained on the transformed data.
    - Classification report on the validation data.
    """

    # Standardize the dataset
    scaler = StandardScaler()
    X_train_standardized = scaler.fit_transform(X_train)
    X_validation_standardized = scaler.transform(X_validation)

    # Apply PCA
    pca = PCA(n_components=n_components)
    X_train_pca = pca.fit_transform(X_train_standardized)
    X_validation_pca = pca.transform(X_validation_standardized)

    # Train and validate a Random Forest model
    rf = RandomForestClassifier(n_estimators=100, random_state=42)
    rf.fit(X_train_pca, y_train)
    y_val_pred = rf.predict(X_validation_pca)

    accuracy = accuracy_score(y_validation, y_val_pred)
    classification_rep = classification_report(y_validation,
       y_val_pred)
```

```
        return rf, accuracy, classification_rep

# You can call the function and print out the results as follows:
rf_model, acc, class_rep = apply_pca(X_train, y_train,
    X_validation, y_validation)

print("Random Forest Model Performance on Validation Data (PCA):")
print("Accuracy:", acc)
print("Classification Report:")
print(class_rep)

# Output
Accuracy: 0.9029760464553593
Classification Report:
               precision    recall  f1-score   support
     fighting       0.88      0.61      0.72        23
      grazing       0.91      0.89      0.90      1121
        lying       0.84      0.70      0.77       295
      running       0.95      0.97      0.96        97
scratch_biting      1.00      0.18      0.30        28
      shaking       1.00      0.80      0.89         5
     standing       0.88      0.94      0.91      1809
     trotting       0.96      0.93      0.95       122
      walking       0.96      0.92      0.94       633

     accuracy                           0.90      4133
    macro avg       0.93      0.77      0.81      4133
 weighted avg       0.90      0.90      0.90      4133
```

Here is what each part of the code does:

1. Import PCA: The PCA class from sklearn.decomposition is imported.
2. Define apply_pca Function: A function is defined to apply PCA to the training and validation datasets before using these transformed datasets to train and validate a Random Forest model. The function parameters allow for specifying the datasets, the number of PCA components (or the amount of variance that needs to be explained by the components), and the model parameters.
3. Standardize the Dataset: The training and validation datasets are standardized using StandardScaler.
4. Apply PCA: PCA is applied to the standardized training data to transform it into a set of linearly uncorrelated components. The n_components = 0.95 parameter means that the number of components chosen by PCA will explain 95% of the variance in the data. The same transformation is applied to the validation dataset.

5. Train and Validate Random Forest Model: A Random Forest classifier is then trained on the PCA-transformed training data. After training, the model makes predictions on the PCA-transformed validation data.
6. Evaluate Model Performance: The model's performance is evaluated using the accuracy_score and classification_report from sklearn.metrics.
7. Function Call and Results Printing: The apply_pca function is called with the original training and validation datasets. The Random Forest Model's performance is printed.

Upon applying PCA and evaluating the Random Forest model, the performance on the validation data yields an accuracy of approximately 90.30%. While precision, recall, and F1-scores across various classes show competent performance, there is a notable difference compared to using all features directly, especially in classes like 'scratch_biting' which shows a significant drop in recall. This illustrates that while PCA can effectively reduce dimensionality and still capture essential patterns in data, there may be a trade-off in model sensitivity, particularly for classes represented by subtle or complex feature interactions.

It is crucial to mention that the effectiveness of PCA is more pronounced in datasets with very high dimensionality. In our case, with a dataset containing only 20 features, the reduction might not be as necessary or impactful as it would be for datasets with hundreds or thousands of features. Nonetheless, exploring PCA provides valuable insights into how dimensionality reduction techniques can influence model performance and offers a perspective on balancing model complexity with computational efficiency.

## Hyperparameter Tuning and Model Evaluation

In this section, we will focus on tuning the parameters of the Random Forest model, which we have identified as the model of our choice due to its superior performance on our dataset. This process involves adjusting the model's hyperparameters to optimize its performance.

For the Random Forest model, several hyperparameters can significantly influence its accuracy and efficiency.

Here is a structured approach to hyperparameter tuning for Random Forest models:

1. Understanding Key Hyperparameters: Before starting the tuning process, it is important to understand the key hyperparameters associated with Random Forest:
    - n_estimators: The total trees in the forest.

- max_depth: The tree's maximum depth.
- min_samples_split: The minimum sample count needed for splitting a node.
- min_samples_leaf: The minimum sample count a leaf node must have.
- max_features: The feature count evaluated for the optimal split.
- bootstrap: Indicates if bootstrap sampling is utilized in tree construction.

2. Hyperparameter selection: Several techniques can be used for this task:
   - Manual Search: Utilize initial settings based on experience to establish a performance baseline.
   - Grid Search: Grid search is a comprehensive search approach where you define a range of values for various hyperparameters. The algorithm will try all possible combinations of hyperparameters.
   - Random Search: Instead of trying out every possible combination like in grid search, random search sets a number of random combinations to try. This can be more efficient and still yield a good set of hyperparameters.
   - Bayesian Optimization: This is a probabilistic model-based optimization approach. Tools like Hyperopt or Optuna can be used for this. Bayesian optimization is more efficient than grid and random search as it builds a probabilistic model of the objective function and uses it to select the most promising hyperparameters to evaluate in the true objective function.
3. Evaluate on Validation Set: After obtaining a set of hyperparameters, evaluate the model's performance on a validation set. This helps in ensuring the model is not overfitting and can generalize well to unseen data.
4. Iterative Process: Hyperparameter tuning is often iterative. Based on results and domain knowledge, you might need to refine the hyperparameter space and perform another search.
5. Final Model Evaluation: Once you have identified the best hyperparameters, train the model using these hyperparameters on the complete training dataset. Then evaluate the model on the test dataset to get an unbiased estimate of its performance.

In our example, we will focus on implementing Bayesian optimization for hyperparameter tuning, leveraging the power of Optuna library. This approach allows us to efficiently search through the hyperparameter space to find the settings that maximize our model's performance.

Below, is a step-by-step process:

### Step: 1 Install and Import Necessary Libraries

```
# Install and Import Necessary Libraries

# !pip install optuna

import optuna
from sklearn.model_selection import cross_val_score
```

First, we need to ensure that Optuna is available in our environment. If it is not already installed, you can do so by running !pip install optuna. After installing Optuna, we import it along with cross_val_score from sklearn.model_selection.

### Step 2: Define Objective Function for Optimization

In this step, we will define the hyperparameter space and train a Random Forest model. We will use *cross-validation* to evaluate the performance of the model. *Cross-validation*, which will be detailed in Chapter 8, is a technique used to assess how well a model will generalize to an independent dataset by dividing the data into multiple parts, training the model on some segments and validating it on others.

```
def objective(trial):
    # Define hyperparameter space
    n_estimators = trial.suggest_int('n_estimators', 50, 300)
    max_depth = trial.suggest_int('max_depth', 10, 40, log=True)
    min_samples_split = trial.suggest_int('min_samples_split', 2, 15)
    min_samples_leaf = trial.suggest_int('min_samples_leaf', 1, 10)
    max_features = trial.suggest_categorical('max_features',
       ['auto', 'sqrt'])
    bootstrap = trial.suggest_categorical('bootstrap', [True, False])

    # Initialize and train classifier
    clf = RandomForestClassifier(
        n_estimators=n_estimators,
        max_depth=max_depth,
        min_samples_split=min_samples_split,
        min_samples_leaf=min_samples_leaf,
        max_features=max_features,
        bootstrap=bootstrap,
        n_jobs=-1
    )
    # Use cross-validation to evaluate model performance
    return cross_val_score(clf, X_train_n, y_train, n_jobs=-1,
       cv=5).mean()
```

This code defines an objective function for use with Optuna focusing on tuning a Random Forest classifier. Here is how it works and why specific numbers are chosen for the hyperparameters:

1. Hyperparameter Space Definition:
    - n_estimators: The suggested number of estimators is between 50 and 300. This range is chosen to give enough variability to see how increasing the number of trees affects performance without going to extremes that might only offer diminishing returns or excessive computation.
    - max_depth: The suggested max depth is between 10 and 40, with a logarithmic scale (log = True). The log scale is used to explore the lower end of the range more finely, where smaller changes can influence the model complexity and performance.
    - min_samples_split: The suggested min samples split is between 2 and 15. This range enables the model to investigate both small and relatively larger node sizes before deciding on a split.
    - min_samples_leaf: The suggested min samples leaf is between 1 and 10. This parameter helps restrict overfitting by preventing the model from learning overly intricate patterns.
    - max_features: To determine the number of features to be evaluated during the search for the optimal split, there are two options. The options are to either use all features (auto) or a square root of the number of features (sqrt). This choice affects how diverse each tree in the forest is and can influence model accuracy and overfitting.
    - bootstrap: The options of bootstrap are True (bootstrap sampling) or False (the entire dataset is used to build each tree). Bootstrapping introduces randomness into the model, which can help improve robustness and accuracy.
2. Random Forest Classifier Initialization and Training: The classifier is initialized with the suggested hyperparameters and set to use all available CPU cores (n_jobs = –1) for faster training.
3. Model Evaluation with Cross-Validation: The cross_val_score function evaluates the classifier's performance using cross-validation with 5 folds (cv = 5), meaning the training set is divided into five parts, with the model trained on four and validated on the fifth, rotating until each part has been used for validation. The function returns the mean accuracy across all folds, which serves as the objective value Optuna which we seek to maximize.

This approach allows Optuna to systematically explore the hyperparameter space, using the objective function's return value to guide the search towards combinations that yield the best cross-validation accuracy.

### Step 3: Run Optimization

The following code initiates and executes a hyperparameter optimization process using Optuna, aiming to maximize the cross-validation score of our model by finding its optimal hyperparameters.

```
study = optuna.create_study(direction='maximize')  # We want to
maximize the cross-validation score
study.optimize(objective, n_trials=100)  # Number of iterations

best_params = study.best_params
best_score = study.best_value

# Output:
best_params:
{'n_estimators': 223,
 'max_depth': 24,
 'min_samples_split': 2,
 'min_samples_leaf': 1,
 'max_features': 'auto',
 'bootstrap': False}

best_score: 0.941725499462175
```

Breakdown of each part of the code:

1. Creating an Optuna Study:
    - optuna.create_study(direction = 'maximize'): A study in Optuna is an optimization task, and here, the direction = 'maximize' argument indicates that the goal is to maximize the objective function, which in this context is the mean cross-validation result yielded by the objective function defined earlier.
2. Optimizing the Study:
    - study.optimize(objective, n_trials = 100): This method starts the optimization process by calling the objective function multiple times (100 trials in this case). For each trial, Optuna proposes a set of hyperparameters that are passed to the objective function, which trains a Random Forest model with those parameters and returns the mean cross-validation score. Optuna uses these scores to guide the search towards hyperparameter values that maximize the performance of the model.
3. Retrieving the Best Parameters and Score:
    - best_params = study.best_params: After completing the specified number of trials, the best hyperparameters identified during the optimization are stored in study.best_params.

- best_score = study.best_value: Similarly, study.best_value holds the highest cross-validation score achieved during the optimization.

By the end of this process, you have a set of optimized hyperparameters (best_params) and an understanding of how well the model can perform with those parameters (best_score), providing a solid foundation for building a high-performing Random Forest classifier.

### Step 4: Evaluate on the Validation Set

After obtaining the best hyperparameters from Bayesian optimization, the model is trained using the training dataset and then assess its performance on the validation set.

```
best_rf = RandomForestClassifier(**best_params, n_jobs=-1)
best_rf.fit(X_train_n, y_train)

predictions = best_rf.predict(X_validation_n)

print(f"Accuracy: {accuracy_score(y_val, predictions) * 100:.2f}%")
print("\nClassification Report:\n", classification_report(y_val,
predictions))

# Output:
Accuracy: 94.82%
Classification Report:

               precision    recall  f1-score   support
     fighting       0.91      0.87      0.89        23
      grazing       0.93      0.95      0.94      1121
        lying       0.95      0.88      0.91       295
      running       0.98      0.99      0.98        97
scratch_biting      1.00      0.32      0.49        28
      shaking       1.00      1.00      1.00         5
     standing       0.95      0.97      0.96      1809
     trotting       0.97      0.94      0.96       122
      walking       0.96      0.95      0.95       633

     accuracy                           0.95      4133
    macro avg       0.96      0.87      0.90      4133
 weighted avg       0.95      0.95      0.95      4133
```

Code breakdown:

1. best_rf = RandomForestClassifier(**best_params, n_jobs = –1) and best_rf.fit(X_train_n, y_train): the best_params are applied to construct and train a new model, best_rf, on our normalized training dataset (X_train_n).

2. predictions = best_rf.predict(X_validation_n): The model is then evaluated on a normalized validation dataset (X_validation_n).

The results demonstrate an improvement in model performance. For example, accuracy has increased to 94.82% from the previously observed 93.90%. Precision, recall, and F1-scores across most classes have shown improvements or remained strong. For instance, the precision and recall for 'fighting' have remained consistent, but we see an improvement in the model's recall for 'lying', from 0.85 to 0.88, and an increase in precision for 'grazing' from 0.92 to 0.93. Comparing these results with those obtained before hyperparameter optimization, it is clear that the systematic search for the optimal settings has yielded a more effective model.

### Step 5: Final Model Training and Evaluation

```
# Train the Random Forest classifier with the best hyperparameters
final_rf = RandomForestClassifier(
    n_estimators=223,
    max_depth=24,
    min_samples_split=2,
    min_samples_leaf=1,
    max_features='auto',
    bootstrap=False,
    n_jobs=-1
)

# Fit the model to the training data
final_rf.fit(X_train_n, y_train)

# Predict on the test set
test_predictions = final_rf.predict(X_test_n)

# Evaluate the predictions
print(f"Accuracy on Test Set: {accuracy_score(y_test,
    test_predictions) * 100:.2f}%")
print("\nClassification Report on Test Set:\n", classification_
    report(y_test, test_predictions))

# Output:
Accuracy on Test Set: 95.11%
Classification Report on Test Set:
              precision    recall  f1-score   support
    fighting       0.85      0.77      0.81        22
     grazing       0.95      0.95      0.95      1121
       lying       0.95      0.85      0.89       295
     running       0.97      0.96      0.96        98
```

```
scratch_biting       0.83      0.36      0.50        28
       shaking       0.83      1.00      0.91         5
      standing       0.95      0.98      0.96      1810
      trotting       0.93      0.94      0.94       122
       walking       0.97      0.96      0.97       633

      accuracy                           0.95      4134
     macro avg       0.92      0.86      0.88      4134
  weighted avg       0.95      0.95      0.95      4134
```

Code breakdown:

1. Initialize the Final Model: The RandomForestClassifier is initialized with the optimal hyperparameters discovered earlier: n_estimators = 223, max_depth = 24, min_samples_split = 2, min_samples_leaf = 1, max_features = 'auto', and bootstrap = False. The n_jobs = –1 parameter allows the classifier to leverage all available CPU cores to accelerate the training process.
2. Model Training: The final model, final_rf, is then trained (fit) on the normalized dataset (X_train_n and y_train).
3. Making Predictions: After training the model, predictions are generated on the normalized test set (X_test_n), which simulates how the model would perform when making predictions on new, unseen data.
4. Mode Evaluation: accuracy_score and classification_report are called to give insights into the overall accuracy and detailed performance metrics for each class.

From the output we can see that the model achieves an accuracy of 95.11% on the test set, indicating a high predictive performance level.

This marks the conclusion of our journey through applying machine learning to accelerometer data of sheep and goats. From preprocessing and exploratory data analysis to feature selection, hyperparameter tuning, and final evaluation, we have seen how systematic approaches can refine a model to achieve high accuracy. This example highlights the practical steps in developing a ML model.

## Summary

In Chapter 3, we tackled the practical side of machine learning by developing a Python-based project to classify the behaviors of sheep and goats. Our approach is hands-on, focusing on the actual steps needed to carry out a machine learning project. We started with the basics: loading and preprocessing data, followed by exploratory data analysis using histograms, correlation matrix, and boxplots to understand the dataset better. We then introduced the feature extraction technique using windowing methods, which is crucial for preparing the data for modeling.

The subsequent steps involved selecting features and tuning the model's hyperparameters to optimize performance. These stages are key to refining the model and enhancing its accuracy. Finally, we demonstrated how to evaluate the model's effectiveness in a real-world scenario, emphasizing the importance of a systematic assessment to ensure reliability. Throughout this chapter, our goal was to illustrate the essential components of a machine learning pipeline, rather than delving into every technical detail. By simplifying some of the content, we aimed to maintain a clear and coherent narrative that makes the core processes accessible and understandable, avoiding the potential for confusion that might arise from a more complex presentation.

CHAPTER 4

# Sensors, Data Collection and Annotation

Over the years, methods for studying and interpreting animal behavior have become increasingly sophisticated, especially with advancements in technology. One of the crucial aspects is the evolution of data collection techniques, which have significantly transformed the way we approach this task.

Animal behavior studies center on the accurate collection, preservation, and analysis of multifaceted data. This involves an animal's movements, its vocalizations, interactions, and various other behavioral attributes. The goal is to adopt an understanding of the animal's physical activity, and also its emotional inclination and its interactions within its environment and social groups.

The field of data collection is massive. A plethora of sensors, each designed for specific observational needs, exist to researchers and professionals. The choice of a particular data collection method is based on many variables. The species of the animal under observation, the behavior allocated for study, and the related conditions during data collection, are instrumental in guiding this choice.

### Deciding on the Data's Relevance

To start a study or project requires a decision on the type of data that would serve the aim. It is not just about having data; it is about having the right data. The decision-making process begins by describing the objectives of the study. What behaviors are we most interested in? Are we looking at long-term patterns or short, episodic bursts of activity? Such questions guide the selection of sensors and data collection methodologies.

## Overview of Data Collection Methods

Data collection for animal behavior studies involves capturing, storing, and analyzing various types of information about an animal's movements, sounds, interactions, and other aspects of its behavior. The primary purpose is to gain insights into the animal's physical state, emotional state, and social interactions.

In the early stages of Precision Livestock Farming (PLF), data collection was mostly manual. Researchers and farmers would engage in direct observations, making use of basic tools and their own senses to evaluate animal behavior. Charts, simple logs, and visual cues served as the means to record observed behaviors. As time progressed, the need for more accurate and granular data drove innovations in data collection methodologies. Manual observations began to be complemented, and in many cases replaced, by mechanical and eventually electronic sensors.

The modern era of PLF has witnessed the smooth integration of technology. The arrival of microelectronics, wireless communication, and advanced software algorithms has revolutionized the way we collect and analyze animal behavior data. These advancements enable real-time monitoring and offer capabilities like remote monitoring, predictive analytics, and automated interventions.

## *The Importance of Good Quality Data*

Collecting high-quality data significantly influences the success of any machine learning project. In the context of animal behavior studies, the data collected via these sensors serves as the basis for training machine learning models. Therefore, it is essential to ensure the data is accurate, relevant, and representative of the behaviors we want our models to learn.

### Defining 'Good' Quality Data

To explore large collections of data and assign what qualifies as 'good', it becomes essential to establish certain benchmarks:

- **Precision:** Precision features the consistency of the data. In the context of animal behavior studies, precision translates to obtaining similar or identical data points under unchanged or repeated conditions.
- **Accuracy:** This denotes the closeness of a data point to its actual or true value. For instance, if a sensor is meant to track an animal's movement across a pasture, its readings should correctly reflect the actual distances traveled.
- **Reliability:** Reliability requires the dependability of data over time and across varied conditions. A reliable sensor, regardless of external conditions or time, will capture data consistently.
- **Relevance:** Data should be relevant to the research objectives. Gathering data that is extraneous or irrelevant might divert the analysis, weakening the eventual insights

### Implications of Poor Quality Data

Compromised data quality can lead to the following issues:

1. **Misinterpretations:** Faulty data can misguide machine learning algorithms,

causing them to misinterpret patterns or associations. For instance, if a sensor erroneously records excessive movement during an animal's rest phase, the algorithm might mistakenly classify that period as active, leading to biased behavioral profiles.

2. **Flawed Insights:** Decisions or conclusions based on poor-quality data are fundamentally faulty. They can misrepresent realities, posing challenges for farmers and researchers.
3. **Potential Risks:** In precision livestock farming, inaccurate data interpretations can risk animal well-being. Misreading signs of distress or disease due to poor data might delay necessary interventions, escalating health risks for animals.

The core of machine learning resides in the quality of data it gets. Ensuring this data is robust in terms of accuracy, precision, reliability, and relevance is a universal exercise and a fundamental necessity.

### Deciding on the Right Data Type

Making informed decisions regarding the appropriate data type is dependent on the understanding of the study's goals, an appreciation of the environmental factors, and a thorough assessment of available resources.

An interpretation of the study's main objectives is at the center of the decision-making process. Each research work in animal behavior influences its set of aims, whether it is mapping the travelling patterns of a herd, decoding the communication signals among a flock, or analyzing the metabolic rates under varied conditions.

A study towards understanding the spatial dynamics of a herd might find GPS data more appropriate. In contrast, an investigation into an animal's response to varying temperatures might rely heavily on thermal sensors. Thus, setting clear objectives allows researchers to align their goals with the most fitting data types, ensuring both relevance and precision.

## *Considerations for Data Collection*

The data collection requires an alignment with the study's goals but also a thorough consideration of various influential factors:

1. **Environmental Factors:** The environment in which the study is conducted always influences the data type and collection methodology. A study in dry and hot fields might need robust sensors resistant to dust and heat, while a marine study would require waterproof equipment. Additionally, environmental conditions like lighting, temperature variations, and seasonal changes can impact sensor readings and should be considered during data type decisions.
2. **Animal Specifics:** The species, size, habits, and physiology of the animals under study can greatly influence the choice of data type. For instance, tracking nocturnal activities might necessitate the use of infrared sensors,

while monitoring the heartbeat of a small bird could require highly sensitive equipment.

3. **Equipment Availability and Suitability:** The availability of specific equipment and its compatibility with the study's needs is vital. Researchers must weigh the pros and cons of available tools, deciding on aspects like battery life, data storage capacities, range, and durability.
4. **Ethical Considerations:** Ensuring the welfare of the animals being monitored is crucial. Any equipment or sensors used should not harm, stress, or overly worry the animals. This requires choosing lightweight, non-invasive tools, and regularly monitoring the animals for signs of discomfort.

## Exploration of Selected Sensors

In this book, our focus will be on the most commonly used sensors that have proven to be impactful in animal activity recognition. These sensors have been selected based on their extensive application and the richness of the data they provide. The following discussion will be related to accelerometers, gyroscopes, cameras, and GPS systems, offering a comprehensive overview of their working principles, data types, and use-cases in the field of PLF.

### *Accelerometers and Gyroscopes*

#### Accelerometers

Accelerometers are devices that measure proper acceleration, commonly referred to as g-force. Proper acceleration is different than coordinate acceleration. For instance, resting accelerometer in the surface of the Earth will provide an acceleration measure of $g = 9.81\text{m/s}^2$. In contrast, an accelerometer in free fall towards the Earth will measure zero.

In the context of animal behavior, accelerometers can provide insights into the movement patterns of animals. For example, they can detect when a cow stands up or lies down, giving clues about its health and well-being.

**Working Principle:** An accelerometer measures the changes in velocity along one or more axes. Modern accelerometers are typically micro-electromechanical systems (MEMS) and are small, affordable, and energy efficient, making them ideal for integration into wearable animal tags.

**Data Types:**

- **Raw Acceleration Data:** The fundamental form of data collected by accelerometers, consisting of acceleration readings along three axes ($x$, $y$, $z$). This tri-axial data provides a comprehensive view of movement in three-dimensional space.

- **Processed Data:** Post-collection, the raw data can be processed to compute various metrics. One common metric is the overall dynamic body acceleration (ODBA), widely used to estimate energy expenditure in animals.

**Capture Methods:** Attaching accelerometers to animals requires a balance between securing comprehensive data and ensuring the welfare and natural behavior of the animal. The methods vary significantly depending on the species:

- For terrestrial animals, accelerometers are often attached via collars or harnesses.
- For aquatic species, direct attachment to the body or fins is common.
- For birds, lightweight backpacks or harnesses are used to minimize impact on flight.

**Use Cases:** Monitoring the activity levels of animals, detecting falls or abnormal movements, understanding grazing patterns, etc.

### Gyroscopes

Gyroscopes are devices used to measure angular velocity. In simpler terms, they detect how fast something is spinning around an axis.

**Working Principle:** Gyroscopes rely on the principles of angular momentum. When an external force is applied, it produces a behavior known as precession, which is the change in the orientation of the rotational axis.

**Data Types:** Gyroscopes provide the rate of rotation around the device's X, Y, and Z axis.

**Capture Methods:** Same as accelerometers.

**Use Cases:** In animal behavior monitoring, gyroscopes can help determine rotational movements, like when an animal is shaking its head or rolling.

Figure 4.1 illustrates the acceleration data captured from an animal over the span of one minute. The data is separated into X, Y, and Z directions. The distinct lines represent acceleration along the X, Y, and Z axes, respectively. The variations in the graph portray the dynamic movements of the animal, with each axis representing a different spatial direction.

Displayed in Figure 4.2 is the gyroscope data showcasing the angular velocity of the animal over a 60-second interval. The lines represent the angular rotational movements about each respective axis. This provides insights into the orientation and turning behaviors of the animal during the observation period.

**The distinction between accelerometers and gyroscopes—accelerometers monitor linear movement, while gyroscopes track rotational dynamics.**

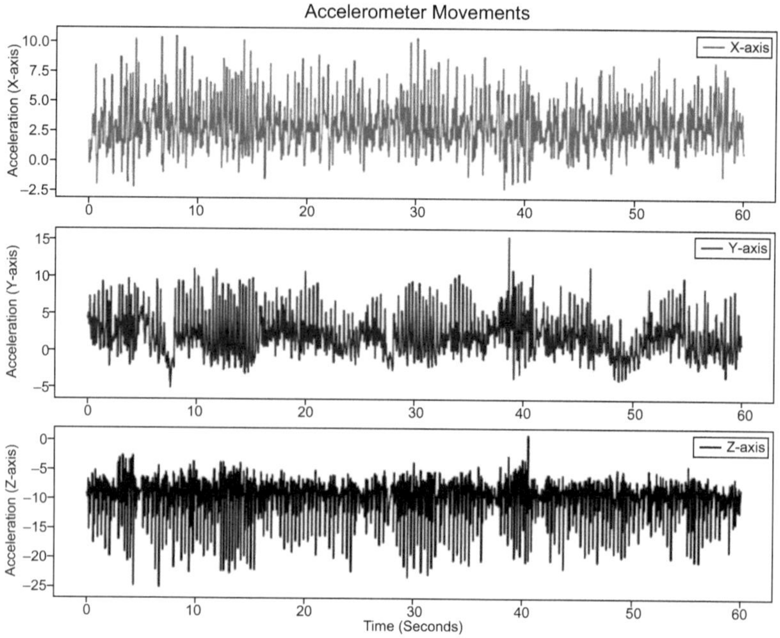


**Figure 4.1: Acceleration data showing movements in X, Y, and Z directions over one minute.**

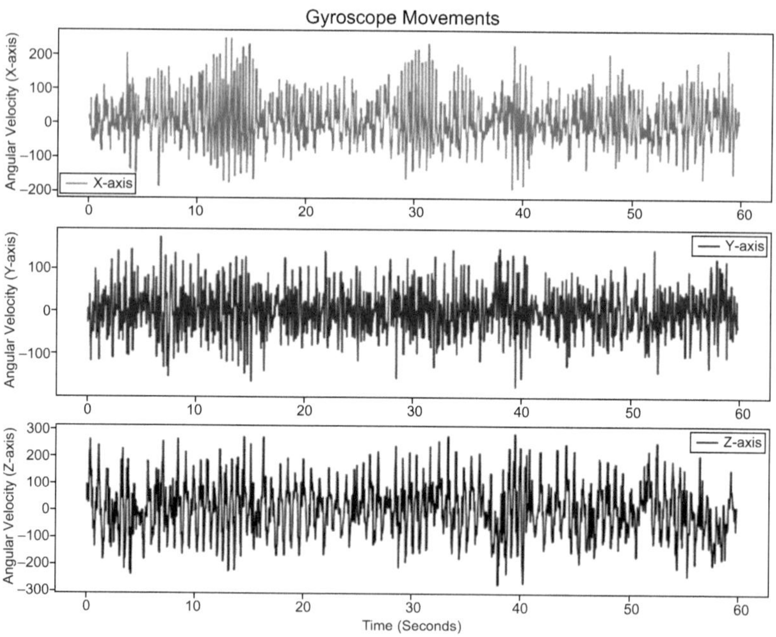


**Figure 4.2: Gyroscope data showing angular velocity in X, Y, and Z directions over one minute.**

## *Cameras*

Cameras, ranging from traditional to thermal and UAV-based, play a pivotal role in capturing visual data from the field.

### Types

1. **Traditional Cameras:** These devices capture visual details by receiving light through a lens, translating it into video or still imagery. They are most effective under sufficiently brightened conditions, making them suitable for daytime observations.
2. **Thermal Cameras:** These take advantage of the principle of infrared radiation detection. By sensing and visualizing temperature differences, they produce images, making nocturnal observations feasible. Thermal variations can indicate potential illnesses, positioning these cameras as vital tools for preventive health checks.
3. **UAV-based Cameras**: Drones or UAVs, equipped with cameras, offer aerial visuals, ideal for monitoring large animal gatherings or expansive terrains. Their primary advantage is mobility, capturing imagery from diverse altitudes and perspectives.

**Working Principle:** Cameras capture light through a lens, which hits a sensor that converts this light into an electronic signal to produce an image.

**Data Types**: Video footage, still images.

**Use Cases:** Monitoring herd movement, detecting heat signatures of animals (especially useful during the night or in dense forests), aerial surveillance of large herds.

## *GPS*

The Global Positioning System (GPS) is a satellite-based navigation system that plays an essential role in animal behavior studies. GPS technology is instrumental in tracking and understanding the spatial movements of animals, offering invaluable insights into their grazing patterns, territory utilization, and migratory habits.

**Operational Mechanism:** GPS operates by triangulating signals from a constellation of satellites orbiting the Earth. Each satellite transmits a signal that includes its location and the exact time the signal was transmitted. A GPS receiver, such as a collar worn by an animal, calculates its precise location by measuring the time delay between the transmission and reception of signals from multiple satellites.

**Data Types:** Coordinates (Longitude, Latitude), Altitude, Time stamp.

Applications in Animal Behavior Studies:

- **Territory Mapping:** GPS collars can map the territories and ranges of animals, providing insights into their habitat utilization.
- **Grazing Pattern Analysis:** For grazing animals like cattle or sheep, GPS data can reveal patterns in grazing behavior, helping to optimize pasture management.
- **Migration Tracking:** The technology is also vital in studying migratory patterns, particularly in wildlife conservation efforts.

**Use Cases:** Tracking animal migration patterns, monitoring the movement of herds over vast pastures, ensuring animals stay within designated boundaries, studying wildlife in their natural habitat.

## Data Collection Process

### *Planning and Setup*

The data collection process starts long before any actual data is captured. In the field of animal activity recognition, thorough planning is crucial. Understanding the objectives of your study will help in narrowing down the most appropriate sensors to be deployed. For instance, if your research aims to track an animal's movement patterns, GPS sensors may be more suitable. On the other hand, understanding an animal's intricate behaviors might require accelerometers, gyroscopes, or cameras. The process requires careful planning, dedication, attention to detail, and often physical labor. In many ways, the quality of this initial stage will state the quality of everything that follows.

#### Choosing the Right Sensors

Given the numerous sensors available, it can seem overwhelming to choose the right one. Begin by understanding the nature of the data you need:

**Type of Data:** Are you looking for visual, motion, location, or thermal data?

**Granularity of Data:** Do you need detailed, high-frequency data or is a broader overview sufficient?

**Environmental Conditions:** Will the data collection occur in a controlled environment or in the wild where conditions can be unpredictable?

#### Visiting the Site

Before any data can be collected, it is essential to visit the site, whether it is a farm, a wildlife reserve, or any other location. This visit allows researchers to familiarize themselves with the environment, understand the daily routines of the animals, and identify potential challenges, such as areas with weak signal reception or possible limitations to equipment placement.

### Handling the Animals

Handling and outfitting animals with sensors is a delicate operation that typically requires the assistance of an expert or veterinarian. It is critical to ensure the animals' safety and comfort during this procedure. An inappropriately placed sensor might yield inaccurate data and could also harm or distress the animal. For instance, when placing collars on sheep, one must consider the weight of the equipment and ensure that it does not restrict the animal's movement or cause discomfort.

**Any sensor attached to animals must be secure and without sharp edges or components that might cause harm or discomfort.**

### Adjustment Period

Once the equipment is attached, animals need time to adjust. Introducing foreign objects, like sensors or collars, might initially alter an animal's behavior. It is vital to provide an adjustment period, allowing the animals to return to their natural behaviors before actual data collection begins. This period can range from half an hour to several hours, depending on the animal and the nature of the equipment.

### Programming the Sensors

Modern sensors often come with a range of settings and options. Researchers must program these sensors according to the specifics of their study. This could involve setting the frequency of data collection, determining whether the collection should be continuous or periodic, and other vital parameters.

### Preparing the Observational Location

For researchers aiming to verify sensor data with visual confirmations, setting up an observational post is necessary. This could involve setting up cameras to record animals, providing a *ground truth* reference for behaviors detected by the sensors. Proper positioning is essential to ensure comprehensive coverage and clear, unobstructed views.

### Attention to Detail

Given that this stage lays the groundwork for all subsequent analysis, researchers must be persistent. A missed detail here could lead to hours of wasted effort later. For example, sensors that are not calibrated correctly might produce data that is consistently off, leading to inaccurate machine learning predictions.

### Setting up Equipment

Once the right sensors are chosen, the next step is setting them up.

**Placement**: Depending on the type of animal and the behavior you are studying, the sensor's placement is critical. For instance, accelerometers placed on an animal's leg will provide data different from those placed on its back. Cameras should be positioned to capture the full range of the animal's activities without obstructions.

**Calibration**: All sensors require calibration to ensure the data they produce is accurate. This step is especially crucial for sensors like accelerometers and gyroscopes. Calibration processes may vary across sensors, but they often involve setting them up in a known state or condition and adjusting them until their outputs match the expected values.

## *Data Acquisition*

With everything set up, the next phase is data acquisition, which involves the actual process of gathering data from the sensors.

### Periodic vs. Continuous Data Collection

1. **Periodic Data Collection:** In this approach, data is collected at specified intervals. For example, a GPS sensor might be configured to record an animal's location every ten minutes. While this method can conserve storage and battery life, it may overlook some short-duration events.
2. **Continuous Data Collection:** This approach involves capturing data without any gaps. It is particularly useful when studying behaviors that can occur unpredictably and spontaneously. However, it demands more storage and energy.

### Data Storage

Given the large volume of data these sensors can produce, especially in continuous collection mode, having an effective storage solution is critical.

1. **On-device Storage:** Many sensors come with built-in storage capabilities. This is convenient but is often limited in capacity.
2. **Cloud Storage:** Some modern sensors can transmit data in real-time to cloud storage solutions. This provides virtually unlimited storage capacity but requires the sensors to be within the network coverage.
3. **Local Storage Solutions:** In scenarios where real-time data transmission is not possible or when on-device storage is insufficient, data can be periodically offloaded to local storage devices like laptops or external hard drives.

In essence, the planning and setup phase is arguably the most labor-intensive and critical stage of the entire process. It is not purely about attaching sensors to animals; it is about ensuring the data that will be collected is as accurate and representative as possible. A well-laid foundation at this stage will undoubtedly benefit the subsequent phases, especially when it is time to analyze the data and draw meaningful conclusions.

## *Post-collection Data Processing*

Once the data has been collected, it is essential to shift the focus towards preparing the collected data for the subsequent phases of analysis and modeling. This step is often termed post-collection data processing. Even though modern sensors are sophisticated and technologically advanced, the data they capture is not always immediately ready for analysis. Post-collection processing is, therefore, a critical bridge between raw data collection and its utilization for visualization or model training.

### Cleaning and Formatting

Raw data collected from sensors can sometimes be noisy, incomplete, or may contain erroneous values. For example, if a sensor malfunctioned momentarily or an animal interacted with the device in an unforeseen manner, false readings might have been recorded. Cleaning involves identifying and fixing these anomalies, ensuring that the dataset is free from outliers or corrupted values that could distort the analysis. This process ensures that the data is coherent and represents true and accurate animal behavior.

Formatting, on the other hand, is about structuring the data into a more consistent and standardized format suitable for analysis. This may involve tasks like aligning timestamps, converting units, or restructuring data tables to be more compatible with the analysis tools and methods we plan to employ.

It is vital to note that while we touch upon these processes in this chapter, a detailed and methodical approach to cleaning and formatting will be extensively covered in Chapter 5, dedicated to data preprocessing and feature extraction.

### Preliminary Analysis

After the initial cleaning and formatting, it is beneficial to conduct a preliminary analysis of the data. This does not involve in-depth modeling or prediction tasks but serves to provide an initial understanding of the data's nature. During this stage, researchers may choose to plot the data to visualize trends, examine the distribution of various readings, or simply check for patterns or anomalies that were not caught during the cleaning phase. This analysis helps in understanding the general characteristics of the data, which, in turn, can inform more advanced analyses later on.

To conclude, post-collection data processing is a preparatory step, ensuring that data is in its best form before looking into analysis or model training. It is a demonstration of the fact that in machine learning and data science, the quality of input (data) often dictates the quality of the output (results or predictions). A thorough approach here sets the stage for optimal results in subsequent phases.

## Data Annotation

Data annotation is a vital step in preparing data for machine learning models, especially for supervised learning. By labeling or annotating the data, we provide context to the raw data, transforming it into informative samples from which the machine learning model can learn.

### *Importance of Annotated Data*

In supervised learning, the model learns to predict or classify based on annotated or labeled examples. This means that each sample in the dataset is paired with the correct outcome or label.

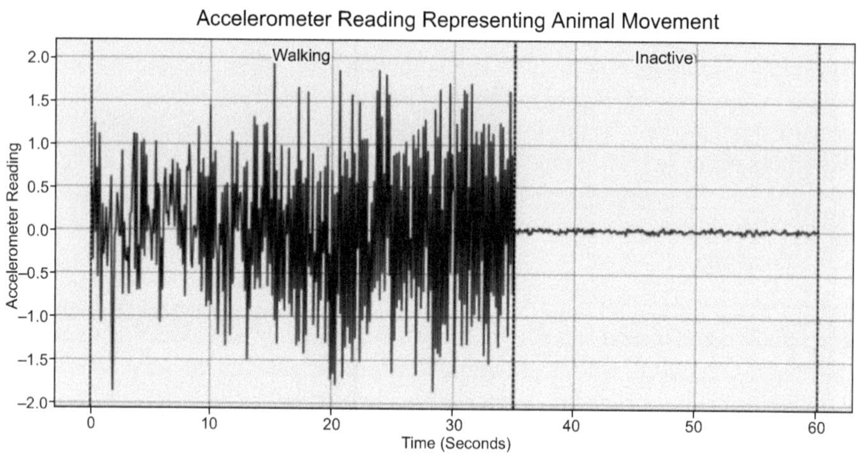


**Figure 4.3: Accelerometer data – A representation of walking and inactivity phases.**

Figure 4.3 offers an example of how accelerometer data can be annotated to gain insights into specific animal behaviors. The graph showcases raw accelerometer readings over a span of one minute. In the first 35 seconds, marked by vertical dashed lines, the pronounced spikes in the data suggest that the animal was actively walking. This is further confirmed by the annotated label '*walking*'. Following this active phase, the accelerometer readings become more uniform and moderated, indicating a period of inactivity for the animal, as denoted by the second annotated segment.

### *Methods of Annotation*

There are various methods through which data can be annotated:

**Manual:** Manual annotation is a labor-intensive process where human annotators label data points individually. This method is highly accurate, as it relies on human judgment and understanding. It is particularly useful for complex tasks where detailed understanding is crucial, such as image segmentation or sentiment analysis. However, manual annotation is time-consuming and can be expensive, especially for large datasets.

**Semi-automatic:** Semi-manual annotation combines human oversight with automated tools. In this approach, algorithms pre-process the data, performing basic annotations or suggesting labels. Human annotators then review and refine these automated suggestions. This method strikes a balance between accuracy and efficiency, allowing for quicker annotation of large datasets while maintaining a high quality standard.

**Automatic:** Automatic annotation is entirely machine-driven, using algorithms and Artificial Intelligence models to label data. This method is the fastest and most scalable, capable of processing vast datasets in a fraction of the time humans would require. Automatic annotation is ideal for well-defined, repetitive tasks, such as tagging objects in images where the context is straightforward. However, it may struggle with complex or subjective tasks.

## *Accelerometer Data Annotation*

Annotating accelerometer data is a process that involves labeling the sensor readings with the corresponding activities or behaviors of interest. Typically, accelerometer data are time-series signals that reflect the intensity of movement along different axes. The annotation process for this type of data can be broken down into several steps:

### Step 1: Data Visualization

Before annotation can begin, accelerometer data must be visualized to understand the patterns and to determine the thresholds for different behaviors. Software tools like MATLAB, Python with libraries such as Matplotlib, or domain-specific tools can be used for this purpose.

### Step 2: Segment Identification

By visual inspection or using automated algorithms, periods where a specific behavior occurs are identified. For instance, a period of grazing might be characterized by a consistent, low-frequency movement pattern.

### Step 3: Labeling

Once segments have been identified, they can be labeled with the corresponding behavior. This can be done manually, where a researcher reviews the segments and assigns labels, or semi-automatically, where thresholds are set to categorize the data based on signal magnitude or frequency.

### Step 4: Review and Refinement

The labeled data should be reviewed to ensure accuracy. It may require multiple iterations to refine the labeling, especially if the behavior is complex or if there is a significant overlap between behaviors.

### Step 5: Export and Usage

Finally, the annotated data are exported in a structure appropriate for training machine learning models. The annotations serve as the ground truth against which the predictions of the model are compared during the training process.

Annotation accuracy is critical since any errors in labeling can lead to poor model performance. Researchers should be trained to recognize the patterns associated with different behaviors and understand how to use the annotation software effectively.

### *Image and Video Annotation*

In animal behavior research, image and video annotation serve as crucial tools for extracting detailed information from visual data. Through the process of annotation, researchers can label specific behaviors, interactions, and even emotional states captured in images and videos, transforming raw footage into a rich dataset ready for analysis.

Image annotation involves tagging still images with labels or markers to highlight features of interest, such as specific animal postures or interactions with the environment. Video annotation goes a step further by tracking these behaviors over time, providing a dynamic view of animal actions and reactions within their habitats.

Despite their importance, the intricate methodologies, tools, and techniques involved in image and video annotation extend beyond the main focus of this book. As such, we will not explore these topics in further detail. Our discussion remains centered on the broader aspects of utilizing sensor technologies in the study of animal behavior, aiming to equip readers with foundational knowledge rather than delve into the specialized domain of visual data annotation.

## Summary

In Chapter 4, we looked at sensor technologies that are revolutionizing the study of animal behavior. We discussed tools, including accelerometers, gyroscopes, GPS trackers, and video cameras, which open up new avenues for understanding the complex lives of animals. We talked about how these devices capture the physical movements of animals and offer information on their social interactions, emotional experiences, and adaptations to their environments. We addressed the critical aspects of data collection, emphasizing the significance of selecting the right sensors for our research objectives to ensure the collection of high-quality, relevant data. Through this process, we introduced challenges of working in diverse field conditions and the need for a thorough process of sensor calibration. Our discussion extended to the involved task of data annotation, where we looked at techniques necessary to transform raw sensor data into valuable insights.

CHAPTER 5

# Preprocessing and Feature Extraction for Animal Behavior Research

This chapter focuses on accelerometer data, emphasizing the critical steps of data preprocessing and feature extraction, and their essential roles in behavioral research. Accelerometer data obtained from animal gait is a valuable resource, offering deep insights into animal behavior, habitat usage, and health. However, transforming raw accelerometer data into meaningful analysis is a complex task. Raw data often comes with noise and complexity, requiring thorough preprocessing to become a useful tool for research.

Preprocessing accelerometer data involves several key steps. These include cleaning the data to remove inaccuracies and normalizing it to ensure uniformity and comparability. These steps are crucial in turning the raw, often chaotic data into an organized and analyzable format. Following preprocessing, the focus shifts to feature extraction. This stage involves isolating the most informative and relevant elements from the processed data. By extracting features in both time and frequency domains, researchers can turn raw measurements into actionable insights.

The chapter presents the theoretical foundations of these techniques and also provides practical applications using Python. These examples are designed to show how filtering and feature extraction can be effectively implemented in real-world research scenarios, with a particular focus on animal behavior. Feature scaling examples using Python are not included in this Chapter.

## Understanding Accelerometer Data

The deployment of accelerometer technology in the study of animal behavior represents a significant increase in the capability to observe and quantify animal movements in their natural habitats. As we mentioned in Chapter 4, accelerometers measure acceleration forces, which include both static forces, such as gravity, and dynamic forces resulting from movement. In animal behavior studies, these devices are instrumental for recording multi-dimensional data, revealing insights into the animal's movements and activities.

## *Importance of Sample Frequency*

Sample frequency, or the rate at which the accelerometer records data, is a critical consideration in study design. This frequency determines the resolution of the movement data captured and has implications for the type of behavior that can be analysed:

- High Sampling Rates.
- Low Sampling Rates.

The choice of sampling rate is influenced by the specific research objectives, the behavior of interest, and practical considerations like device battery life and data storage capacity.

## *Low Sampling Rate vs. High Sampling Rate*

When capturing accelerometer data, the sampling rate impacts the level of detail and the volume of collected data. A low sampling rate captures fewer data points per second, which can miss finer details of the animal's movement but is more manageable in terms of data volume and battery life. Conversely, a high sampling rate captures more data points per second, providing a more detailed picture of the movement but at the cost of generating larger data volumes and increased battery consumption.

In the context of an animal walking, a low sampling rate might capture the general rhythm of the steps but may miss subtle variations in movement. A high sampling rate, however, can capture these details, including slight changes in speed or gait.

The following figure (Figure 5.1) illustrates these differences. They simulate an animal walking, with the $x$-axis representing the movement. Peaks in the graph indicate steps taken by the animal.

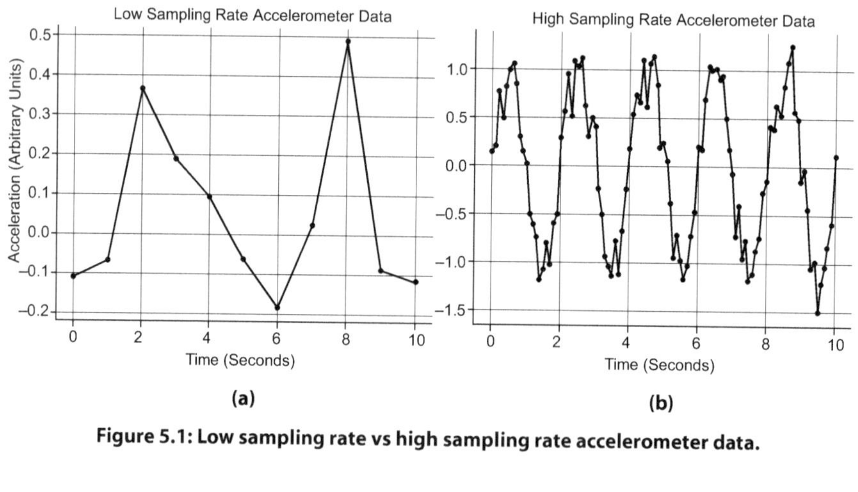


**Figure 5.1: Low sampling rate vs high sampling rate accelerometer data.**

**Figure 5.1(a): Low Sampling Rate Accelerometer:**

This graph shows data captured at a low sampling rate (1 sample per second). The points are broader and less frequent, providing a general view of the walking pattern. However, due to the lower frequency of data points, finer details of the movement are not captured. This could result in missing subtle variations in the animal's gait or speed.

**Figure 5.1(b): High Sampling Rate Accelerometer Data:**

This graph represents data captured at a high sampling rate (10 samples per second). The increased frequency of data points offers a more detailed view of the walking pattern. It captures the details and variations in the movement, including the smaller fluctuations that occur between steps. This level of detail is crucial for studies requiring a detailed understanding of animal behavior.

These figures demonstrate how the choice of sampling rate can significantly impact the data captured by accelerometers. The choice of sampling rate should be carefully considered based on the specific requirements and constraints of each research study.

## Data Preprocessing

Data preprocessing stands as a critical phase where raw data is converted into an appropriate format for detailed analysis. This section covers the essential aspects of data preprocessing, including strategies for data cleaning, scaling, normalization, and filtering techniques. Each topic is explored, providing insights into their importance and practical application.

### *Data Cleaning*

Data cleaning is a preliminary step in preparing accelerometer data for analysis. It involves identifying and correcting (or removing) errors and inconsistencies in the data to improve its quality.

Handling Missing Data:

1. **Identification:** The first step is identifying missing values in the dataset. This can occur due to various reasons like sensor malfunction or interruptions during data transmission.
2. **Imputation Techniques:** Depending on the context, different imputation methods can be used to fill in missing data. Simple methods include using the mean or median of nearby points, while more complex approaches might involve predictive modelling.
3. **Deletion:** In cases where imputation may not be appropriate or feasible, especially if the missing data is substantial, deletion might be the chosen strategy.

Addressing Noise and Outliers:

1. **Noise Reduction:** Techniques such as smoothing filters or rolling averages can help reduce noise.
2. **Outlier Detection and Treatment:** Outliers can skew the analysis and lead to incorrect interpretations. They can be detected using statistical methods like the Z-score and can either be removed or corrected.

## *Data Scaling and Normalization*

Scaling and normalization are processes that adjust the data on a consistent scale and are crucial for accurate analysis, particularly when comparing datasets from different sources or sensors.

### Techniques and Significance

1. **Standardization (Z-score Scaling):** Involves scaling the data to have a mean and standard deviation of 0 and 1, respectively.

   Formula:

   $$Z = \frac{X - \mu}{\sigma}.$$

   Here, $X$ is the original value, $\mu$ is the mean of the data, and $\sigma$ is the standard deviation.

   **Use Case:** Standardization is particularly useful when the data needs to be compared across different conditions or sensors. It is also a common practice in many machine learning algorithms.

> **Standardization assumes that the data follows a Gaussian distribution. If the data is not normally distributed, this method might not be appropriate.**

Figure 5.2 illustrates the accelerometer readings along the $x$-axis for the first 10 seconds. The original data is shown in dark gray, exhibiting the raw sensor readings with natural variations. The standardized data, displayed in light gray, provides a standardized view that is essential for certain analytical comparisons. The time is marked on the $x$-axis, and the acceleration values are on the $y$-axis, offering a clear visual representation of the effect of standardization on accelerometer data.

2. **Min-Max Scaling (also known as Normalization):** Transforms the data to fit within a specific range, 0 to 1, that can be useful in certain types of analyses where relative differences are more important than absolute values.

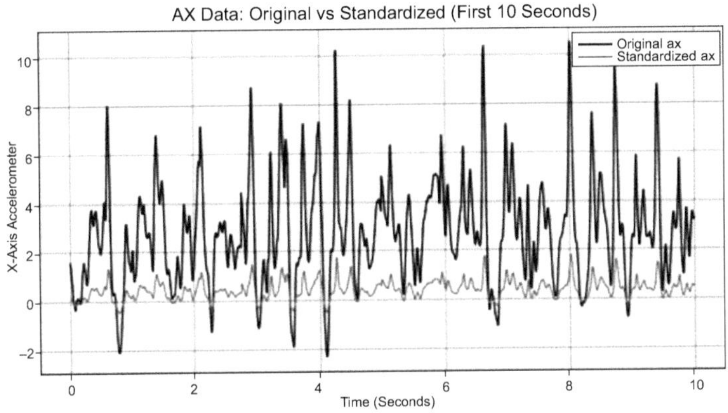


**Figure 5.2: Comparison of original and standardized accelerometer data.**

Formula:

$$X_{minmax} = \frac{X - X_{min}}{|X_{max} - X_{min}|}.$$

In this formula, $X_{min}$ and $X_{max}$ are the minimum and maximum values in the data, respectively.

**Use Case:** Min-Max scaling is beneficial when the dataset needs to be normalized within a specific range while preserving the distinctions in value ranges.

### Potential Issues with Min-Max Scaling

1. **Sensitivity to Outliers:**
   - Min-Max scaling is highly sensitive to outliers in the data. Since it involves the minimum and maximum values, outliers can skew these values significantly, resulting in a distorted scale where most "normal" data points are transformed to a narrow range in the middle of the scale.
   - It is advisable to handle outliers before applying Min-Max scaling. Outlier detection and treatment should be the first step in your data preprocessing pipeline if you plan to use Min-Max scaling.
2. **Not Suitable for Non-Uniform Data:**
   - If the data is not uniformly distributed, Min-Max scaling might not be the best choice. It can compress the majority of the data into a small interval, which might not be desirable for certain analyses.
   - Analyze the data distribution first. If the data is not roughly uniformly distributed, consider using other scaling techniques like standardization, especially if your subsequent analysis involves algorithms that assume normally distributed data.

3. **Loss of Information:**
   - In scenarios where the maximum and minimum values carry important information (e.g., extreme but meaningful fluctuations), Min-Max scaling might lead to a loss of this information.
   - Carefully consider the context of your data. If the extreme values are meaningful and should be preserved, explore alternative scaling techniques, or modify the range of Min-Max scaling accordingly.

> **Although Min-Max scaling effectively constrains data within a predefined range, it is vulnerable to outliers. A single outlier has the potential to significantly skew the scaled data.**

### Applying Min-Max Scaling

After ensuring that the data is suitable for Min-Max scaling and outliers have been appropriately handled, you can apply the scaling.

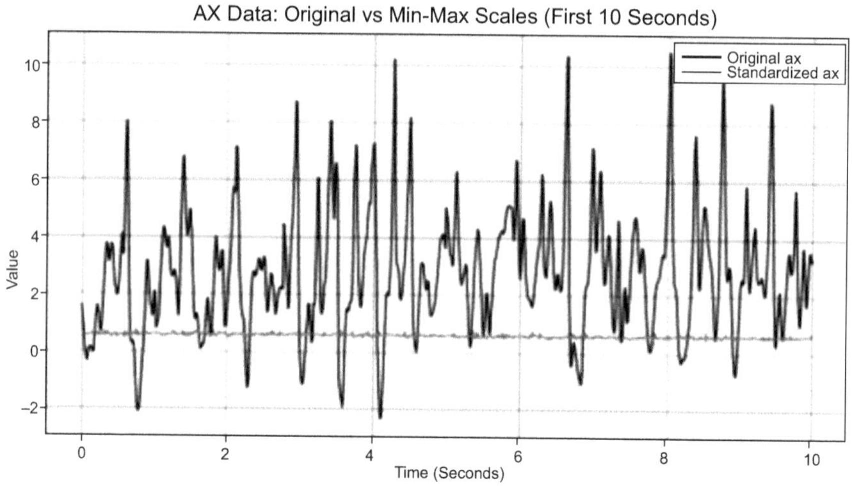


**Figure 5.3: Impact of outliers on Min-Max scaling.**

Figure 5.3 illustrates a 10 second block from a large dataset alongside its min-max scaled version. Notably, the scaled data is tightly clustered within a small range, indicating a possible influence of outliers on the normalization process. Such clustering highlights the necessity of addressing outliers before applying Min-Max scaling to ensure a more evenly distributed range of transformed data.

## *Filtering Techniques*

Filtering plays a crucial role in data preprocessing, particularly with accelerometer data, where it is vital to isolate the frequencies of interest and reduce noise.

Different types of filters serve distinct purposes, depending on the nature of the data and the analysis goals.

It is important to note that filtering is not always necessary; its application depends on the specific requirements and nature of the data being analyzed. The decision to use filters should be based on the scientist's assessment of the data quality and the objectives of the study. When deemed appropriate, filtering can significantly enhance the signal quality by reducing noise, thereby facilitating more accurate activity recognition.

### *Python Code Availability*

For practical implementation, the Python code for each of these filtering techniques has been made available on https://github.com/nkcAna/WSDpython. You can find these examples under the name Chapter_5_data_filtering_examples.ipynb. The data used to show the examples is retrieved from (Kamminga et al., 2017). If interested readers would like to use this dataset for research, they must cite the authors.

### *The Selective Use of Filtering in Accelerometer Data*

The necessity of filtering in accelerometer data relies on various aspects, including the presence of noise and the desired outcomes of the analysis. Filtering can be crucial for separating the signal from noise, but in some scenarios, the raw data in its unfiltered state may provide the needed information.

### Overview of Filtering Techniques

1. **Low-Pass Filtering (LPF):** LPFs allow frequencies below a certain cutoff frequency to pass through while attenuating (reducing) frequencies above this cutoff. They are used to remove high-frequency noise from the data.

   Python Example:

```
# 1. Low-pass filtering
from scipy.signal import butter, filtfilt

def apply_low_pass_filter(data, cutoff_frequency, sampling_rate,
order=5):
    nyquist = 0.5 * sampling_rate
    normal_cutoff = cutoff_frequency / nyquist
    b, a = butter(order, normal_cutoff, btype='low', analog=False)
    return filtfilt(b, a, data)

# Applying the filter (example parameters)
cutoff_freq = 10  # Cutoff frequency in Hz
sampling_rate = 200  # Sampling rate in Hz
filtered_data_lpf = apply_low_pass_filter(selected_data['ax'],
    cutoff_freq, sampling_rate)
```

Code Explanation:

- Starting from scipy.signal import butter, filtfilt: This line imports the butter and filtfilt functions from the scipy.signal module. butter is used for creating the filter coefficients, and filtfilt is a function that applies the filter to a data sequence.
- def apply_low_pass_filter(data, cutoff_frequency, sampling_rate, order = 5): This function, apply_low_pass_filter, is designed to apply a low-pass filter to the input data. It takes four parameters: the data to be filtered, the cutoff frequency (cutoff_frequency), the sampling rate of the data (sampling_rate), and the order of the filter (order), which defaults to 5 if not specified.
- nyquist = 0.5 * sampling_rate: The Nyquist frequency is calculated as half of the sampling rate. It represents the highest frequency that can be effectively captured by the sampling process.
- normal_cutoff = cutoff_frequency / nyquist: The cutoff frequency is normalized by dividing it by the Nyquist frequency. This normalization is necessary because the butter function expects the cutoff frequency in units relative to the Nyquist frequency.
- b, a = butter(order, normal_cutoff, btype = 'low', analog = False): The butter function designs a low-pass filter with the specified order and normalized cutoff frequency. The btype = 'low' argument specifies it as a low-pass filter. The function returns two arrays, b and a, which represent the filter's coefficients.
- return filtfilt(b, a, data): The filtfilt function applies the filter to the data. It uses the filter coefficients b and a to process the input data. This function is used for zero-phase filtering, indicating it does not introduce any phase shift in the filtered data.
- cutoff_freq = 10 # Cutoff frequency in Hz
  sampling_rate = 200 # Sampling rate in Hz
  filtered_data_lpf = apply_low_pass_filter (selected_data['ax'], cutoff_freq, sampling_rate): Finally, the apply_low_pass_filter function is called with actual parameters: a cutoff frequency of 10 Hz, a sampling rate of 200 Hz, and the data to be filtered (in this case, selected_data['ax'], which might represent the *x*-axis readings of an accelerometer).

In practice, this function is very effective for processing accelerometer data in scenarios where high-frequency noise needs to be removed to analyze the underlying trends or movements. By adjusting the cutoff frequency, the filter can be tailored to the unique attributes of the data and the requirements of the analysis.

2. **High-Pass Filtering (HPF):** HPFs are used to remove low-frequency components from the data, allowing only frequencies higher than a certain threshold. This can be useful for eliminating drift or offset in the accelerometer.

   Python example:

```
# 2. High-pass filter
def apply_high_pass_filter(data, cutoff_frequency, sampling_rate,
order=5):
    nyquist = 0.5 * sampling_rate
    normal_cutoff = cutoff_frequency / nyquist
    b, a = butter(order, normal_cutoff, btype='high', analog=False)
    return filtfilt(b, a, data)

# Applying the filter
cutoff_freq = 0.5  # Cutoff frequency in Hz
filtered_data_hpf = apply_high_pass_filter(selected_data['ax'],
    cutoff_freq, sampling_rate)
```

   Code Explanation:

   - def apply_high_pass_filter(data, cutoff_frequency, sampling_rate, order = 5): Similar to the low-pass filter function, this function, apply_high_pass_filter, applies a high-pass filter to the input data. The parameters are the same: the data to be filtered, the cutoff frequency, the sampling rate, and an optional filter order.
   - The Nyquist frequency calculation (nyquist = 0.5 * sampling_rate) and the normalization of the cutoff frequency (normal_cutoff = cutoff_frequency / nyquist) are identical to the low-pass filter code. These steps are standard in preparing the parameters for filter design in digital signal processing.
   - b, a = butter(order, normal_cutoff, btype = high, analog = False): This line uses the butter function to design a high-pass filter. The only change from the low-pass filter is the btype = 'high' argument, indicating that it's a high-pass filter. The function returns the filter coefficients b and a.
   - The application of the filter with filtfilt(b, a, data) is the same as in the low-pass filter code. filtfilt is used for zero-phase filtering, ensuring no phase shift in the output signal.
   - cutoff_freq = 0.5 # Cutoff frequency in Hz
     filtered_data_hpf = apply_high_pass_filter(selected_data['ax'], cutoff_freq, sampling_rate): The high-pass filter is applied to the data (selected_data['ax']) with a specified cutoff frequency (0.5 Hz in this case) and the sampling rate of the data.

The high-pass filter is particularly useful in scenarios where low-frequency components, such as gravitational effects or sensor drift, need to be removed from accelerometer data. By setting an appropriate cutoff frequency, the filter allows only frequencies higher than the cutoff to pass, effectively removing these low-frequency components. This can be crucial in analyses focusing on dynamic movements or changes in acceleration rather than static or slow-moving signals.

3. **Band-pass Filtering:** Band-pass filters retain frequencies within a certain range, filtering out both high and low frequencies outside this range. Ideal for isolating the frequency bands that are most representative of certain types of animal behaviors, such as walking, running, or specific repetitive movements.

   Python Example:

```
# 3. Band-pass filter
def apply_band_pass_filter(data, lowcut, highcut, sampling_rate,
order=5):
    nyquist = 0.5 * sampling_rate
    low = lowcut / nyquist
    high = highcut / nyquist
    b, a = butter(order, [low, high], btype='band', analog=False)
    return filtfilt(b, a, data)

# Applying the filter
lowcut = 0.5  # Low cutoff frequency in Hz
highcut = 10  # High cutoff frequency in Hz
filtered_data_bpf = apply_band_pass_filter(selected_data['ax'],
   lowcut, highcut, sampling_rate)
```

The above code demonstrates the implementation of a band-pass filter, a technique that allows signals within a certain frequency range (between lowcut and highcut) to pass through while attenuating signals outside this range. Band-pass filtering is particularly useful in applications where only a specific range of frequencies is of interest, such as isolating types of animal movements from accelerometer data.

Code Explanation:

- The function apply_band_pass_filter is defined to apply a band-pass filter to the given data. Parameters include data (the signal to be filtered), lowcut (the lower bound of the frequency range), highcut (the upper bound of the frequency range), sampling_rate (the rate at which data is sampled), and order (the complexity of the filter).
- Filter Design:
  - The Nyquist frequency is calculated (nyquist = 0.5 * sampling_rate),

and the low and high cutoff frequencies are normalized relative to the Nyquist frequency.

  - The butter function designs a band-pass filter (btype = 'band') with the specified order and frequency range, returning the filter coefficients b and a.

- The filtfilt function applies the filter to the data, ensuring zero phase distortion.
- The band-pass filter is applied to a segment of accelerometer data (selected_data['ax']) with defined lowcut and highcut frequencies.

4. **Moving Average Filter:** Smoothens data by averaging subsets of data points, effective for reducing random noise.

   Python example:

```
# 4. Moving average filter
def apply_moving_average_filter(data, window_size=20):
    return np.convolve(data, np.ones(window_size) / window_size,
    mode='valid')

# Applying the filter
window_size = 20  # Number of samples to average over
filtered_data_maf = apply_moving_average_filter(selected_data['ax']
    ,window_size)
```

   Code Explanation:

   - apply_moving_average_filter:
     - The function takes in a dataset (data) and a window_size, indicating the number of samples to be averaged over.
     - The default window size is set to 20, meaning that by default, each point in the filtered data will be the average of 20 consecutive points in the original data.
   - Applying the Filter:
     - The function uses np.convolve to apply a moving average filter. This function convolves the data with a window of specified size, effectively computing a running average.
     - The window is created using np.ones(window_size) / window_size, which generates an array of ones (of length window_size) divided by the window size, ensuring that the sum of the window's elements is 1.
     - The mode = 'valid' argument ensures that the output is only computed for points where the window fits entirely within the signal boundaries,

leading to a slight reduction in the size of the output array compared to the input.

- The moving average filter is applied to the accelerometer data (selected_data['ax']) with a specified window size.

5. **Savitzky-Golay Filter:** Smooths the data while preserving features of the signal such as relative maxima, minima, and width, which can be lost with other types of smoothing. It is particularly useful when the shape of the accelerometer signal is important, such as in gesture recognition or detailed movement analysis.

   Python Example:

```
# 5. Savitzky golay filter
from scipy.signal import savgol_filter

def apply_savitzky_golay_filter(data, window_size=5, polynomial_
    order=2): return savgol_filter(data, window_size, polynomial_
    order)

# Applying the filter
window_size = 29  # Must be odd
polynomial_order = 2  # The order of the polynomial fit
filtered_data_sgf = apply_savitzky_golay_filter(selected_data['ax'],
    window_size, polynomial_order)
```

Code Explanation:

- apply_savitzky_golay_filter:
  - This function applies the Savitzky-Golay filter to the provided data.
  - Parameters: the data, the size of the smoothing window (window_size), and the order of the polynomial used in the filter (polynomial_order).
- Applying the Savitzky-Golay Filter:
  - The savgol_filter function from scipy.signal is used to apply the filter. It performs a polynomial regression on a sliding window of data points, effectively smoothing the data.
  - The window_size must be odd because the filter works by fitting a polynomial to a symmetric window of data around each point. This ensures that each point has an equal number of neighbouring data points on both sides, which is essential for maintaining the symmetry of the regression fit.

- Filter Parameters:
  - window_size is set to 29, which must be an odd number for the reasons explained above.
  - polynomial_order is set to 2, indicating that a quadratic polynomial is used for the fit within each window.
- Finally, the function is used to apply the Savitzky-Golay filter to a segment of accelerometer data (selected_data['ax']).

This technique smooths the data while preserving key features like peaks and troughs, making it easier to detect meaningful patterns in the animal's movement.

6. **Kalman Filtering:** A sophisticated technique that combines measurements over time to produce estimates of unknown variables.
   - It is useful in settings where the data is affected by uncertainty and various types of noise.
   - The Kalman filter is an iterative filter that estimates the state of a dynamic system from a series of incomplete and noisy measurements.
   - It is widely used in applications like tracking and navigation systems, where accuracy and efficiency are crucial.
   - The Kalman filter works in two steps: prediction and update. In the prediction step, the filter produces estimates of the current state variables, along with their uncertainties. During the update step, the filter refines its predictions based on new measurements.

   Python Example:

```
# 6. Kalman filter

from filterpy.kalman import KalmanFilter
import numpy as np

def apply_kalman_filter(data, sampling_rate):
    # Initialize the Kalman Filter
    kf = KalmanFilter(dim_x=1, dim_z=1)

    # Define initial state and state transition matrix
    kf.x = np.array([0])  # initial state
    kf.F = np.array([[1]])  # state transition matrix

    # Define measurement function
    kf.H = np.array([[1]])

    # Define initial covariance matrix
    kf.P *= 1000
```

```
    # Define measurement noise (this might need to be adjusted)
    kf.R = np.array([[0.5]])  # Assuming 0.5 g^2 as measurement
      noise variance

    # Define process noise (also might need adjustment)
    process_noise_std = 0.5  # Assuming 0.5 g^2 as process noise
      variance
    kf.Q = np.array([[process_noise_std**2]])

    # Kalman Filter Operation
    filtered_data = []
    for measurement in data:
        kf.predict()
        kf.update(np.array([measurement]))
        filtered_data.append(kf.x[0])
    return np.array(filtered_data)

# Applying the filter to your accelerometer data
filtered_data_kf = apply_kalman_filter(selected_data['ax'], 200)
# 200 Hz sampling rate
```

Kalman Filter Function:

- Function apply_kalman_filter:
  - The function initializes and applies a Kalman filter to a given set of data, assuming a certain sampling_rate.
  - It returns the filtered data, which represents the estimated true state of the system.
- Initialization and Configuration:
  - KalmanFilter(dim_x = 1, dim_z = 1): Initializes a Kalman filter object for a 1D state (dim_x) and 1D measurement (dim_z).
  - kf.x: Sets the initial state estimate. Here, it's initialized to 0, assuming the system starts from a neutral state.
  - kf.F: The state transition matrix. Set to [[1]] for a simple 1D model, assuming the state does not change much between measurements (e.g., constant velocity model).
  - kf.H: The measurement function, relating the state to the measurement. Also, a simple 1D identity matrix, assuming direct observation of the state.
- Noise and Covariance:
  - kf.P: The initial estimate of the state covariance. A high value (1000) reflects initial uncertainty about the state.
  - kf.R: The measurement noise covariance. Set to [[0.5]], which represents the variance of the measurement noise (assumed to be 0.5 $g^2$). This

adjustment should be made according to the specific noise attributes of the accelerometer.

- kf.Q: The process noise covariance, reflecting uncertainty in the model. Set based on an assumed standard deviation of 0.5 g. This value might need adjustment based on the dynamics of the animal's movement.

- Kalman Filter Operation:
  - The for loop iterates through each measurement, applying the predict and update steps of the Kalman filter. The predict step estimates the next state based on the current state and model, while update adjusts this prediction based on the new measurement.
  - The estimated state (kf.x[0]) is stored in filtered_data for each measurement.

Applying the Filter:

- The function is applied to a segment of accelerometer data (selected_data['ax']).

Reasons for Parameter Choices:

- Initial State (kf.x): The choice of zero as the initial state is a common practice when there is no prior knowledge about the system's state.
- State Transition Matrix (kf.F): The simple identity matrix reflects a model where the state does not change dramatically between measurements, which is a reasonable assumption for many physical systems over short time intervals.
- Measurement Noise (kf.R): The assumption of 0.5 $g^2$ noise variance is a starting point, and it should be refined based on the actual noise characteristics of the accelerometer used.
- Process Noise (kf.Q): The assumed standard deviation of 0.5 g for process noise is a guess. In practice, this should be based on how much you expect the system (animal movement) to deviate from the model predictions between measurements.

Sometimes, the filtered data and the original data might exhibit minor or no differences. The reasons for similarity between Original and Kalman Filtered Data are the following:

- Low Noise Levels in Original Data:
  - If the original accelerometer data is relatively clean with low levels of noise, the Kalman filter may not make significant alterations to the data. Kalman filters are most effective when there is substantial measurement noise that they can help mitigate.

- Filter Parameters:
  - The effectiveness of a Kalman filter greatly depends on the tuning of its parameters, including the process noise covariance (kf.Q) and measurement noise covariance (kf.R).
  - If the measurement noise covariance (kf.R) is set too low, the filter might place too much trust in the measurements and not enough in its own model predictions, leading to filtered data that closely follows the original measurements.
  - Equally, if the process noise covariance (kf.Q) is set too low, it implies that the system is expected to change very little, leading the filter to make only minor adjustments to the data.
- Simple State Model:
  - The example provided uses a very basic 1D state model, represented by the state transition matrix (kf.F) and measurement matrix (kf.H). This simple model might not be adequate to capture the complexities of animal movement, leading to minimal adjustments by the filter.
  - In real-world applications, especially with complex movements, a more sophisticated model that includes aspects like velocity and possibly acceleration would be more effective.
- Nature of Animal Movements:
  - If the animal's movements are relatively smooth and predictable, with minimal sharp changes or erratic behavior, the Kalman filter's impact might be less noticeable, especially if it is configured with a simple state model.

Key Considerations:

- Kalman filtering is best suited for scenarios with significant noise or where integrating data over time is crucial for accurate estimation.
- The effectiveness of the filter depends on careful tuning and, often, a deep understanding of the system being modelled. This tuning involves setting the right parameters and possibly designing a more complex system model.
- In cases where the original data is already of high quality, or if the system dynamics do not align well with the model of the filter, the impact of Kalman filtering may be subtle.

Practical Advice:

- Experiment with the filter parameters and consider more complex models for the state and measurement processes.
- Understand the characteristics of the accelerometer data and the nature of the movements being captured.

- Use visualizations and statistical analyses to assess the effectiveness of the filter in different scenarios.

### *Summing up*

Applying filters to accelerometer data for animal activity recognition requires careful consideration of several factors, including the nature of the signal, the characteristics of the specific behaviors you are interested in, and the sampling rate of the accelerometer. Here is a simple guide on how to apply these filters and what to keep in mind:

Understanding Your Signal

- Characteristics of Animal Movement:
  - Identify the typical frequencies associated with the animal behaviors of interest. For instance, the frequency range of walking might differ significantly from running or resting.
- Noise Characteristics:
  - Understand the source of noise in your data. Is it high-frequency sensor noise, low-frequency drifts, or transient disturbances?

Sampling Rate Considerations

- Nyquist Theorem: Ensure your sampling rate is at least twice the highest frequency component you wish to capture (Nyquist rate). For instance, if the maximum frequency of interest is 50 Hz, you should sample at least at 100 Hz.
- Aliasing: If the sampling rate is too low, you risk aliasing, where high-frequency components are misrepresented in your data.

Applying Filters

- Low-Pass Filter:
  - When to Use: If your data contains high-frequency noise that is not characteristic of the animal's movement.
  - Setting Cutoff Frequency: The cutoff frequency should be set just above the highest frequency of interest for the animal behavior.
- High-Pass Filter:
  - When to Use: To remove bias or drift, especially if the accelerometer captures gravitational effects.
  - Cutoff Frequency: Set it below the typical frequency range of the behaviors but high enough to eliminate drift.
- Band-Pass Filter:
  - When to Use: To isolate frequencies that correspond to specific animal

behaviors, filtering out both high-frequency noise and low-frequency drift.

  - Setting Range: Define the lower and upper cutoff frequencies based on the behavior's frequency characteristics.

- Moving Average and Savitzky-Golay Filters:
  - When to Use: For smoothing the data while preserving important trends.
  - Window Size: Larger windows smooth over a longer period, which might obscure short-duration behaviors.

As we conclude our discussion of various filtering techniques for preprocessing accelerometer data, it is important to reflect on the significance of these methods in enhancing the quality and interpretability of our data. Through the application of low-pass, high-pass, band-pass, moving average, Savitzky-Golay, and Kalman filters, we have prepared ourselves with a robust toolkit to address a wide range of noise and signal distortion issues commonly encountered in accelerometer data, particularly in the context of animal behavior research.

Now, having established a solid foundation in data preprocessing, we will transition to the next critical phase: Feature Extraction. In the upcoming section, we look into the techniques and methodologies to extract meaningful features from our processed accelerometer data.

## Feature Extraction

Feature extraction is another important phase in the analysis of accelerometer data. It involves transforming raw accelerometer signals into a set of representative and informative metrics that can be used for further analysis and interpretation. In this section, we will discuss *time-domain* and *frequency-domain* features.

Time-domain features are extracted directly from the raw accelerometer data without any transformation to the frequency domain. These features often include statistical measures that capture the central tendency, dispersion, and shape of the signal over time. They are intuitive and often straightforward to compute, making them a common option in numerous applications. Examples include mean, variance, standard deviation, and other statistical moments.

Frequency-domain features, on the other hand, involve transforming the time-series data into the frequency domain using techniques such as Fourier and Wavelet Transforms [ref]. These features are crucial in identifying the dominant frequencies and periodicities in the data, which can be particularly informative in understanding periodic or repetitive movements in animal behavior. Spectral density, power spectral density, and specific frequency bands' energy are typical examples of frequency-domain features.

### Windowing in Feature Extraction

An important aspect of feature extraction from accelerometer data is the concept of *windowing*. Instead of extracting features from each data point individually, we segment the data into overlapping or non-overlapping windows and extract features from these segments. This approach is essential because it captures temporal patterns and dependencies in the data that are significant in understanding dynamic movements. The size of the window and the sampling rate are critical factors in this process, as they determine the resolution and the type of movements that can be captured.

As we discuss specific time-domain and frequency-domain features, we will consider how these features can be applied to accelerometer data to get insights into animal behavior. This will include practical considerations for choosing the right features and understanding their implications in the context of animal movement and behavior studies.

A range of time-domain and frequency-domain features will be discussed in the upcoming sections, providing detailed explanations and discussions on their significance and applications in accelerometer data analysis. This will equip you with the necessary tools and knowledge to effectively extract meaningful information from accelerometer datasets.

### Time-domain and Frequency-domain Features in Animal Behaviour Studies

Over recent years, Deep Learning (DL) has surpassed traditional Machine Learning (ML) methods in performance. DL acquires a significant advantage in its capacity to directly learn features from raw data during the training phase, eliminating the need for a separate feature extraction process in preprocessing. However, for DL to excel and achieve optimal results, it often requires large datasets. This is a significant challenge, particularly in fields like Animal Activity Recognition (AAR), where acquiring large enough datasets can be time-consuming and not always feasible. Consequently, ML methods remain a relevant choice, and feature extraction continues to be a crucial step, especially in classification problems.

In the context of continuous accelerometer measurements, unlike some other types of data, a single value is not sufficient to characterize activity patterns. This necessity underscores the importance of feature extraction. Various techniques have been proposed to represent information in raw accelerometer signals for use with ML algorithms in classifying gait activities. Prior research has explored a wide range of features in both time and frequency domains. The advantage of time-domain features lies in their straightforward computation. However, they are prone to errors from measurement and calibration. In contrast, frequency-domain features typically need additional preprocessing steps like windowing, filtering, and transformations. Although these features robustly represent signal information, they typically require more computational effort compared to time-domain features.

### Time-domain Features

Following is a list of time-domain features extracted from accelerometer signals ($x$, $y$, $z$) in animal behavior studies:

#### *Mean*

The mean feature of accelerometer data is a fundamental statistical measure used to describe the central tendency of the acceleration values along each axis ($x$, $y$, $z$) of the accelerometer. This feature is particularly useful in understanding the average behavior over a period of time.

If we have a set of accelerometer readings along a single axis (let's say the $x$-axis) recorded over a period of time, the mean (average) of these readings can be calculated as follows:

Let $x_1, x_2, \ldots, x_n$ be the accelerometer readings along the $x$-axis over $n$ samples.

The mean ($\bar{x}$) of these readings is given by:

$$\overline{x} = \frac{1}{n} \sum_{i=1}^{n} x_i$$

where,

- $\bar{x}$ is the mean of the accelerometer readings along the $x$-axis.
- $x$ is the $i$th accelerometer reading.
- $n$ is the total number of samples.

The formula sums up all the accelerometer readings along the $x$-axis and then divides this total by the number of readings to find the average value.

Application to Accelerometer Data:

- The same formula can be applied independently to the $y$-axis and $z$-axis readings to get the mean values for those axes.
- In the context of accelerometer data, calculating the mean for each axis helps in understanding the average acceleration or movement in each direction.
- This can be particularly important in scenarios where you are interested in the overall trend or bias in the movement.

We can also calculate the average of all axes:

$$combined\ mean = \frac{\overline{x}+\overline{y}+\overline{z}}{3}.$$

#### *Standard Deviation (SD)*

The standard deviation is a statistical measure that quantifies the amount of variation or dispersion in a set of data values. For accelerometer data the standard deviation can be calculated for each axis to understand the variability in acceleration measurements along that axis.

The standard deviation for each axis is calculated using the formula:

$$SD_{axis} = \sqrt{\frac{1}{n-1} \sum_{i=1}^{n} \Sigma \left(axis_i - \overline{axis}\right)^2}$$

where,

- $SD_{axis}$ is the standard deviation for the specified axis.
- $axis_i$ is the $i$th reading of the accelerometer on that axis.
- $\overline{axis}$ is the mean of the accelerometer readings on that axis.
- $n$ is the number of readings.

The standard deviation for each axis provides insights into the variability or consistency of movement along that axis. A higher standard deviation indicates greater variability in acceleration, which could be indicative of more dynamic or erratic movements. Conversely, a lower standard deviation suggests more uniform or steady movements.

In the context of accelerometer data analysis, especially for applications like gait analysis or activity recognition, understanding the standard deviation along each axis is crucial for interpreting the nature and intensity of movements.

### *Variance*

Variance is a statistical measure that quantifies the degree of spread or dispersion of a set of data points. In the context of accelerometer data, which typically includes readings along $x$, $y$, and $z$ axes, variance can be calculated for each axis to understand the extent of variation in acceleration measurements.

The variance for each axis is calculated using the following formula:

$$Variance_{axis} = SD_{axis}^{2}.$$

Understanding Variance in Accelerometer Data:

Variance measures how much the acceleration readings deviate from their average value (mean).

In accelerometer data analysis, particularly in studies of animal movement or gait analysis, understanding the variance along each axis can provide insights into the nature of the movements, such as their regularity, predictability, or intensity.

The variance is a foundational statistical measure that, along with the mean and standard deviation, offers a comprehensive view of the distribution and variability of accelerometer data.

### *Inverse Coefficient of Variation (ICV)*

The Inverse Coefficient of Variation (ICV) is a statistical measure that provides insights into the relative variability of a dataset in relation to its mean. It is particularly useful when comparing the variability of datasets with different mean values.

The ICV is defined as the ratio of the mean to the standard deviation of a dataset. The formula for ICV is:

$$ICV = \frac{\overline{x}}{SD}.$$

The ICV is essentially the reciprocal of the coefficient of variation (CV), which is a measure of relative dispersion. A higher ICV indicates lower relative variability, meaning the data points are more consistent in relation to the average value. A lower ICV suggests higher relative variability, implying that the data points are more spread out in comparison to the mean.

Application in Accelerometer Data:

In the context of accelerometer data, the ICV can be calculated for each axis ($x$, $y$, $z$) to assess the consistency of the acceleration measurements. This is particularly useful in applications where it is important to understand the magnitude of the variability (as indicated by the standard deviation) and how that variability relates to the overall level of the signal (the mean).

For example, in gait analysis or animal behavior studies, the ICV can help in distinguishing between different types of movements or activities. Steady, regular movements might exhibit a higher ICV (lower relative variability), while erratic or irregular movements could result in a lower ICV (higher relative variability). This insight can be valuable in identifying patterns, classifying behaviors, or detecting anomalies.

### *Median*

The median is a statistical measure representing the middle value in a sorted list of numbers. In the context of accelerometer data, the median is calculated separately for each axis ($x$, $y$, $z$). It provides a central value of the acceleration readings that divides the dataset into two equal halves.

Median:

- To find the median for a given axis, the acceleration readings along that axis are first sorted in ascending order.
- For the $x$-axis, if $n$ (the number of readings) is odd, the median ($Median_x$) is the middle value: $\frac{x_{n+1}}{2}$.
- If $n$ is even, the median is the mean of the two middle values.
- The same method is applied to find the median for the $y$ and $z$ axes.

Context in Accelerometer Data Analysis:

Central Tendency: The median provides a measure of central tendency that is less sensitive to outliers and extreme values than the mean. This can be particularly useful in scenarios where the accelerometer data may include sudden spikes or drops.

Typical Behavior: In many applications, such as monitoring animal movements or human activities, the median can give a more representative value of typical behavior, especially in the presence of non-uniform data.

Data Distribution: Analyzing the median alongside other measures like mean can offer insights into the distribution of the accelerometer data. A significant difference between these measurements might indicate a skewed or non-normal distribution.

By assessing the median of accelerometer data, one can gain insights into the central tendency of the movements being measured, while minimizing the influence of atypical readings. This is especially valuable in fields like biomechanics, physical therapy, and animal behavior research, where understanding the typical motion patterns is essential.

### *Minimum and Maximum*

In the context of accelerometer data, the minimum and maximum values are crucial metrics for understanding the range of motion captured by the accelerometer. These values are often examined for each axis ($x$, $y$, and $z$) independently.

Context in Accelerometer Data Analysis:

- Range of Motion: The minimum and maximum values are often used together to understand the overall range of motion that the accelerometer has captured. This range can be crucial in applications like animal gait analysis, where the extent of movement is a key factor in identifying various types of activities or behaviors.
- Identifying Extremes: These values can help in identifying extreme movements, which might be indicative of specific behaviors or events, such as a sudden acceleration or deceleration.
- Calibration and Sensor Limits: Understanding the minimum and maximum values can also be important for assessing the calibration of the accelerometer and ensuring that the sensor's operational limits are not exceeded during data collection.

### *Pairwise Correlation*

Pairwise correlation in the context of accelerometer data refers to the statistical relationship between two different axes of the accelerometer readings. It measures the degree to which the acceleration on one axis is linearly related to the acceleration on another axis.

Computing Pairwise Correlation:

The pairwise correlation is typically calculated using Pearson's correlation coefficient. The formula for Pearson's correlation coefficient between two axes, say $x$ and $y$, is given by:

$$r_{xy} = \frac{\sum_{i=1}^{n}(x_i - \overline{x})(y_i - \overline{y})}{\sum_{i=1}^{n}(x_i - \overline{x})^2(y_i - \overline{y})^2}$$

where,

- $r_{xy}$ is the correlation coefficient between the $x$ and $y$ axes.
- $x_i$ and $y_i$ are individual readings from the $x$ and $y$ axes, respectively.
- $\overline{x}$ and $\overline{y}$ are the means of the $x$ and $y$ axes readings.
- $n$ is the number of paired readings.

The correlation coefficient ranges from –1 to 1. A value close to 1 implies a strong positive linear relationship, a value close to –1 implies a strong negative linear relationship, and a value around 0 implies nonlinear relationship.

Context in Accelerometer Data Analysis:

- Inter-axis Dynamics: Pairwise correlation helps in understanding how movements along different axes are related. For instance, in gait analysis, certain types of gaits might show a higher correlation between horizontal and vertical movements.
- Movement Coordination: High correlation values could indicate coordinated movements or coupled motion patterns, while low correlation values could suggest independent or uncoupled movements.
- Data Redundancy: In some cases, high correlation might indicate redundancy in data, which could be leveraged in data reduction or feature selection processes.

### *Mean Distance Between Axes*

The mean distance between axes in accelerometer data is a measure used to quantify the average difference in acceleration readings between two axes. This metric can provide insights into how the movements in one direction can be compared to the movements in another.

Calculating Mean Distance Between Axes:

The mean distance between two axes, say $x$ and $y$, is calculated by taking the average of the absolute differences between their respective values at each point in time.

Formula:

$$Mean\ Distance_{xy} = \frac{1}{n}\sum_{i=1}^{n}|x_i - y_i|.$$

The absolute value ensures that the distance is always a non-negative number. Similarly, you can calculate the mean distances for other pairs of axes, *Mean Distance*$_{xz}$, *Mean Distance*$_{yz}$.

Context in Accelerometer Data Analysis:

- Comparing Movement Patterns: This measure can be particularly useful in studies where the relationship or disparity between movements in different planes is of interest. For example, in gait analysis, comparing vertical and horizontal movements could provide valuable insights.
- Movement Synchronization: The mean distance can help assess the synchronization or divergence in movement between different axes. A smaller mean distance might indicate synchronized movements, while a larger mean distance could suggest divergent or independent movements across axes.
- Activity Characterization: Different activities may exhibit characteristic patterns in the mean distances between axes. For instance, activities like walking or running might show distinct mean distance patterns compared to stationary activities.

### *Signal Magnitude (SM)*

Signal Magnitude (SM) is a key metric used in accelerometer data analysis that represents the overall magnitude of the acceleration vector at each point in time. It is a comprehensive measure that combines the acceleration data from all three axes ($x$, $y$, and $z$) to provide a singular value representing the total acceleration.

Calculating SM:

The SM is calculated by determining the magnitude of the acceleration vector at each individual data point and then aggregating these values as needed (e.g., averaging over time).

The formula for the magnitude of the acceleration vector at the $i$th data point is:

$$SM_i = \sqrt{x_i^2 + y_i^2 + z_i^2}$$

where, $x_i$, $y_i$ and $z_i$ are the acceleration readings at the $i$th point in time for the $x$, $y$, $z$ axes, respectively. Unlike the Average Signal Magnitude, which involves averaging these magnitudes, SM is often considered for each individual data point or used in other forms of aggregation.

Context in Accelerometer Data Analysis:

- Overall Acceleration: SM provides a direct measure of the total acceleration experienced by the sensor at each point in time, accounting for movements in all directions.
- Activity Intensity: In applications such as physical activity monitoring or gait analysis, SM is a valuable tool for assessing the intensity of movements. It

can help distinguish between different levels of activity or different types of motion.

- Aggregated Measures: While SM itself is a point-wise measure, it can be aggregated over time to create features like mean SM, sum of SMs over a window, or even used as a basis for more complex features like energy or entropy.

### *Movement Intensity/Average Signal Magnitude*

Movement Intensity (MI) or Average Signal Magnitude is a measure used in accelerometer data analysis to quantify the overall intensity of physical activity or movement. It is calculated by combining the magnitudes of acceleration across all three axes ($x$, $y$, and $z$) to form a single composite measure.

Calculating MI:

The MI is calculated by first computing the magnitude of the acceleration vector at each point in time and then taking the average of these magnitudes over a specified window or the entire dataset.

The MI over $n$ points is:

$$\mathrm{MI} = \frac{1}{n}\sum_{i=1}^{n} SM_i.$$

Context in Accelerometer Data Analysis:

- Overall Activity Level: MI provides a summary measure of the total activity level captured by the accelerometer, encompassing movement in all directions.
- Physical Activity Intensity: In fields like sports science, physical therapy, or animal behavior studies, MI is useful for quantifying the intensity of physical activity or detecting changes in activity levels over time.
- Robustness to Direction: Since MI considers all axes, it is not biased towards movements in any specific direction, making it a comprehensive measure of motion intensity.

### *Signal Magnitude Area (SMA)*

Signal Magnitude Area (SMA) is a metric used to quantify the cumulative magnitude of acceleration over a period of time. It reflects the overall physical activity, or the intensity of movements captured by the accelerometer.

Calculating SMA:

SMA is calculated as the sum of the absolute values of the accelerometer readings across all three axes ($x$, $y$, and $z$) over a specified time window. This approach captures the total 'activity' represented by the accelerometer data.

The formula for SMA is:

$$SMA = \frac{1}{n}\left(\sum_{i=1}^{n}|x_i| + \sum_{i=1}^{n}|y_i| + \sum_{i=1}^{n}|z_i|\right).$$

Context in Accelerometer Data Analysis:

- Assessing Overall Activity Level: SMA provides a straightforward measure of the overall level of activity, considering the intensity of movements in all directions. It is particularly useful in scenarios where the total amount of movement is more important than the specific direction or pattern of movement.
- Useful in Various Applications: SMA is widely used in fields like health monitoring, sports science, and physical therapy, where it helps in quantifying physical activity levels, monitoring rehabilitation progress, or assessing the overall intensity of exercises.
- Window-based Analysis: The calculation of SMA typically involves dividing the data into windows (e.g., seconds, minutes) and then computing the SMA for each window. This window-based approach allows for the analysis of changes in activity levels over time.

The Signal Magnitude Area is an effective way to summarize accelerometer data, providing a single metric that captures the aggregate motion detected by the sensor. It is especially valuable in applications that require a simple yet comprehensive measure of physical activity or movement intensity.

### *Skewness and Kurtosis*

Skewness and Kurtosis are statistical measures that describe the shape and characteristics of the distribution of data points.

Skewness:

Skewness measures the asymmetry of the data distribution around the mean. It indicates whether the data points are skewed to the left (negative skewness) or to the right (positive skewness) of the mean.

Formula: The skewness can be calculated as:

$$skewness = \frac{n}{(n-1)(n-2)}\left(\sum_{i=1}^{n}\frac{(x_i-\overline{x}_i)^3}{SD^3}\right).$$

In accelerometer data, skewness can indicate whether the majority of the movements are concentrated above or below the average value, which can be useful in identifying biased movements or predominant directions.

Kurtosis:

Kurtosis measures the 'tailedness' of the data distribution. It describes the extremity of deviations from the mean, compared to a normal distribution.

Formula: Kurtosis is typically calculated as:

$$kurtosis = \frac{n(n+1)}{(n-1)(n-2)(n-3)}\left(\sum_{i=1}^{n}\frac{(x_i-\overline{x}_i)^4}{SD^4}\right) - \frac{3(n-1)^2}{(n-2)(n-3)}.$$

High kurtosis in accelerometer data could imply more frequent extreme movements, while low kurtosis could suggest a flatter distribution of movements. This can be particularly relevant in studies where the presence of extreme movements or outliers is of interest.

Application in Accelerometer Data Analysis:

Skewness and Kurtosis offer a nuanced view of the accelerometer data's distribution, providing information that goes beyond basic measures like mean and standard deviation. Analyzing these metrics can help in understanding the underlying characteristics of the movement, such as the presence of sporadic, intense activities or a tendency towards certain types of motion.

### *Root Mean Square (RMS)*

RMS is a statistical measure used to quantify the magnitude of a varying quantity. In the context of accelerometer data, RMS is used to represent the average magnitude of the acceleration readings, providing an overall measure of the intensity of motion.

Calculating RMS:

The RMS value for accelerometer data is calculated by squaring each reading, averaging these squared values, and then taking the square root of the average.

The formula for RMS along a single axis (say, the $x$-axis) is:

$$RMS_x = \sqrt{\frac{1}{n}\sum_{i=1}^{n} x_i^2}.$$

This calculation is typically done separately for each axis ($x$, $y$, and $z$) to obtain the RMS values for each directional component of the acceleration.

Context in Accelerometer Data Analysis:

- Overall Intensity: RMS provides a comprehensive measure of the overall intensity of the movement captured by the accelerometer. It effectively combines the amplitude and frequency components of the acceleration signal into a single value.
- Steady-state and Vibrational Analysis: In applications such as vibration analysis or monitoring steady-state activities, RMS is particularly valuable as it reflects both the amplitude and the consistency of the movements or vibrations.
- Comparison Across Axes: By calculating RMS for each axis separately, it is possible to compare the intensity of movements in different directions, which can be insightful in understanding the nature of physical activities or behaviors.

### *Zero Crossing Rate (per window)*

The Zero Crossing Rate (ZCR) is a measure used in signal processing that quantifies the rate at which a signal changes from positive to negative or vice versa. In the context of accelerometer data, it represents how often the acceleration readings cross zero, indicating changes in the direction of movement.

Calculating Zero Crossing Rate using Signum Function:

The ZCR is calculated by counting the number of times the signal crosses the zero line within a given window of data. For a discrete signal represented by a sequence of readings, a zero crossing is said to occur if consecutive readings change in sign. The ZCR can be calculated using the signum (sgn) function. The signum function, denoted as sgn, returns –1 for negative numbers, 0 for zero, and 1 for positive numbers.

A simple way to calculate the ZCR for a window of data is as follows:

$$ZCR = \frac{1}{n-1}\sum_{i=1}^{n-1}\left[\frac{1}{2}\left|sgn(x_{i+1}) - sgn(x_i)\right|\right].$$

Context in Accelerometer Data Analysis:

- Movement Pattern Analysis: The ZCR can provide insights into the frequency and nature of directional changes in movement.
- Dynamic vs. Static Activities: A higher ZCR may indicate more dynamic or variable activities, while a lower ZCR can be an indication of more static or steady activities.

### *Peak-to-Peak*

The Peak-to-Peak feature in signal processing, including accelerometer data analysis, represents the difference between the maximum and minimum values of a signal within a given window. It is a measure of the signal's overall range during that window, capturing the extent of variability or fluctuation.

Calculating Peak-to-Peak Feature:

The Peak-to-Peak value for a set of accelerometer readings within a window is calculated by finding the maximum and minimum values of the readings and then computing the difference between these two values.

The formula can be applied to each axis $(x, y, z)$ of accelerometer data to determine the Peak-to-Peak value for that specific axis in each window.

The formula for Peak-to-Peak:

$$Peak\text{-}to\text{-}Peak = max\,(x) - min(x)$$

where,

- $max(x)$ is the maximum value of the accelerometer readings in the window.
- $min(x)$ is the minimum value of the accelerometer readings in the window.

Context in Accelerometer Data Analysis:

Range of Motion: The Peak-to-Peak feature is useful for understanding the range of motion or the extent of movement variability within each window of data.

- Activity Characterization: This measure can help distinguish between different types of animal activities or behaviors, especially those that involve varying degrees of movement and amplitude.
- Signal Fluctuation: It provides insights into the fluctuation levels of the signal, which can be crucial in applications like gait analysis, gait assessment in animals' medical diagnostics, or activity classification.

### *Integrals and Squared Integrals*

In the analysis of accelerometer data, the concepts of integrals and squared integrals play a crucial role in extracting meaningful information from raw acceleration readings. These measures are particularly useful in understanding the overall movement dynamics and the energy characteristics of the motion being measured.

Integrals in Accelerometer Data:

The integral of accelerometer data refers to the cumulative sum of acceleration readings over time. This measure provides insight into the overall movement magnitude or displacement (when double-integrated) of an object.

Formula:

For discrete accelerometer data, where readings are taken at regular intervals, the integral over a time period can be approximated as the sum of the absolute values of the accelerometer readings across all three axes ($x$, $y$, and $z$), multiplied by the time interval between these readings.

Integrals can be defined:

$$Integrals = \int_{t=0}^{T} |x_i(t)|\ dt + \int_{t=0}^{T} |y_i(t)|\ dt + \int_{t=0}^{T} |z_i(t)|\ dt.$$

This integral is a measure of the total 'activity' or motion detected by the accelerometer, considering all three dimensions.

Squared Integrals in Accelerometer Data:

The squared integral of accelerometer data involves summing the squares of the acceleration readings. This calculation emphasizes larger values and can be indicative of the energy or power-like quantities associated with the motion.

$$Squared\ Integrals = \left( \int_{t=0}^{T} |x(t)|\ dt \right)^2 + \left( \int_{t=0}^{T} |y(t)|\ dt \right)^2 + \left( \int_{t=0}^{T} |z(t)|\ dt \right)^2.$$

Both the integral and squared integral provide distinct yet complementary views of accelerometer data.

### *Energy*

The Energy feature typically refers to a measure that captures the intensity or power of movements recorded by the accelerometer. This feature is often derived from the magnitude of the acceleration signal and is indicative of the vibrational or dynamic energy present in the motion.

Calculation of Energy:

Energy in accelerometer data can be quantified in several ways, but a common approach is to calculate the sum of the squared accelerometer readings, which is analogous to computing the signal's power. For discrete accelerometer data, this can be expressed as:

$$Energy = \sum_{i=1}^{n} (x_i^2 + y_i^2 + z_i^2).$$

In some applications, it might be useful to normalize this energy measure, either by dividing by the number of readings (to get an average energy per reading) or by applying another form of normalization suitable to the context of the data.

The energy feature is particularly relevant in areas like animal activity monitoring, where it can help differentiate between low-intensity and high-intensity activities.

The selection of the energy calculation process may rely on the particular requirements of the application, such as whether absolute energy levels or relative changes in energy are of interest. In some cases, additional processing steps like filtering might be applied to the accelerometer data before calculating the energy, especially to isolate specific frequency bands of interest.

### *Entropy*

Entropy, a fundamental principle in information theory, measures a signal's uncertainty, complexity, or randomness. In accelerometer data analysis for animal activity recognition, entropy can be calculated either for each axis independently or for a combination of all axes, providing insights into the movement dynamics.

Entropy per axis:

This method involves calculating entropy separately for each axis ($x$, $y$, $z$). It is frequently used when the directional components of motion are of interest or when the activities being studied have distinct characteristics in specific axes. For example, vertical movement might be more pronounced in certain activities, making the $z$-axis entropy particularly informative.

Formula (per axis):

For each axis ($x, y, z$), Shannon's entropy is calculated as:

$$Entropy_{axis} = -\sum_{i=1}^{m} p_{axis,i} \log_2(p_{axis,i})$$

where, $p$ is the probability of the $i$th measure of an axis.

Entropy for Combined Axes:

This method involves combining the $x$, $y$, and $z$ data into a single dataset and then calculating entropy. It is often used when a general view of motion is needed. This approach is suitable for studies where the overall complexity or randomness of the motion is more relevant.

Formula (Combined Axes):

For the combined data from all axes, entropy is calculated by merging the $x$, $y$, and $z$ readings:

$$Entropy_{combined} = -\sum_{i=1}^{m} p_i \log_2(p_i).$$

In accelerometer data analysis, especially in the context of animal activity recognition or human motion analysis, you will find examples of both approaches: calculating entropy per axis and calculating entropy for the combined data from all axes. The choice between these methods often relies on the particular goals of the study and the characteristics of the data.

## Frequency-domain Features

As we move from time-domain features to frequency-domain features in accelerometer data analysis, we examine a different aspect of understanding animal behavior and movement. While time-domain features provide insights based on the raw accelerometer readings over time, frequency-domain features offer a perspective based on the frequency content of these signals. This shift allows us to uncover patterns and characteristics that are not immediately apparent in the time-domain.

Frequency-domain analysis involves transforming the time-based accelerometer signals into the frequency domain. This transformation reveals the signal's frequency components, highlighting how the signal's power is distributed across different frequencies.

Why Frequency-domain Features?

- Capturing Repetitive Patterns: Many animal activities, such as walking, running, or specific behavioral patterns, exhibit repetitive motions. These patterns are more apparent in the frequency domain, where they manifest as distinct peaks or characteristic distributions.

- Filtering Noise: Frequency-domain analysis allows for effective separation of signal and noise. By focusing on key frequency bands, one can isolate the essential parts of the signal related to animal behavior.
- Understanding Energy Distribution: Analyzing how the energy of the signal is distributed across various frequencies can reveal insights into the intensity and nature of different activities.

### *The Process: Time-domain to Frequency-domain Transformation*

To analyze frequency-domain features, we first need to transform our time-domain data into the frequency domain. This is typically done using the Fast Fourier Transform (FFT), an algorithm for efficiently computing the Discrete Fourier Transform (DFT).

### *Understanding FFT and Fourier Transform*

The FFT is an algorithm to compute the DFT efficiently. Understanding FFT requires a grasp of the Fourier Transform, a fundamental concept in signal processing.

### *Fourier Transform*

The Fourier Transform is a mathematical transformation used to analyze the frequencies contained in a signal. It decomposes a time function into its basic frequencies.

Mathematical Formula:

The continuous Fourier Transform $F(\omega)$ of a continuous time-domain signal $f(t)$ is given by:

$$F(\omega) = \int_{-\infty}^{+\infty} f(t)e^{-j\omega t}dt$$

where,

- $F(\omega)$ is the Fourier Transform of $f(t)$.
- $\omega$ is the angular frequency (in radians per second).
- $t$ is time.
- $e^{-j\omega t}$ is a complex exponential function, where $j$ is the imaginary unit.

The result $F(\omega)$ is a function of frequency with complex values. The magnitude of $F(\omega)$ gives the amplitude of each frequency component in a signal, while the phase $F(\omega)$ of provides the phase shift of each frequency component.

**Discrete Fourier Transform (DFT)**

In digital signal processing, we often work with discrete signals. The DFT is the version of the Fourier Transform used for sequences of discrete values.

Mathematical Formula:

For a discrete signal $x[n]$ with $N$ samples, the DFT $X[k]$ is defined as:

$$X[k] = \sum_{n=0}^{N-1} x[n]e^{-j\frac{2\pi}{N}kn}$$

where,

- $X[k]$ is the $k^{th}$ element of the of the frequency domain representation of the sequence where $k$ corresponds to the frequency bin.
- $x[n]$ is the $n^{th}$ sample in the time-domain sequence.
- $N$ is the total number of samples.
- $j$ is the imaginary unit ($\sqrt{-1}$).

The DFT decomposes the signal into a sum of sinusoids of different frequencies, each with its own amplitude and phase.

**In the context of DFT, amplitude represents the magnitude of a frequency component within a signal, indicating its strength, while phase indicates the position of the wave relative to a reference point, describing the shift of the frequency component in time.**

### *Fast Fourier Transform (FFT)*

The FFT is a method for calculating the DFT and its inverse efficiently. FFT does not have a single, distinct formula in the same way that the DFT does because the FFT is an algorithmic approach to compute the DFT of a sequence. The essence of the FFT lies in its method of breaking down the DFT of a sequence of length $N$. FFT algorithm handles this summation by dividing the DFT into smaller parts, using properties such as the even-odd decomposition or the Cooley-Tukey algorithm, which is among the most commonly used FFT algorithms. The Cooley-Tukey algorithm, for example, recursively divides the DFT into two halves: one for the even-indexed elements and one for the odd-indexed elements, and then combines these results to produce the final DFT. This method significantly reduces the number of computations needed from $O(N^2)$ (as in the direct computation of DFT) to $O(NlogN)$, assuming $N$ is a power of 2.

FFT is widely used in accelerometer data analysis to transform time-domain signals into the frequency domain. This transformation helps in identifying the dominant frequencies in the motion and understanding the frequency content of different animal activities. The FFT output can be used to extract various frequency-domain features, such as spectral energy, peak frequency, and power spectral density, which are crucial for activity recognition and behavior analysis.

### *Optimal Frequency and Window Size Selection for FFT in Accelerometer Data Analysis*

This section provides guidance on selecting the optimal frequency and window size for FFT, ensuring an efficient and meaningful analysis of accelerometer data.

Understanding the Importance of Sampling Rate and Window Size:

- Sampling Rate Considerations:
  - The sampling rate at which accelerometer data is collected dictates the frequency range that can be accurately analyzed using FFT.
- Window Size for FFT:
  - FFT algorithms are most efficient when the window size is a power of 2.
  - The smallest window size for FFT that still captures the essence of the signal depends on the sampling rate. For example, at a 12 Hz sampling rate, a window size of 16 ($2^4$) can be the starting point as it is the smallest power of 2 greater than 12. However, this only covers around one second of data.
  - For instance, for a sampling rate of 12 Hz, if you require a window to span at least 2 or 3 seconds, you will need at least 24 to 36 samples (since we have 12 samples per second $\times 2$ or $\times 3$ seconds). The nearest power of 2 that accommodates this requirement is 64, which covers approximately 5.33 seconds (64/12 seconds) of data.
- Frequency Resolution:
  - The frequency resolution which is the smallest distinguishable frequency difference, is determined by the window size and the sampling rate. A larger window size offers better frequency resolution but less temporal resolution. For a 12 Hz sampling rate and a window size of 64, the frequency resolution is $12/64 = 0.1875$ Hz.

When applying FFT to accelerometer data, consider these factors to choose an appropriate window size and understand the implications on frequency resolution and computational efficiency. The goal is to balance capturing enough data for meaningful analysis while maintaining computational speed.

Following are some key frequency-domain features:

- **Spectral Energy**

  Spectral Energy is a fundamental feature in the frequency-domain analysis of accelerometer data, especially in the context of animal activity recognition. This feature provides insight into the total energy present across various frequency bands of the accelerometer signal.

  Definition and Calculation of Spectral Energy:

Spectral Energy quantifies the cumulative sum of the squared magnitudes of the frequency components of a signal. It reflects the overall 'power' or intensity contained within the signal's frequency spectrum.

Calculation: For a given frequency spectrum obtained from FFT, the Spectral Energy is calculated as:

$$Spectral\ Energy = \sum_{i=1}^{N} |X(f_i)|^2$$

where,

- $|X(f_i)|$ is the magnitude of the FFT at the frequency $f_i$.
- $N$ is the total number of frequency bins in the spectrum.
- The sum runs over all frequency bins, summing the squared magnitudes.

Application in Activity Recognition:

- Energy Distribution: Spectral Energy helps in understanding how the energy of the accelerometer signal is distributed across different frequencies. This is particularly useful in distinguishing between different types of animal activities, as various activities often exhibit unique energy distributions.
- Intensity of Motion: Higher spectral energy values typically correspond to more intense motion activities, while lower values may indicate more subtle movements.

- **Power Spectral Density (PSD)**

  PSD is a crucial frequency-domain feature that shows how the power of a signal is distributed across various frequencies. Peaks in the PSD can identify dominant rhythms or cycles in the activity. It is particularly useful in identifying how much of the signal's power lies within specific frequency bands.

  **Calculation of PSD**

  PSD is calculated from FFT of the signal. It involves squaring the magnitude of the FFT to get the power at each frequency component and normalizing it appropriately.

  Formula: If $X(f)$ represents the FFT of the signal at frequency $f$, the $PSD(f)$ is given by:

$$PSD(f) = \frac{1}{N}|X(f)|^2$$

where, $N$ is the normalization factor, often the length of the time-domain signal or the FFT window.

Application in Activity Recognition:

- Dominant Rhythms: Peaks in the PSD indicate dominant rhythms or cycles in the activity, which can be characteristic of specific types of movements like trotting or running.
- Comparative Analysis: PSD allows for comparison of frequency content across different activities, aiding in understanding and distinguishing between them.

- **Dominant Frequency**

  Dominant Frequency is a crucial feature in frequency-domain analysis of accelerometer data. This feature identifies the frequency component that has the highest energy or amplitude in the frequency spectrum.

  Definition: The Dominant Frequency is the frequency at which the amplitude of the signal's spectrum reaches its maximum. This frequency represents the most significant periodic component within the window of the accelerometer data.

  Steps for Calculating Dominant Frequency considering Energy Content:

  - Calculate PSD
  - Identify Significant Frequencies: Instead of just finding the peak, analyze the PSD to identify frequencies that contribute significantly to the total power. This can be done in several ways:
    a. Thresholding: Set a threshold (e.g., a percentage of the maximum PSD value) to identify significant peaks in the PSD.
    b. Energy Concentration: Identify frequencies or bands where a substantial portion of the signal's total energy is concentrated.
  - Account for Complex Signals: In cases where the signal is complex with multiple significant frequencies, you might consider identifying multiple dominant frequencies or even defining dominant bands.

- **Spectral Entropy**

  Spectral entropy measures the complexity or randomness in the frequency distribution, indicating the unpredictability of the frequency patterns.

  Calculation of Spectral Entropy:

  Concept: Spectral Entropy is calculated from the PSD. It treats the normalized PSD values as a probability distribution and calculates the entropy of this distribution.

  Formula: If $PSD(f)$ is the normalised PSD, the Spectral Entropy is calculated as:

$$H = -\sum_{i=1}^{N} PSD(f_i) \log_2 PSD(f_i)$$

where, $PSD(f_i)$ is the normalised PSD at frequency $f_i$, and $N$ is the total number of frequency bins.

Application in Activity Recognition:

- Complexity Analysis: Higher Spectral Entropy indicates a more complex or less predictable frequency pattern, which might be characteristic of certain types of activities.
- Distinguishing Between Activities: Different activities may exhibit distinct patterns of Spectral Entropy, aiding in their classification and recognition.

- **Peak Frequency**

  Peak Frequency is a specific frequency-domain feature used in accelerometer data analysis. It refers to the frequency within a given window of data that has the highest amplitude in the frequency spectrum.

  Difference from Dominant Frequency:

  Peak Frequency vs. Dominant Frequency: While "Peak Frequency" and "Dominant Frequency" might sound similar and are sometimes used interchangeably, however there can be subtle differences based on the context or specific implementation:

  - Peak Frequency typically refers to the frequency with the highest peak (maximum amplitude) in the spectrum, regardless of the total energy content across the spectrum.
  - Dominant Frequency often implies the frequency or frequencies that contribute most significantly to the signal, which can be interpreted in terms of energy (spectral energy) rather than just amplitude.

  Context-Dependent Interpretation: In some cases, especially when the spectrum has a clear single peak, both Peak Frequency and Dominant Frequency might refer to the same frequency. However, in a more complex spectrum with multiple peaks, the "dominant" aspect might consider additional factors like the width or energy of the peak.

  Calculation of Peak Frequency:

  FFT Analysis: First, perform a FFT on the accelerometer data.

  Identifying the Peak: The Peak Frequency is determined by finding the frequency at which the FFT amplitude (magnitude) is maximum.

- **Spectral Centroid**

  The Spectral Centroid is a measure that indicates the *center of mass* of the frequency spectrum. It shows where the bulk of the energy in the spectrum is concentrated, often used to characterize the sharpness of a signal.

The Spectral Centroid is calculated as a weighted average of the frequencies present in the signal, weighted by their amplitudes.

Formula:

$$Spectral\ Centroid = \frac{\sum f_i \times a_i}{\sum a_i}$$

where, $f_i$ is the frequency and $a_i$ is the amplitude of the $i$th bin in the FFT.

## Python Example for Feature Extraction

In this section, we will look into a Python code that demonstrates the extraction of time and frequency domain features from accelerometer data. The code provided acts as a practical guide to implementing the concepts discussed earlier. The initial steps involve importing required libraries and creating a synthetic dataset to simulate accelerometer data. This synthetic dataset is instrumental in illustrating how the code functions and can be adapted to real-world datasets. Subsequent to dataset creation, the code introduces various helper functions. Each of these functions are defined to calculate a specific feature from the accelerometer data. These features include basic statistical measures, zero-crossing rate, entropy, spectral energy, and more.

The final part of the code brings everything together, demonstrating how to apply these functions to extract meaningful features from the dataset. The code example can be found in the GitHub repository containing this book (https://github.com/nkcAna/WSDpython) under the name *Chapter_5_Feature_Extraction.ipynb* in *Chapter_5* folder.

Let's look at the python code:

```
# Importing the libraries
import numpy as np
import pandas as pd
from scipy import stats
from statistics import mode

# Firstly, we create a dataset with 1000 rows and 3 columns (for x,
y, z axes)
np.random.seed(0)
data = np.random.rand(1000,3)

# We now create a dataframe from the data
columns = ['x', 'y', 'z']
df = pd.DataFrame(data, columns = columns)
```

```
# For our example, we manually assign the labels of our dataframe
labels = ['walking']*300 + ['standing']*400 +['walking']*300

# We then add the label column to the dataframe
df['label'] = labels

# Display the dataframe
print(df)

# Output
            x         y         z    label
0    0.548814  0.715189  0.602763  walking
1    0.544883  0.423655  0.645894  walking
2    0.437587  0.891773  0.963663  walking
3    0.383442  0.791725  0.528895  walking
4    0.568045  0.925597  0.071036  walking
..        ...       ...       ...      ...
995  0.698630  0.503697  0.025738  walking
996  0.774353  0.560374  0.082494  walking
997  0.475214  0.287293  0.879682  walking
998  0.284927  0.941687  0.546133  walking
999  0.323614  0.813545  0.697400  walking

[1000 rows x 4 columns]
```

- Importing numpy, pandas, and scipy.stats and statistics.mode.
- A synthetic dataset is created to simulate accelerometer data, representing the x, y, and z axes. This dataset is then converted into a pandas DataFrame for ease of manipulation.

Functions for zero crossing rate, and entropy features:

```
# Function to calculate zero crossing rate
def zero_crossing_rate_sgn(window_data):
    # Define the signum function
    sgn = np.vectorize(lambda x: -1 if x < 0 else (1 if x > 0 else
    0))

    # Apply the signum function to window data
    sign_data = sgn(window_data)

    # Calculate the ZCR using the sgn function
    zcr = np.sum(np.abs(np.diff(sign_data))) / (2 * (len(window_
    data) - 1))
    return zcr

# Function to calculate entropy
def calculate_entropy(data):
```

```
histogram, bin_edges = np.histogram(data, bins='auto',
density=True)
probabilities = histogram * np.diff(bin_edges)
probabilities = probabilities[probabilities > 0]
entropy = -np.sum(probabilities * np.log2(probabilities))
return entropy
```

### Function to Calculate Zero Crossing Rate (ZCR)

- Function Definition:
  - zero_crossing_rate_sgn(window_data): Calculates the zero-crossing rate of a given segment of data (window_data). This is typically applied to a window of time-series data.
- Signum Function Definition:
  - A lambda function is defined to act as a signum function (sgn), which returns –1 for negative values, +1 for positive values, and 0 for zero values. This is vectorized using np.vectorize for efficient application to arrays.
- Application of Signum Function:
  - The signum function is applied to the window_data to create sign_data, which consists of –1, 0, and +1, representing the sign of each data point.
- Calculating ZCR:
  - ZCR is calculated by finding the absolute difference between consecutive points in sign_data using np.diff(sign_data), then summing these differences and normalizing by (2 * (len(window_data) –1)).
  - This normalization accounts for the fact that each pair of consecutive data points contributes to two comparisons in the dataset.
  - ZCR is indicative of the frequency at which the signal changes its sign. It is useful in detecting the rate of change in the time-series data.

### Function to Calculate Entropy

- Function Definition:
  - calculate_entropy(data): Computes the entropy of a dataset, providing a measure of its randomness or unpredictability.
- Histogram Computation:
  - The function first computes a histogram of the data using np.histogram, with the number of bins determined automatically (bins = 'auto'). This histogram represents the distribution of data values.

- The parameter density = True ensures that the histogram is normalized to form a probability density.

- Probability Calculation:
  - The probabilities for each bin are calculated by multiplying the histogram values by the width of each bin (np.diff(bin_edges)).
  - The probabilities are then filtered to exclude any non-positive values since log of zero or negative numbers is undefined.
- Entropy Computation:
  - Entropy is calculated using the formula –np.sum(probabilities * np.log2(probabilities)). This formula is based on the Shannon entropy, a fundamental concept in information theory that quantifies the amount of uncertainty or surprise associated with random variables.
  - Higher entropy values indicate a more random or unpredictable dataset, while lower values suggest more predictability.

Python code for other helper functions:

```
# Function to calculate spectral energy
def calculate_spectral_energy(fft_values):
    spectral_energy = np.sum(np.abs(fft_values)**2)
    return spectral_energy

# Function to calculate dominant frequency
def calculate_dominant_frequency(fft_vals, fft_freq):
    # Filter to keep only the positive frequencies
    positive_indices = fft_freq > 0
    positive_fft_vals = fft_vals[positive_indices]
    positive_fft_freq = fft_freq[positive_indices]

    # Calculate the Power Spectral Density (PSD) for positive
    frequencies
    positive_psd = np.abs(positive_fft_vals)**2

    # Identify significant frequencies
    threshold = 0.1 * np.max(positive_psd)  # Example: 10% of the
      max PSD value
    significant_freqs_indices = np.where(positive_psd >= threshold)
      [0]
    significant_freqs = positive_fft_freq[significant_freqs_indices]

    # Calculate the mean or median of these significant frequencies
    dominant_freq = np.mean(significant_freqs) if len(significant_
      freqs) > 0 else 0
    return dominant_freq
```

```
# Function to calculate psd
def calculate_psd(fft_vals, N):
    psd = np.abs(fft_vals)**2 / N
    return psd

# Function to calculate spectral entropy
def calculate_spectral_entropy(psd):
    normalized_psd = psd / np.sum(psd)
    spectral_entropy = -np.sum(normalized_psd * np.log2(normalized_
      psd))
    return spectral_entropy

# Function to calculate peak frequency
def calculate_peak_frequency(fft_vals, fft_freq):
    # Calculate the magnitude of the FFT
    magnitude = np.abs(fft_vals)

    # Find the index of the maximum magnitude
    idx_peak = np.argmax(magnitude)

    # Return the corresponding frequency
    peak_frequency = fft_freq[idx_peak]
    return peak_frequency
```

### Function to Calculate Spectral Energy

- Purpose: This function calculates the total energy in the frequency spectrum of a signal.
- How It Works:
  - fft_values: The input to the function, which are the Fourier coefficients obtained from applying FFT on a time-series signal.
  - np.abs(fft_values)**2: Computes the square of the absolute value of each FFT coefficient. This represents the energy at each frequency component.
  - np.sum(...): The sum of these energies across all frequency components gives the total spectral energy.

### Function to Calculate Dominant Frequency

- Purpose: Identifies the frequency that contributes the most to the signal, in terms of energy.
- How It Works:
  - Filters out only the positive frequencies.
  - Computes the Power Spectral Density (PSD) for these frequencies.

- Sets a threshold to identify significant frequencies (those with high energy).
- The dominant frequency is then calculated as the mean of these significant frequencies.

### Function to Calculate Power Spectral Density (PSD)

- Purpose: Calculates the power distribution across frequencies.
- How It Works:
  - N: The total number of samples in the original time series.
  - np.abs(fft_vals)**2 / N: Computes the normalized power at each frequency component.

### Function to Calculate Spectral Entropy

- Purpose: Measures the disorder or complexity in the frequency distribution of a signal.
- How It Works:
  - Normalizes the PSD to make it a probability distribution.
  - Calculates the entropy of this distribution using the Shannon entropy formula.

### Function to Calculate Peak Frequency

- Purpose: Identifies the frequency at which the signal's amplitude is highest.
- How It Works:
  - Calculates the magnitude of each FFT coefficient.
  - Finds the index of the coefficient with the maximum magnitude and returns the corresponding frequency as the peak frequency.

Function to calculate spectral features:

```
# Function to calculate spectral features
def calculate_spectral_features(fft_vals, fft_freq):
    # Compute power spectrum
    psd = np.abs(fft_vals)**2
    total_power = np.sum(psd)
    psd_normalized = psd / total_power

    # Spectral Centroid
    spectral_centroid = np.sum(fft_freq * psd_normalized)
```

```
    # Spectral Spread
    spectral_spread = np.sqrt(np.sum(((fft_freq - spectral_
       centroid)**2) * psd_normalized))

    # Spectral Skewness and Kurtosis
    spectral_skewness = np.sum(((fft_freq - spectral_centroid)**3)
       * psd_normalized) / (spectral_spread**3)

    spectral_kurtosis = np.sum(((fft_freq - spectral_centroid)**4)
       * psd_normalized) / (spectral_spread**4)

    # Spectral Flatness
    spectral_flatness = np.exp(np.mean(np.log(psd + 1e-10))) /
       np.mean(psd + 1e-10)

    return     spectral_centroid, spectral_spread, spectral_skewness,
               spectral_kurtosis, spectral_flatness
```

- calculate_spectral_features: Defines the function calculate_spectral_features with two parameters: fft_vals (FFT values of the signal) and fft_freq (corresponding frequencies).
- psd = np.abs(fft_vals)**2: Calculates the Power Spectral Density (PSD) by squaring the absolute values of the FFT coefficients (fft_vals).
- total_power = np.sum(psd): Computes the total power of the signal by summing all the PSD values.
- psd_normalized = psd / total_power: Normalizes the PSD by dividing each value by the total power, ensuring the PSD sums to 1.
- spectral_centroid = np.sum(fft_freq * psd_normalized): Calculates the spectral centroid, which is the weighted mean of the frequencies present in the signal. It is weighted by the normalized PSD, representing the 'center of mass' of the spectrum.
- spectral_spread = np.sqrt (np.sum(((fft_freq – spectral_centroid)**2) * psd_normalized)): Computes the spectral spread, which measures the width of the spectrum around the spectral centroid.
- spectral_skewness = np.sum(((fft_freq – spectral_centroid)**3) * psd_normalized) / (spectral_spread**3): Calculates the spectral skewness, a measure of the asymmetry of the frequency distribution around the spectral centroid.
- spectral_kurtosis = np.sum(((fft_freq – spectral_centroid)**4) * psd_normalized) / (spectral_spread**4): Computes the spectral kurtosis, indicating how 'peaked' the frequency distribution is compared to a normal distribution.

- spectral_flatness = np.exp(np.mean(np.log(psd + 1e–10))) / np.mean(psd + 1e–10): Calculates the spectral flatness, which measures how 'flat' or 'peaky' the spectrum is. A higher value indicates a flatter spectrum. The 1e–10 is added to avoid logarithm of zero.
- return spectral_centroid, spectral_spread, spectral_skewness, spectral_kurtosis, spectral_flatness: The return statement returns the calculated spectral features: centroid, spread, skewness, kurtosis, and flatness.

The following function is created to extract time and frequency domain features from the dataset:

```
# Create a function to extract time-domain features from the
dataframe

def calculate_features(data, delta_t):
    # We initialize a dictionary to store the features
    features = {}

# Basic statistical measures
    for col in data.columns:
        features[f'{col}_mean'] = data[col].mean()
        features[f'{col}_std_dev'] = data[col].std()
        features[f'{col}_variance'] = data[col].var()
        features[f'{col}_icv'] = features[f'{col}_mean'] /
                                 features[f'{col}_std_dev']
        features[f'{col}_median'] = data[col].median()
        features[f'{col}_minimum'] = data[col].min()
        features[f'{col}_maximum'] = data[col].max()
        features[f'{col}_skewness'] = data[col].skew()
        features[f'{col}_kurtosis'] = data[col].kurt()
        # Calculate Interquartile Range
        features[f'{col}_iqr'] = stats.iqr(data[col])
    features[f'{col}_zcr'] = zero_crossing_rate_sgn(data[col])
    features[f'{col}_peak_to_peak'] = np.max(data[col]) -
      np.min(data[col])
# Calculate Entropy for each axis
    features[f'{col}_entropy'] = calculate_entropy(data[col])

# Paiwise Correlation
    corr_matrix = data.corr().values
    for i, col1 in enumerate(data.columns):
        for j, col2 in enumerate(data.columns):
            if i<j: # to avoid duplicate pairs
                features[f'correlation_{col1}_{col2}'] = corr_
                      matrix[i,j]
```

```
# Signal Magnitude Area
    features['sma'] = data.abs().sum().sum() / len(data)

# Energy calculation
    energy = np.sum(data['x']**2 + data['y']**2 + data['z']**2)
    features['energy'] = energy

# Combined entropy
    combined_data = np.hstack((data['x'], data['y'], data['z']))
    features['combined_entropy'] = calculate_entropy(combined_data)

# Calculate Frequency-Domain Features
    fs = 1 / delta_t  # Frequency based on delta_t
    N = len(data)

    for col in ['x', 'y', 'z']:
        # Perform FFT
        fft_vals = np.fft.fft(data[col])
        fft_freq = np.fft.fftfreq(N, 1/fs)

# Filter to keep only the positive frequencies
    positive_indices = fft_freq > 0
    positive_fft_vals = fft_vals[positive_indices]
    positive_fft_freq = fft_freq[positive_indices]

# Spectral Energy
    features[f'{col}_spectral_energy'] = calculate_spectral_
    energy(fft_vals)

# Dominant Frequency
    features[f'{col}_dominant_frequency'] =
    calculate_dominant_frequency (positive_fft_vals,
    positive_fft_freq)

# Calculate PSD and Spectral Entropy
    psd = calculate_psd(fft_vals, N)
    features[f'{col}_max_psd'] = np.max(psd)  # Storing PSD
    values

    spectral_entropy = calculate_spectral_entropy(psd)
        features[f'{col}_spectral_entropy'] = spectral_entropy

# Calculate Peak Frequency
    features[f'{col}_peak_frequency'] =
    calculate_peak_frequency(positive_fft_vals, positive_fft_freq)

# Spectral Features Calculation
    spectral_centroid, spectral_spread, spectral_skewness,
```

```
    spectral_kurtosis, spectral_flatness =
    calculate_spectral_features(positive_fft_vals, positive_fft_freq)

    features[f'{col}_spectral_centroid'] = spectral_centroid
    features[f'{col}_spectral_spread']   = spectral_spread
    features[f'{col}_spectral_skewness'] = spectral_skewness
    features[f'{col}_spectral_kurtosis'] = spectral_kurtosis
    features[f'{col}_spectral_flatness'] = spectral_flatness

    return features
```

Below is the explanation of the calculate_features function line by line, focusing on the Python operations and constructs:

- Function Definition: Defines calculate_features with parameters data (a pandas DataFrame) and delta_t (a time interval).
- Initializing Feature Dictionary: An empty dictionary named 'features' is created to store computed feature values.
- Basic Statistical Measures Loop: Iterates over each column in the DataFrame data.
  - Calculates mean, standard deviation, variance, inverse coefficient of variation (mean divided by standard deviation), median, minimum, maximum, skewness, kurtosis, interquartile range (using stats.iqr), zero-crossing rate (using a custom function zero_crossing_rate_sgn), and peak-to-peak distance for each column.
- Entropy Calculation: Computes entropy for each axis/column using a custom calculate_entropy function.
- Pairwise Correlation Calculation:
  - rates over column pairs to calculate pairwise correlations, avoiding duplicate pairs.
- Signal Magnitude Area:
  - Calculates the SMA as the sum of absolute values divided by the length of the DataFrame.
- Energy Calculation:
  - Computes the total energy of the signal by summing the squares of 'x', 'y', and 'z' values.
- Combined Entropy Calculation:
  - Concatenates 'x', 'y', and 'z' columns using np.hstack and calculates their combined entropy.

- Frequency-domain Feature Calculations Loop: Iterates over 'x', 'y', and 'z' columns.
  - Performs FFT and calculates frequency using np.fft.fft and np.fft.fftfreq.
  - Filters to keep only positive frequencies.
  - Calculates spectral energy, dominant frequency, PSD, spectral entropy, peak frequency, and various spectral features (centroid, spread, skewness, kurtosis, flatness) using custom functions.
- Return Statement: Returns the features dictionary containing all the calculated features.

This function is comprehensive, covering a wide range of time-domain and frequency-domain features, making it highly valuable for detailed signal analysis.

Now, the following function is defined to extract the features from the accelerometer data using sliding windows:

```
# Function to extract features using sliding windows
def feature_extraction_with_windows(data, window_size, step_size,
    calculate_features_func, delta_t):
    num_samples = len(data)
    features_list = []

    for start in range(0, num_samples-window_size+1, step_size):
        end = start + window_size
        window_data = data.iloc[start:end]
        window_features = calculate_features_func(window_data.
        drop(columns=['label']), delta_t)

# Get the most frequent label in the window
    try:
    window_label = mode(window_data['label'])
    except:
    # In case there is no unique mode
    # Use the first label in case of a tie
        window_label = window_data['label'].value_counts().index[0]

    window_features['label'] = window_label
    features_list.append(window_features)

    return pd.DataFrame(features_list)
```

Here is an explanation of the feature_extraction_with_windows function:

- Function Definition: A function named feature_extraction_with_windows is defined to extract features from a DataFrame over sliding windows.

- Calculating Number of Samples: The total number of samples in the DataFrame data is determined.
- Initializing Feature List: An empty list features_list is created to store features from each window.
- Sliding Window Loop:
  - A loop starts, which iterates from the beginning to the end of the DataFrame, using a step size of step_size. This loop creates the sliding windows.
  - For each iteration, the start and end indices of the window are determined.
  - The subset of data for the current window is extracted using iloc.
- Feature Calculation for Each Window:
  - The function calculate_features_func is called with the current window data (excluding the 'label' column) and delta_t to calculate features for this window.
- Label Determination for Each Window:
  - A try-except block is used to handle the label assignment for the window.
  - The most frequent label in the current window is determined using the mode function.
  - If there is no unique mode (i.e., a tie in label frequency), the first label in the frequency sorted list is chosen (value_counts().index[0]).
- Appending Window Features to List:
  - The label for the window is added to the window_features dictionary.
  - The window_features dictionary is appended to the features_list.
- Return Statement:
  - The function returns a pandas DataFrame created from features_list, where each row represents the features extracted from a window.

This function is useful for time-series analysis where behaviors or patterns are observed over intervals. By analyzing data in windows, it captures temporal dynamics and variations within the data.

In the following code, we use the created functions to extract the features from our dataset:

```
# Example

window_size = 64 # Number of samples in each window
step_size = 10 # Step size for the moving window
```

```
delta_t = 1/12  # Example: 12 Hz sampling rate

preprocessed_data = feature_extraction_with_windows(df,
                                                    window_size,
                                                    step_size,
                                                    calculate_features, delta_t)
# Check the first five rows of the data
preprocessed_data.head()
```

The provided code demonstrates how to apply the feature_extraction_with_windows function to a DataFrame df to extract the time-domain and frequency-domain features from time-series data.

- Setting Parameters:
  - window_size = 64: Defines the number of samples in each window. Each window will contain 64 data points (5.333 seconds).
  - step_size = 10: The window will move 10 samples at a time. This setting introduces overlap between consecutive windows.
  - delta_t = 1/12: Sets the time interval between samples, corresponding to a sampling rate of 12 Hz.
- Feature Extraction:
  - The function feature_extraction_with_windows is called with the DataFrame df, the window size, step size, the feature calculation function calculate_features, and the time interval delta_t.
  - The result, preprocessed_data, is a new DataFrame where each row contains features extracted from a sliding window of the original data.
- Checking the First Five Rows:
  - preprocessed_data.head(): Shows the first five rows of the preprocessed_data.

The Output:

- The output is a DataFrame with 94 rows, and 76 feature columns including the label column.
- Each column corresponds to a specific feature.
- Each row is a feature set for a window of data, capturing both time-domain and frequency-domain characteristics.

## Summary

In Chapter 5 we looked into the critical stages of preprocessing and feature extraction in the context of animal behavior research using accelerometer data. We talked about the fundamental principles underlying accelerometer data, setting a foundation for understanding how these datasets can be interpreted and utilized.

Then we introduced the various steps of data preprocessing. Starting with data cleaning, followed by a thorough discussion on data scaling and normalization, highlighting the importance of transforming data into a standard format for consistency and comparability across different datasets.

We further discussed filtering techniques. Filtering helps in reducing noise and improving the signal quality, which is crucial for accurate analysis and interpretation of animal movements and behaviors. The section provided a detailed overview of different filtering methods and their practical applications.

Moving into the core of the chapter, feature extraction is extensively covered. We discussed windowing in feature extraction, a technique that involves dividing the continuous stream of data into manageable segments or 'windows'. This approach is significant in capturing the temporal dynamics of animal behavior.

The chapter delves into the discussion on time-domain and frequency-domain features, providing an understanding of how these features can be extracted and their relevance in animal behavior studies. The time-domain features offer insights into the basic statistical properties of the data, while the frequency-domain features shed light on the periodic nature of the animal movements.

Next, we presented a practical Python example for feature extraction, offering a hands-on approach to applying the theoretical concepts discussed. The section is particularly valuable for readers looking to implement these techniques in real-world scenarios.

CHAPTER 6

# Feature Selection Techniques

Following the exploration of data preprocessing and feature extraction in Chapter 5, Chapter 6 progresses into feature selection techniques. This chapter stands at the crucial intersection where we refine the raw power of extensive datasets into meaningful insights, focusing on accelerometer data derived from sheep.

Feature selection forms the backbone of effective activity recognition. The vast array of features extracted from datasets often encompasses redundant, irrelevant, or potentially misleading information. Such inappropriate data can negatively impact the efficiency and accuracy of predictive or classification models. While exhaustive search algorithms have their advantage in locating distinct features, their practical use faces significant challenges, especially within large, high-dimensional datasets. To avoid these challenges, an array of feature selection techniques has been employed across various fields to achieve an optimal set of features. These refined features are then utilized in classification or predictive models, improving their performance and interpretability.

In this chapter, we will investigate the most commonly used algorithms for feature selection, categorized into filter, wrapper, and hybrid methods. Each category offers unique approaches and advantages in feature selection, and their detailed exploration will provide a comprehensive understanding of their applications and effectiveness.

We will also present a hands-on Python example using a preprocessed dataset data collected from farm animals. This example will illustrate the practical application of these feature selection techniques. Through this example, readers will gain practical experience and insights into applying these techniques to similar datasets in their respective fields.

## Filter Methods

Filter methods represent a fundamental of feature selection techniques used in machine learning and data preprocessing. These methods are characterized by their use of various statistical measures to evaluate the importance of different features independent of any machine learning algorithm. The key advantage of filter methods lies in their simplicity and efficiency in high-dimensional datasets. They are generally faster and less computationally intensive compared to wrapper and hybrid methods, making them suitable for initial feature reduction in large datasets.

The primary mechanism of filter methods involves assessing each feature's contribution to the model based on proxy measures such as information content, statistical attributes like variance, consistency, and similarity scores. These measures allow for an objective evaluation of each feature's standalone predictive power.

The most commonly used filter methods include:

- Information-based filters: These assess the amount of information or mutual information a feature shares with the target variable.
- Statistical filters: These use statistical tests to evaluate the relationship of features with the target variable.
- Similarity-based filters: Features are evaluated based on their ability to distinguish between samples that are near each other.

Despite their advantages, filter methods have limitations. One significant drawback is their inability to capture feature dependencies and interactions since they treat each feature independently. This univariate approach can sometimes lead to the retention of redundant or correlated features, which might not be beneficial for the model's performance.

Moreover, filter methods are model-agnostic, meaning they do not take into account the model that will eventually be used. This can be both an advantage and a disadvantage. While it allows these methods to be universally applied across different models, it also means they might not select the best feature subset for a specific model type.

In summary, filter methods are an effective first step in feature selection, especially when dealing with large datasets. They provide a quick way to reduce dimensionality and remove irrelevant features. However, it is essential to consider their limitations and, in some cases, to complement them with other feature selection methods that take into account model-specific characteristics.

### *Information Gain*

Information Gain is a technique used in feature selection to determine the importance of features by measuring the reduction in entropy. It is particularly useful in scenarios where we need to understand how much information a feature provides about the class.

## *Chi-square Test*

The Chi-square test is a statistical method used in feature selection to evaluate the independence between two variables. In machine learning, it is often used to determine the relevance of categorical features with respect to categorical labels. The essence of the Chi-square test in feature selection is to assess the dependency between each feature and the target variable: the higher the Chi-square value, the more likely the feature is dependent on the target, making it a potentially good predictor.

### Some Notes on Chi-square

- Categorical Features and Target: The Chi-square test is typically applied for categorical features. It assesses whether the distribution of a sample matches an expected distribution.
- Independence of Samples: The test assumes that the samples are independent.
- Expected Frequency: The expected frequency of each category in each feature should generally be greater than 5. This condition is necessary to ensure the legitimacy of the test outcomes.
- Applicability to Numerical Data: While the Chi-square test is typically used for categorical data, it can be applied to continuous or numerical data by converting these into categorical data (binning). However, this conversion needs to be done carefully, as it may result in information loss.

Given these points, whether Chi-square test can be used on the dataset depends on the nature of the features. If the dataset consists of numerical features (like accelerometer data), you will need to discretize these features into bins or categories to apply the Chi-square test effectively. This binning process can be arbitrary and might not always yield meaningful results, especially if the discretization does not align well with the underlying patterns in the data.

Therefore, while technically possible, using the Chi-square test on a dataset primarily composed of numerical features like accelerometer data might not be the most appropriate choice. Instead, other feature selection methods that cater to continuous data, such as mutual information or feature importance from tree-based models, could be more suitable and informative.

## *Analysis of Variance (ANOVA) F-Value*

ANOVA F-Value is a statistical approach used in feature selection, particularly useful in scenarios involving classification tasks. This method operates under the principle of variance analysis to determine if the means of different groupings are considerably different.

The F-Value in ANOVA is a ratio that compares the variance between different classes to the variance within each class. A higher F-Value for a feature indicates that the feature has a significant distinction between classes, suggesting its potential importance in predictive modeling.

### Advantages of Using ANOVA F-Value

- Model Independence: ANOVA F-Value is a model-agnostic method. It does not need the building of a model and hence is computationally efficient and independent of any learning algorithm biases.
- Scalability: This method can handle high-dimensional datasets efficiently, making it suitable for initial feature selection stages.
- Ease of Interpretation: The statistical basis of the ANOVA F-Value makes it straightforward to interpret and understand.

### Limitations

- Assumption of Normality: ANOVA assumes that the data for each class is normally distributed, which may not always be the case in real-world datasets.
- Equal Variance: ANOVA assumes homogeneity of variance (equal variance across groups), which can be a limitation in certain datasets.

## *Correlation Coefficient*

The correlation coefficient is a statistical metric indicating the degree of linear relationship between two variables. In the context of feature selection, it is used to identify features that have a strong linear relationship among them.

### Importance in Feature Selection

- Identifying Redundant Features: Correlation coefficients can reveal if two features essentially provide the same information. A high correlation between two features suggests redundancy; one of them can be removed without much loss of information.
- Understanding Feature-Feature Relationships: A high correlation between a feature and the other variable indicates that the feature is likely important for predicting the target.

### Types of Correlation Coefficients

- Pearson Correlation: Measures linear correlation between two variables. Values close to 1 or –1 indicate a strong positive or negative linear relationship, respectively.
- Spearman Correlation: A non-parametric measure that assesses how well the relationship between two variables can be described using a monotonic function (whether linear or non-linear).

### Limitations

- Only Linear Relationships: Pearson correlation only captures linear relationships. Non-linear but strong relationships might be overlooked.
- Sensitive to Outliers: Correlation can be heavily influenced by outliers.

## *Mean Absolute Difference (MAD)*

MAD is a statistical measure used in various fields, particularly in feature selection and signal processing. MAD offers an effective way to understand the variability or dispersion in a dataset.

Understanding MAD:

MAD is the average of the absolute differences from the mean of a dataset. It measures the average magnitude of deviations from the dataset's mean, giving a sense of the *spread* of the data.

Formula: For a dataset $X = \{x_1, x_2, \ldots, x_n\}$ with mean μ, MAD is calculated as:

$$MAD = \frac{1}{n}\sum_{i=1}^{n}|x_i - \bar{x}|$$

where, $|x_i - \bar{x}|$ is the absolute difference of each data point from the mean. A higher MAD value indicates greater variability in the dataset, while a lower MAD suggests less dispersion.

Application in Feature Selection:

- Indicator of Variability: MAD can be used to quantify the variability of each feature in a dataset. Features with very low MAD may not be very discriminative, as they do not vary much across observations.
- Filter Method: In feature selection, MAD can serve as a criterion in filter methods. Features with higher MAD are often considered more informative since they exhibit more significant variation.
- Preprocessing Step: It is a useful preprocessing step to identify and possibly remove features that show little to no variation, as these might not contribute to predictive models.

## *Relief and ReliefF*

Relief and ReliefF algorithms are influential feature selection methods, especially useful in scenarios where the understanding of feature interactions is crucial. These methods are effective for problems with complex data structures where features interact in non-linear and dependent ways.

### Relief Algorithm

The Relief algorithm is designed to assign a weight to each feature based on how well the feature can distinguish between instances that are near each other. It works by randomly selecting an instance and then finding its nearest neighbor from the same and opposite classes. If a feature value is different for neighbors of different classes (indicating it is a good feature), its weight is increased; if it is different for neighbors of the same class, its weight is decreased. Relief is good for tasks where interactions between attributes are important. However, it is generally limited to binary classification tasks.

### ReliefF Algorithm

ReliefF is an extension of the Relief algorithm that generalizes it to multiclass problems and is more robust to noisy and incomplete datasets. Unlike Relief, ReliefF considers multiple nearest neighbors rather than just one. It averages the feature weights across multiple neighbors, which makes it more reliable, especially in datasets with more noise and outliers. ReliefF can handle a variety of data types (discrete, continuous, among others) and is suitable for both binary and multiclass classification problems.

## *Variance Threshold*

The Variance Threshold is another method within filter-based feature selection techniques. It operates on a simple principle: Features with low variance are less likely to be informative or discriminative. In other words, if a feature does not vary much within the dataset, it is less likely to impact the predictive power of the model.

The Variance Threshold method involves setting a specific threshold for variance, and only the features that exceed this threshold are retained. This approach is particularly useful for removing constant or quasi-constant features[1], which do not contribute to the model's learning process. By doing so, it helps in reducing the dimensionality of the dataset, making data processing and model training more efficient.

This method is especially effective in scenarios where data contains many redundant or irrelevant features. However, it is crucial to choose the variance threshold wisely, as setting it too high might lead to the loss of potentially useful features, while setting it too low might not effectively reduce dimensionality. In essence, the Variance Threshold is a powerful tool in the feature selection process, especially in the initial stages of data preprocessing to discard features that offer little to no variability and hence, informational value.

---

[1] Quasi-constant features refer to those variables within a dataset that exhibit little to no variance among the observations.

## Wrapper Methods

Wrapper methods in feature selection are a set of techniques that select features based on the performance of a predictive or classification model. Unlike filter methods, which rely on general characteristics of the data, wrapper methods use a specific machine learning model to evaluate the effectiveness of different subsets of features.

The process usually involves training a model on various combinations of features and assessing the model's performance to determine the best set of features. This evaluation can be based on different criteria, such as accuracy, precision, recall, or any relevant performance metric. The key steps in wrapper methods typically include:

- Subset Generation: This step involves creating different combinations of features. These combinations could be generated through different strategies, such as forward selection, backward elimination, or recursive feature elimination.
- Model Training and Evaluation: For each feature subset, a model is trained and evaluated. Commonly used machine learning models in this context include Naïve Bayes, Support Vector Machines (SVM), and Random Forests. The choice of model can significantly impact the selection process, as different models have varying sensitivities to different types of features.
- Performance Assessment: The performance of the model with each subset of features is assessed using a chosen metric. The subset that yields the best performance is considered the optimal set of features.

While wrapper methods can be effective in finding the best subset of features for a given model, they have some limitations. The most significant one is computational expense. Evaluating every possible combination of features can be impractical, especially with large datasets and a high number of features. This often restricts the use of wrapper methods to datasets with a smaller number of features or necessitates the use of more efficient subset generation strategies.

Despite these challenges, wrapper methods are popular due to their model-specific approach, which can lead to better feature selection for the specific predictive task at hand, compared to the more general approach of filter methods.

### *Forward Selection*

Forward Selection is a type of wrapper method used in feature selection. It is an iterative method that starts with having no feature in the model. In each iteration, it adds the feature that provides the most significant improvement to the model until a specified criterion is reached.

Here is how forward selection typically works:

- Start with No Features: Initially, the model starts without any feature.
- Iterate through Features: In each iteration, the algorithm loops through each feature that has not been selected yet and adds it to the model temporarily.
- Evaluate the Model: For each added feature, the model is trained, and its performance is evaluated using a specified metric (such as accuracy, AUC and R-squared).
- Select the Best Feature: After evaluating all the potential new features, the one that results in the highest performance improvement is permanently added to the model.
- Repeat or Stop: This process is repeated until adding a new feature does not improve the model performance beyond a specified threshold, or until a predefined number of features is reached.

Forward selection is useful when dealing with a large set of features. It allows for a more manageable, step-wise approach to model building and feature selection. However, it can be computationally expensive as it involves repeatedly training the model.

### *Backward Elimination*

Backward elimination is another wrapper method. Unlike forward selection, backward elimination starts with all features included and iteratively removes the least significant feature until a specified stopping criterion is met.

The general process of backward elimination:

- Start with All Features: Initially, the model includes all available features.
- Evaluate and Remove: In each iteration, the model is trained, and its performance is evaluated. The least significant feature (the one whose removal most improves model performance or least deteriorates it) is then removed from the set.
- Iterate Until Criteria Met: This process is repeated until a stopping criterion is met, which could be a set number of features, a performance threshold, or a statistical significance level.
- Final Model: The final model contains the subset of features that provides the best performance according to the chosen metric.

Backward elimination can be more efficient than forward selection when the number of features is not excessively high, as it begins with a full model and removes redundant or non-informative features.

### *Recursive Feature Elimination (RFE)*

RFE works by recursively removing features and building a model on the features that remain. RFE uses the model's weights or coefficients to identify which features contribute the least to predicting the target variable and removes them until the desired number of features is reached. RFE is always used in combination with cross-validation. In each iteration it removes one feature at a time.

Recursive Feature Elimination Steps:

- Train the Initial Model: Fit a model using all available features.
- Rank Features: After the model is trained, features are ranked based on their importance to the model. The importance can be determined by coefficients in regression models or feature importances in classification models.
- Remove the Least Important Feature: The least important feature(s) are removed from the current set of features.
- Iterate: Repeat the process with the reduced set of features, retraining the model and removing the least important feature in each iteration.
- Stop When Desired Feature Count is Reached: The process is repeated until the specified number of features is left.

### *Exhaustive Feature Selection (EFS)*

EFS is a thorough approach to feature selection that evaluates all possible feature combinations to identify the best performing subset for a given model. This method falls under the wrapper methods category and is highly comprehensive, but can be computationally expensive, especially for datasets with a large number of features. Here's an overview:

How Exhaustive Feature Selection Works:

- Combination of Features: EFS examines all possible combinations of features. For a dataset with features, it evaluates all $2^N$–1 non-empty subsets.
- Model Training and Evaluation: For each subset, it trains a model and assesses its performance using a specified metric (such as accuracy and F1-score).
- Best Subset Selection: After evaluating all subsets, the combination of features that yields the best performance according to the chosen metric is selected as the optimal feature set.

Considerations and Challenges:

- Computational Cost: The major drawback of EFS is its computational intensity. The number of combinations increases exponentially with the number of features, making it impractical for datasets with many features.

- Overfitting Risk: Due to its exhaustive nature, there is a risk of overfitting, especially if the evaluation metric does not penalize model complexity.
- Time and Resources: EFS requires significant computational resources and time, which might not be feasible in many practical scenarios.

### *Boruta*

Boruta is a specific feature selection algorithm that falls under the category of wrapper methods. It is known for its effectiveness in identifying the most relevant features for a predictive model, particularly when using random forests.

An overview of how Boruta works:

- Shadow Features Creation: Boruta starts by duplicating every feature in the dataset to create *shadow features*. These shadow features are shuffled versions of the real features, which means they should not contain any useful information about the target variable.
- Random Forest Classifier: A random forest classifier is then trained on the extended dataset, including both the real and shadow features. Random forests are chosen for their ability to handle a large number of features and their robustness to overfitting.
- Feature Importance: After training, the algorithm evaluates the importance of each feature.
- Comparison with Shadow Features: The importance of each real feature is compared to the maximum importance among the shadow features. If a real feature is more important than the most important shadow feature, it is deemed relevant.
- Iterative Process: This process is repeated several times. In each iteration, the shadow features are reshuffled, and the model is retrained. Features that consistently outperform the shadow features are kept as relevant, while those that do not are progressively discarded.
- Final Selection: The algorithm ends when all features are either confirmed or rejected, or after a specified number of iterations is reached.

Boruta is useful for datasets where the relevance of features is uncertain, and it provides a robust method to ensure that only features that have a genuine impact on the model's predictive power are selected. It is especially favored in biological and medical datasets where the interpretability of the model and the understanding of which features are truly important is crucial.

One of the key benefits of Boruta is that it takes into account the interaction between variables, which can be missed by simpler filter methods. However, like other wrapper methods, Boruta can be computationally intensive, especially for datasets with a large number of features.

### *Genetic Algorithms (GA)*

GA are generally categorized under *heuristic search methods* in feature selection. They are often considered a subset of wrapper methods, though they can also be seen as a distinct category due to their unique approach.

An Overview of GAs:

Genetic algorithms are inspired by the process of natural selection in biological evolution. They use mechanisms such as mutation, crossover, and selection to evolve a set of solutions towards the best possible solution over generations.

Application in Feature Selection:

- Initial Population: GAs start with a randomly generated population of potential solutions (feature subsets in this context).
- Fitness Function: Each subset is evaluated using a fitness function, often based on the performance of a predictive model.
- Evolution: Subsets are evolved using genetic operators like crossover (combining parts of two subsets) and mutation (randomly altering a subset).
- Selection: The best-performing subsets are selected for the next generation, gradually moving towards an optimal or near-optimal set of features.

Characteristics of Genetic Algorithms:

- Global Search Capability: Unlike methods that make decisions based on local information (like stepwise selection), GAs can explore a wide range of the search space, reducing the risk of getting stuck in local optima.
- Flexibility: GAs can be adapted to various kinds of optimization problems, including feature selection with different types of data and models.
- Parallelization: GAs can evaluate multiple solutions simultaneously, making them suitable for parallel computing environments.
- Customization: The fitness function, genetic operators, and other parameters of GAs can be tailored to specific problems.

Challenges:

While GAs can be more efficient than exhaustive search, they can still be computationally demanding, especially for large datasets with many features. Also, GAs require careful tuning of parameters like population size, mutation rate, and number of generations. As with other heuristic methods, GAs offer no guarantee of finding the absolute best solution, though they often find good solutions in practice.

Practical Use:

Genetic algorithms are used in feature selection when the dataset is complex, and there is a need for a robust method that can explore a large search space efficiently. They are useful in situations where traditional methods might fail to capture the global structure of the problem or when there is a need to balance exploration and exploitation in search strategy.

## Embedded and Hybrid Methods

Embedded and Hybrid Feature Selection Methods are two approaches in machine learning that optimize the process of identifying the most relevant features for a model. Each method offers distinct advantages and is suited for different scenarios.

### *Embedded Methods*

Embedded methods stand out due to their integration of feature selection directly into the model training process. This approach ensures that feature selection is specifically tailored to the model being trained.

How They Work: These methods often utilize learning algorithms that have built-in mechanisms for feature selection. Classical examples are regularization techniques in Lasso and Ridge regression, where the former can shrink the coefficients of certain features to zero, effectively removing them from the model.

Advantages: The primary benefit of embedded methods is their efficiency. By combining feature selection with model training, they often yield more effective models. This is because the selection process is optimized for the specific model, potentially leading to improved performance.

#### L1 (Lasso) Regularization

Lasso, or L1 regularization, plays a pivotal role in embedded methods by adding a penalty equal to the absolute value of coefficients. This approach encourages sparsity in the model's coefficients, effectively performing feature selection by reducing some coefficients to zero. Lasso's utility spans into both regression and classification tasks, aiding in high-dimensional data by simplifying model complexity and mitigating overfitting.

#### L2 (Ridge) Regularization

Ridge regularization, or L2, adds a penalty based on the squared magnitude of the coefficients. This technique is valuable in scenarios of multicollinearity or when the number of predictors exceeds the number of observations, helping to maintain all features in the model but with minimized coefficients.

#### Elastic Net

Combining the strengths of L1 and L2 regularization, Elastic Net applies both penalties to the loss function. This hybrid approach is particularly effective in dealing with correlated features, offering a versatile solution that adjusts the balance between Lasso and Ridge benefits through its control parameters, $\alpha$ and $\lambda$.

### Decision Trees and Random Forests

Decision Trees and Random Forests are powerful and popular ML algorithms used in both classification and regression tasks. They are part of the ensemble learning methodology, where multiple models are combined to solve a single problem.

#### *Embedded Feature Selection with Decision Trees and Random Forests*

In the context of embedded feature selection, both Decision Trees and Random Forests can be used to identify important features. These algorithms inherently perform feature selection as they assign importances to each feature based on how useful they are at predicting the target variable.

Feature Importance in Decision Trees: Decision Trees provide a straightforward method to understand feature importance. It calculates the importance based on the reduction in impurity (such as Gini impurity or entropy) that each feature brings to the tree.

Feature Importance in Random Forests: In Random Forests, the importance of a feature is measured by how much the tree nodes, which use that feature, reduce impurity on average across all trees in the forest. The reduction is typically measured by a decrease in Gini impurity or mean squared error.

### Gradient Boosting Machines (GBM)

GBM are powerful machine learning algorithms that build models in a stage-wise fashion. They are widely used due to their effectiveness in handling various types of data and their ability to improve model performance by reducing overfitting.

In the context of feature selection, GBMs, like Random Forests, offer a way to assess feature importance. This capability stems from the nature of how these models are constructed. GBM builds an ensemble of decision trees, each attempting to correct the errors of the previous ones. The importance of a feature is often measured by how much it contributes to reducing the prediction error across all trees.

### Regularized Trees (e.g., XGBoost, LightGBM)

Regularized Trees, such as eXtreme Gradient Boosting (XGBoost) and Light Gradient Boosting Machine (LightGBM), are advanced gradient boosting frameworks that incorporate regularization techniques to prevent overfitting. This makes them highly effective for both prediction and feature selection. Regularization in these models adds a penalty on the model's complexity, encouraging simpler models that generalize better. XGBoost and LightGBM both offer fast training speeds and high model performance. They are widely used in various machine learning tasks.

### *Hybrid Methods*

Hybrid methods are a strategic combination of filter and wrapper or embedded methods, designed to leverage the strengths of both to optimize feature selection.

How They Work: Typically, a hybrid method begins with a filter approach to reduce the feature space by removing less relevant features. This is followed by a wrapper or embedded method that further refines the selection in the reduced space.

Advantages: Hybrid methods strike a balance between the computational efficiency of filter methods and the effectiveness of wrapper methods. They are particularly useful for large feature sets, where they can reduce overall complexity and execution time.

Common Examples: A typical example would be using a variance threshold filter followed by recursive feature elimination. This two-step process first reduces the number of features to a manageable size and then applies a more computationally intensive wrapper method to select the most effective features.

In conclusion, the choice between embedded, hybrid, or other feature selection methods depends on the specific requirements of the dataset and the problem at hand. Embedded methods are integral to the learning algorithm and are efficient for models with built-in feature selection capabilities. In contrast, hybrid methods combine the initial simplicity of filter methods with the targeted effectiveness of wrapper methods, offering a balanced approach suitable for large datasets.

## Python in Feature Selection

In this section, we focus on feature selection. Firstly, we will load the dataset, which contains pre-extracted features, from a CSV file ('gsdata.csv') and then we examine the features. This dataset has been specifically created to serve as a resource for feature selection examples that will be demonstrated, allowing for a practical understanding of how to identify and select relevant features for machine learning tasks in activity recognition.

The dataset contains 85 features, such as the mean, standard deviation, variance, ICV, median, minimum, maximum, skewness, kurtosis, interquartile range (IQR), peak-to-peak distance, entropy, correlation between axes, signal magnitude area (SMA), signal vector magnitude, average signal vector magnitude, orientation angles (pitch, roll, inclination), integral and squared integral of the signal, energy, combined entropy, and a variety of spectral features including spectral energy, dominant frequency, maximum PSD, spectral entropy, peak frequency, spectral centroid, spectral spread, spectral skewness, kurtosis, flatness, slope, and rolloff across the $x$, $y$, and $z$ axes. Each record in the dataset is labeled with a corresponding activity. The labels in our dataset include different types of animal behaviors: walking, trotting, lying, grazing, and running. However, we have omitted labels such as shaking, fighting, and scratch biting due to insufficient data representation. The script for loading and preparing the dataset for feature selection can be found on our GitHub repository titled Chapter_6_feature_selection_methods-gsdata.ipynb.

An important step in our workflow is splitting the dataset into training and testing sets. We have partitioned our dataset with a test size of 30%. Now we will explore various feature selection methods using Python, applying them to our dataset. We will also introduce some methods that may not be directly applicable to our dataset but are important in the broader context of machine learning. These methods will be demonstrated through Python code examples, providing the reader with practical insights into their implementation and usage.

## *Filter Methods in Python*

### Information Gain

```
# Importing the libraries
from sklearn.feature_selection import mutual_info_classif
import pandas as pd

# Calculate Information Gain for each feature
info_gain = mutual_info_classif(X_train, y_train)
info_gain_series = pd.Series(info_gain, index=X_train.columns)

# Sort the features by Information Gain in descending order
sorted_features = info_gain_series.sort_values(ascending=False)

# Print the sorted features with their Information Gain values
print(sorted_features)

# Plotting the Information Gain of the top 30 features
plt.figure(figsize=(10, 8))
sorted_features.head(30).plot(kind='bar')
plt.title('Information Gain by Feature')
plt.ylabel('Information Gain')
plt.xlabel('Features')
plt.show()
```

In the below python example, we demonstrate how to use Information Gain for feature selection. We use the already split dataset: X_train, X_test, y_train, and y_test.

Information Gain can be calculated using the mutual_info_classif function from sklearn.feature_selection. This function measures the dependency between variables. Based on the calculated Information Gain, we can then select a subset of features that contribute the most to predicting the output.

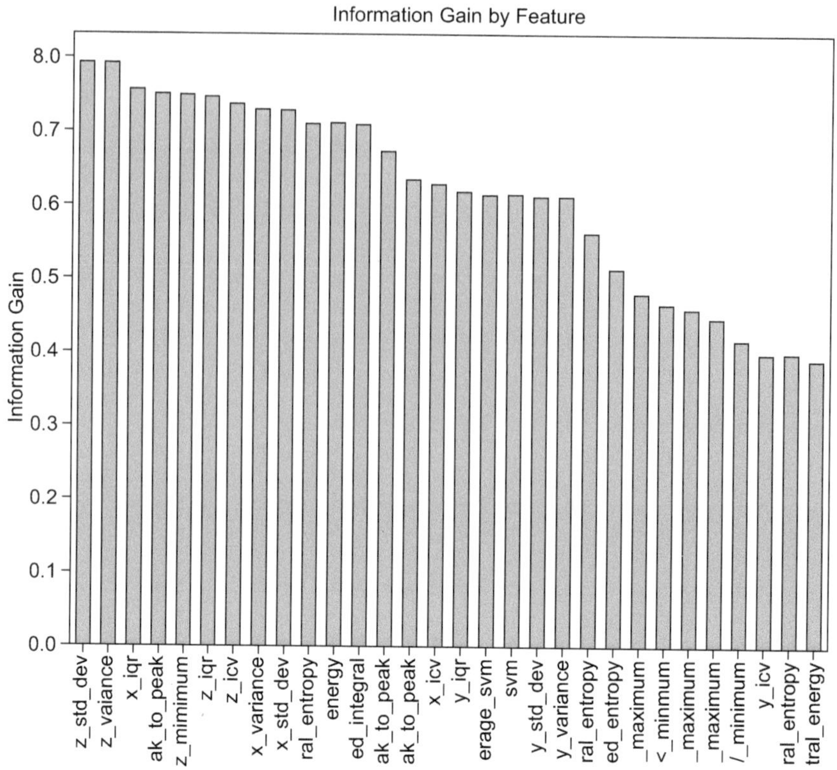


**Figure 6.1: Bar chart of information gain for the top 30 features.**

Figure 6.1 displays a bar chart where each bar represents a feature from the dataset used for animal behavior analysis. The height of each bar indicates the Information Gain associated with that feature, providing a visual representation of feature importance.

## Chi-square Test

For datasets like ours, which predominantly consist of numerical data, the direct application of the Chi-square test is not straightforward. Numerical data would first need to be discretized (binned) into categories, which could lead to loss of information and potentially misleading results. Hence, while we can technically apply the Chi-square test, it might not be the most suitable method for our dataset.

Nevertheless, for the purpose of demonstration and learning, we can briefly go through how one would apply the Chi-square test in Python, bearing in mind the limitations of its applicability to our dataset (this example is not included in our python script since it is not suitable for our dataset).

```
# Importing the libraries
from sklearn.feature_selection import SelectKBest, chi2
from sklearn.preprocessing import MinMaxScaler

# Scaling the features
scaler = MinMaxScaler()
X_train_scaled = scaler.fit_transform(X_train)
X_test_scaled = scaler.transform(X_test)

# Select features based on the Chi-Square scores.
chi2_selector = SelectKBest(chi2, k=20)  # Adjust k as needed
X_train_chi2_selected = chi2_selector.fit_transform(X_train_scaled,
y_train)
X_test_chi2_selected = chi2_selector.transform(X_test_scaled)
```

In the above code we import the necessary libraries SelectKBest, chi2. The Chi-square test requires non-negative values, and hence we scale the dataset using MinMaxScaler appropriately. We then use SelectKBest to select features based on the Chi-square scores.

### ANOVA F-Value

To utilize ANOVA F-Value for feature selection in Python, the scikit-learn library provides a straightforward implementation. Here is an example:

```
# Importing the library
from sklearn.feature_selection import SelectKBest, f_classif

# Applying ANOVA F-Value
selector = SelectKBest(score_func=f_classif, k='all')
X_train_selected = selector.fit_transform(X_train, y_train)

# Getting scores for each feature
scores = selector.scores_

# Creating a Series for the scores
feature_scores = pd.Series(scores, index=X_train.columns)

# Sorting features based on scores
sorted_features = feature_scores.sort_values(ascending=False)

# Displaying the sorted features and their scores
print("Feature Scores:", sorted_features)
# Plotting feature scores (top 20 features)
plt.figure(figsize=(12, 6))
sorted_features.head(20).plot(kind='bar')
```

```
plt.title('Feature Importance based on ANOVA F-Value')
plt.xlabel('Features')
plt.ylabel('F-Value')
plt.show()

# Output
Feature Scores:
z_std_dev          35172.453652
z_iqr              32214.437313
svm                25555.910370
average_svm        25555.910370
z_peak_to_peak     23178.704585
                       ...
x_entropy             66.373729
y_kurtosis            55.642534
z_kurtosis            53.889757
y_skewness            10.330706
x_skewness             9.027106
Length: 85, dtype: float64
```

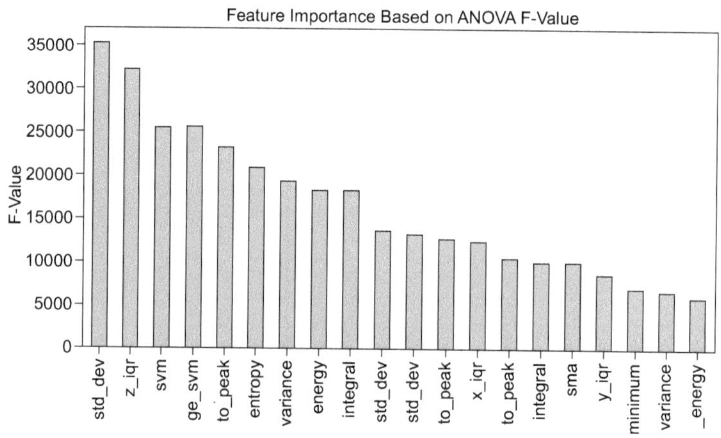


**Figure 6.2: Top 20 features ranked by ANOVA F-value.**

This code snippet demonstrates the process of feature selection using ANOVA F-Value, which is a statistical method for finding the most significant features that contribute to classifying the data. Here is what each part of the code does:

1. Importing Libraries: We first import SelectKBest and f_classif. SelectKBest selects features according to the k highest scores, and f_classif computes the ANOVA F-Value for provided features.
2. Applying ANOVA F-Value: We configure the SelectKBest function to use the f_classif scoring function and set to select all features (k = 'all'). We then

apply it to X_train with the corresponding y_train to determine the importance of each feature.

3. Getting Scores for Each Feature: The scores_ attribute of the selector object contains the F-Values computed for each feature.
4. Creating a Series for the Scores: We create a pandas Series the F-Values, indexed by the feature names from the X_train dataset.
5. Sorting Features Based on Scores: We sort the Series of feature scores in descending order to identify the most important features first.
6. Displaying Sorted Features: The sorted features along with their scores are printed out, providing a clear look at which features are considered most significant by the ANOVA F-test.
7. Plotting Feature Scores: We plot the F-Values of the top 20 features (Figure 6.2).

The output shows that features like z_std_dev, z_iqr, svm, average_svm, and z_peak_to_peak have the highest F-Values, indicating they are potentially the most important features for the model. Features with lower F-Values, such as x_entropy, y_kurtosis, z_kurtosis, y_skewness, and x_skewness, may be less important.

### Correlation Coefficient

In the following python code, we use Pearson's correlation coefficients for feature selection:

```
# Calculating correlation matrix
corr_matrix = X_train.corr()

# Identifying highly correlated features
threshold = 0.6
highly_correlated_features = [column for column in corr_matrix.
columns if any(corr_matrix[column] > threshold)]

print("Highly correlated features: ", highly_correlated_features)
```

In the above code, the objective is to identify features in our training dataset (X_train) that are highly correlated with each other. Here is s a brief explanation of the code:

- Calculating Correlation Matrix: The first line computes the correlation matrix for the X_train dataset using the corr() function. This matrix shows the correlation coefficients between every pair of features in your dataset.
- Identifying Highly Correlated Features:

- A threshold value of 0.6 is set. This value is used to determine what level of correlation is considered *high.*
- A list comprehension is then used to iterate through each column (feature) in the correlation matrix. For each column, it checks if any of its values are greater than the threshold (0.6 in this case). This means we are looking for features that have a high correlation with at least one other feature.

- The features that meet this criterion are added to the list, highly_correlated_features.
- Finally, the code prints out the list of highly correlated features. These are the features in our dataset that have a correlation coefficient greater than 0.6 with at least one other feature.

**Mean Absolute Difference (MAD)**

The calculation of MAD is straightforward.

```
# Calculate the mean for each feature
feature_means = X_train.mean()

# Calculate MAD for each feature manually
mad_values = (X_train - feature_means).abs().mean()

# Displaying the MAD values
print(mad_values)

# Output
x_mean                  4.420863
x_std_dev               0.958864
x_variance              5.853508
x_icv                  17.396949
x_median                4.388141
                          ...
z_spectral_skewness     2.301578
z_spectral_kurtosis    48.428439
z_spectral_flatness     0.019177
z_spectral_slope        0.498021
z_spectral_rolloff      3.398551
Length: 85, dtype: float64
```

In the above code:

- feature_means stores the mean of each feature in X_train.
- To calculate MAD, we subtract the mean of each feature from the feature values (X_train–feature_means), take the absolute value (.abs()), and then compute the mean for each feature (.mean()).
- Then we print out the results.

### Relief and ReliefF

Both Relief and ReliefF are not natively available in popular libraries like scikit-learn but can be accessed through specialized libraries like scikit-rebate.

Here is an example of how to use ReliefF in Python:

```
from skrebate import ReliefF

# Initialize the ReliefF algorithm
fs = ReliefF(n_neighbors=10, n_features_to_select=15)

# Fit the model and select the top features
fs.fit(X_train.values, y_train.values)
print("Feature rankings:", fs.feature_importances_)

# Feature importances from ReliefF
importances = fs.feature_importances_

# Creating a DataFrame for visualization
feature_importances = pd.DataFrame(importances, index=X_train.
    columns, columns=["Importance"])

# Sort the DataFrame by importance
sorted_features = feature_importances.sort_values(by="Importance",
    ascending=False)

# Get the top 'n_features_to_select' features
n_features_to_select = 15
top_features = sorted_features.head(n_features_to_select)

print("Top Features Selected by ReliefF:")
print(top_features)

# Output
Top Features Selected by ReliefF:
                    Importance
x_spectral_entropy    0.232102
y_spectral_entropy    0.216700
combined_entropy      0.198497
z_spectral_entropy    0.191691
roll                  0.171168
z_minimum             0.155979
z_std_dev             0.152093
z_iqr                 0.143996
correlation_x_z       0.134285
y_iqr                 0.133620
z_peak_to_peak        0.123054
x_mean                0.119706
x_std_dev             0.113344
x_median              0.113085
x_iqr                 0.107914
```

In the above code:

- ReliefF is initialized with parameters like n_neighbors (number of nearest neighbors) and n_features_to_select.
- ReliefF might be expecting a NumPy array instead of a pandas DataFrame or Series. Therefore, we convert X_train and y_train to NumPy arrays before passing them to ReliefF.
- The algorithm is fit to the dataset and the feature_importances_ are obtained.
- The feature_importances_ attribute gives the ranking of features according to their importance.

The ReliefF algorithm provides feature importances, but it does not automatically select the features. Instead, it ranks them based on their importance. To identify which features are selected based on the ReliefF rankings, we can use the following approach:

- Sort the Importances: Sort the feature importances returned by ReliefF and get the top n_features_to_select.
- Identify the Feature Names: Map these top importance scores back to the feature names in your dataset.
- Display the top n_features_to_select. You can then use these features for further analysis or model building. The X_train.columns provides the feature names corresponding to each importance score.

### Variance Threshold

To implement this in Python, we can use the VarianceThreshold from the sklearn.feature_selection module.

Here is how we can use it:

```
from sklearn.feature_selection import VarianceThreshold

# Setting up the Variance Threshold
selector = VarianceThreshold(threshold=0.2)

# Fitting and transforming the data
X_selected = selector.fit_transform(X_train)

# Getting the mask of the features selected
features_mask = selector.get_support(indices=True)

# Mapping mask to get feature names
selected_features = X_train.columns[features_mask]

print("Selected Features Based on Variance Threshold:")
print(selected_features, selected_features.shape)
```

Code explanation:

- We import the VarianceThreshold class from sklearn.feature_selection.
- We then set Variance Threshold.
- We apply the fit_transform method to our data.
- Finally, we identify which features have been retained by printing them out.

The selector.get_support(indices = True) line returns the indices of the features that are above the threshold. Using these indices, we can then identify the names of the features that have been retained after applying the threshold. The selected_ features will contain the names of these features. Remember, the choice of the variance threshold (threshold) is crucial. It depends on the nature of your dataset and the specific requirements of your analysis.

## *Wrapper Methods in Python*

As we proceed to illustrate examples of wrapper methods in Python, we must note an important adjustment to our dataset. For the sole purpose of these demonstrations, we have reduced the size of our dataset. This reduction is exclusively for illustrative clarity and to facilitate a smoother learning experience within the scope of this book. This approach allows readers to experiment with the code without facing the significant computational demands that larger datasets entail.

It is critical to understand that this reduction is not a recommendation for handling real-world data. In practical scenarios, especially in comprehensive data science projects, retaining the full dataset is crucial.

We can now explore the implementation of various wrapper methods in Python using our adapted dataset.

### Forward Selection

To implement forward selection in Python, we can use SequentialFeatureSelector from the mlxtend.feature_selection module.

```
from mlxtend.feature_selection import SequentialFeatureSelector as
SFS from sklearn.ensemble import RandomForestClassifier

# Create a random forest classifier
clf = RandomForestClassifier(n_estimators=100, random_state=42)

# Forward Selection
sfs = SFS(clf,
    k_features=10,  # Number of features to select
    forward=True,
    floating=False,
```

```
    scoring='accuracy',
    cv=4)

# Fit SFS to the training data
sfs = sfs.fit(X_sampled, y_sampled)

# Get the selected feature indices
selected_features_indices = sfs.k_feature_idx_

# Get the names of the selected features
selected_feature_names = X_sampled.columns[list(selected_features_
indices)]
print("Selected features:", selected_feature_names)

# Output
Selected features:
Index(['x_maximum', 'y_minimum', 'y_maximum', 'z_std_dev',
    'z_minimum','z_iqr', 'svm', 'squared_integral', 'combined_
    entropy', 'x_spectral_entropy'],
    dtype='object')
```

Python code breakdown:

- SequentialFeatureSelector: SequentialFeatureSelector from mlxtend is used for implementing forward selection.
- Random Forest Classifier: We are using a random forest classifier, but you can choose any other classifier or regressor based on your problem.
- Forward Selection: Set forward = True and floating = False for standard forward selection.
- Scoring and Cross-validation: The scoring parameter defines the performance evaluation metric, and cv sets the number of cross-validation folds.
- Selected Features: sfs.k_feature_idx_ gives us the indices of the selected features.

### Backward Elimination

```
from mlxtend.feature_selection import SequentialFeatureSelector as
SFS from sklearn.ensemble import RandomForestClassifier

# Initialize the classifier
clf = RandomForestClassifier(n_estimators=100, n_jobs=-1, random_
state=42)

# Backward Elimination
sfs = SFS(clf,
```

```
    k_features=10,  # Number of features to select
    forward=False,
    floating=False,
    scoring='accuracy',
    cv=4)

# Fit SFS to the training data
sfs = sfs.fit(X_sampled, y_sampled)

# Get the selected feature indices
selected_features = sfs.k_feature_idx_
print("Selected features:", selected_features)

# Get the names of the selected features
selected_feature_names = X_sampled.columns[list(selected_features)]
print("Selected features:", selected_feature_names)

# Output
Selected features:
Index(['x_minimum', 'y_minimum', 'y_maximum', 'z_minimum', 'svm',
    'average_svm', 'squared_integral', 'combined_entropy',
    'x_spectral_entropy', 'z_spectral_energy'],
    dtype='object')
```

Explanation of the above code:

- SequentialFeatureSelector: This class from the mlxtend library is also used for backward elimination.
- Classifier: The random forest classifier is used again, but the method is model-agnostic.
- Some Python libraries and functions support parallel processing. For instance, RandomForestClassifier from scikit-learn can be set to use multiple cores by adjusting the n_jobs parameter. This can significantly speed up model training, a major component of wrapper methods. Setting n_jobs = –1 tells the classifier to use all available CPU cores
- Backward Elimination: Set forward = False to perform backward elimination.
- Scoring and Cross-Validation: As in forward selection, define the scoring metric and the number of cross-validation folds.
- Selected Features: After fitting, sfs.k_feature_idx_ provides the indices of the final set of features.

### Recursive Feature Elimination

```
from sklearn.feature_selection import RFE
from sklearn.ensemble import RandomForestClassifier
```

```
# Initialize the classifier
clf = RandomForestClassifier(n_estimators=100, n_jobs=-1, random_
    state=42)

# Initialize RFE with the random forest classifier
selector = RFE(clf, n_features_to_select=10, step=1)

# Fit RFE to the training data
selector = selector.fit(X_sampled, y_sampled)

# Get the selected feature indices and print them
selected_features = selector.get_support(indices=True)
print("Selected features:", selected_features)

# Output
Selected features: [ 2  4 12 13 15 16 21 22 27 28]
```

Explanation:

- RFE Constructor: RFE is initialized with the classifier (RandomForestClassifier in this case) and the number of features to select (n_features_to_select).
- Step: The step parameter determines how many features should be eliminated at each iteration.
- Fit the Model: selector.fit trains the model and performs feature elimination.
- Selected Features: selector.get_support(indices = True) provides the indices of the features that are selected as most important.

### Exhaustive Feature Selection

In Python, Exhaustive Feature Selection can be implemented using the mlxtend library.

```
from mlxtend.feature_selection import ExhaustiveFeatureSelector as
EFS
from sklearn.ensemble import RandomForestClassifier

# Assuming X_train and y_train are the training data and labels
clf = RandomForestClassifier(n_estimators=100, n_jobs=-1)

efs = EFS(clf,
    min_features=3,
    max_features=10,
    scoring='accuracy',
    print_progress=True,
    cv=5)
```

```
efs = efs.fit(X_sampled, y_sampled)

# Best feature indices and corresponding score
print('Best features:', efs.best_idx_)
print('Best score:', efs.best_score_)
```

In this example, min_features and max_features define the range of feature numbers to consider, limiting the search space. However, even with these limits, the process can be time-consuming. In practice, EFS is often replaced by more efficient methods (like RFE or feature importance-based selection) when dealing with datasets that have a large number of features.

Note to Readers:

In this section, we have provided practical Python examples for wrapper methods. However, you may notice that we did not include specific code implementations for Boruta and GA. The primary reason for this omission is their computational intensity. Both Boruta and GA are advanced methods that often require substantial computational resources, particularly for large datasets or high-dimensional feature spaces. In the interest of focusing on techniques that are more broadly accessible and less demanding in terms of computational resources, we have chosen to leave these out.

## *Embedded Methods in Python*

### L1 Regularization (Lasso Regression)

```
from sklearn.linear_model import LogisticRegression
from sklearn.feature_selection import SelectFromModel
from sklearn.preprocessing import StandardScaler
from sklearn.pipeline import make_pipeline
from sklearn.preprocessing import LabelEncoder

# Initialize the LabelEncoder
label_encoder = LabelEncoder()

# Fit and transform the labels to numerical form
y_train_encoded = label_encoder.fit_transform(y_train)
y_test_encoded = label_encoder.transform(y_test)

# Standardizing the features and creating logistic regression with
L1 regularization
scaler = StandardScaler()

lasso_clf = LogisticRegression(penalty='l1', solver='liblinear',
C=10, random_state=42, max_iter=1000)
```

```
# Using a pipeline
lasso_pipeline = make_pipeline(scaler, lasso_clf)

# Fit the model
lasso_pipeline.fit(X_train, y_train_encoded)

# Feature selection using SelectFromModel
model = SelectFromModel(lasso_pipeline.named_
steps['logisticregression'], prefit=True)
X_train_lasso = model.transform(X_train)
X_test_lasso = model.transform(X_test)

# Extract the indices of the selected features
selected_feature_idx = model.get_support(indices=True)

# Get the feature names using the indices
selected_features = X_train.columns[selected_feature_idx]
print("Selected Features with Lasso: ", selected_features)
```

Code Breakdown:

- Label Encoding:
  - The LabelEncoder from sklearn.preprocessing is initialized and used to convert the categorical target labels (y_train and y_test) into a numerical form.
- Standardization and Logistic Regression with L1 Regularization:
  - StandardScaler is used for standardizing the features.
  - LogisticRegression is set up with L1 regularization (penalty = 'l1'). The solver liblinear is chosen as it supports L1 regularization. The regularization strength is controlled by C (inverse of regularization strength; smaller values specify stronger regularization).
- Pipeline Creation:
  - A pipeline is created using make_pipeline, combining the scaler and logistic regression model. This ensures that operations are sequentially applied: first scaling, then model fitting.
- Model Fitting:
  - The pipeline is fitted to the training data (X_train, y_train_encoded).
- Feature Selection:
  - SelectFromModel is used to select features based on the importance weights derived from the logistic regression model. The model is already fitted (prefit = True), so it directly accesses the feature importance.

  - It then transforms the training and test datasets to include only the selected features, effectively reducing the feature space.
- Extracting Selected Features:
  - The indices of the selected features are extracted (model.get_support(indices = True)).
  - These indices are used to retrieve the feature names from X_train columns.
- Output:
  - The features deemed important by the Lasso logistic regression model are printed.

### L2 Regularization (Ridge Regression)

In the following Python code, we show how to employ Ridge Regression for feature selection:

```
from sklearn.linear_model import Ridge
from sklearn.feature_selection import SelectFromModel
from sklearn.preprocessing import StandardScaler
from sklearn.pipeline import make_pipeline

# Standardizing the features and creating Ridge regression
scaler = StandardScaler()
ridge = Ridge(alpha=1.0, random_state=42)

# Using a pipeline
ridge_pipeline = make_pipeline(scaler, ridge)

# Fit the model
ridge_pipeline.fit(X_train, y_train_encoded)

# Feature selection using SelectFromModel
# Note: Adjust the threshold as needed
model = SelectFromModel(ridge_pipeline.named_steps['ridge'],
prefit=True, threshold=0.1)
X_train_ridge = model.transform(X_train)
X_test_ridge = model.transform(X_test)

# Extract the indices of the selected features
selected_feature_idx = model.get_support(indices=True)

# Get the feature names using the indices
selected_features = X_train.columns[selected_feature_idx]
print("Selected Features with Ridge: ", selected_features)
```

Code Breakdown:

- We standardize the features using the StandardScaler.
- Ridge Regression:
  - Ridge regression is initialized with alpha = 1.0, which is the regularization strength.
- Creating a Pipeline:
  - The make_pipeline combines the scaler and Ridge regression model.
- Model Fitting:
  - The pipeline is fitted to the X_train, and y_train_encoded.

Feature Selection:

- SelectFromModel is employed for feature selection, using the fitted Ridge regression model within the pipeline (ridge_pipeline.named_steps['ridge']).
- Transforming Data:
  - The transform method of SelectFromModel is then used to reduce X_train to only the features deemed important by the Ridge model.

### L1/L2 Regularization (Elastic Nets)

```
from sklearn.linear_model import ElasticNet
from sklearn.feature_selection import SelectFromModel

# Standardizing the features and creating Elastic Net model
elastic_net = ElasticNet(alpha=1.0, l1_ratio=0.5, max_iter=5000,
random_state=42)

# Using a pipeline
elastic_net_pipeline = make_pipeline(scaler, elastic_net)

# Fit the model
elastic_net_pipeline.fit(X_train, y_train_encoded)

# Feature selection using SelectFromModel
model_elastic = SelectFromModel(elastic_net_pipeline.named_
steps['elasticnet'], prefit=True)
X_train_elastic = model_elastic.transform(X_train)
X_test_elastic = model_elastic.transform(X_test)

# Extract the indices of the selected features
selected_feature_idx_elastic = model_elastic.get_
support(indices=True)
```

```
# Get the feature names using the indices
selected_features_elastic = X_train.columns[selected_feature_idx_
elastic]
print("Selected Features with Elastic Net: ", selected_features_
elastic)
```

Key components of this code:

- Elastic Net Initialization:
  - ElasticNet is initialized with alpha = 1.0 and l1_ratio = 0.5. Here, alpha controls the overall strength of the regularization, while l1_ratio determines the balance between L1 and L2 regularization. A l1_ratio of 0.5 means the regularization is equally split between L1 and L2.
  - max_iter = 5000 sets the maximum number of iterations for the optimization algorithm. Increasing this value can help the algorithm converge, especially for more complex datasets.
- Pipeline with StandardScaler:
  - make_pipeline, combines the StandardScaler (for feature standardization) with the Elastic Net model.
- Model Fitting:
  - The pipeline is fitted to the encoded training data.
- Feature Selection with SelectFromModel:
  - The SelectFromModel is used for feature selection, which works with the Elastic Net model within the pipeline (elastic_net_pipeline.named_steps['elasticnet']).
- Transforming the Data:
  - The transform method of SelectFromModel is applied to reduce X_train to the features considered important by the Elastic Net model. The resulting X_train_elastic (and X_test_elastic) contains a subset of the original features.
- Extracting and Displaying Selected Features:
  - The indices of the selected features are extracted using model_elastic.get_support(indices = True).
  - The feature names corresponding to these indices are then retrieved from X_train.columns to identify which features have been selected by the Elastic Net model.

### Feature Selection Using Random Forest

```
from sklearn.ensemble import RandomForestClassifier
import numpy as np

# Initialize the Random Forest classifier
rf = RandomForestClassifier(n_estimators=100)

# Fit the model
rf.fit(X_train, y_train)

# Get the feature importances
importances = rf.feature_importances_

# Sort the feature importances in descending order and get the
    indices
indices = np.argsort(importances)[::-1]

# Print the feature rankings
print("Feature ranking:")

for f in range(X_train.shape[1]):
    print(f"{f + 1}. feature {indices[f]}
    ({importances[indices[f]]}) - {X_train.columns[indices[f]]}")

# Select features based on importance (you can set a threshold)
#  e.g., the median of the importances
threshold = np.median(importances)
selected_features_rf = X_train.columns[importances > threshold]
print("Selected Features:", selected_features_rf)
```

- Random Forest Initialization and Fitting:
  - A RandomForestClassifier is initialized with n_estimators = 100. Subsequently it is fitted to the X_train and y_train.
- Feature Importances:
  - importances = rf.feature_importances_ extracts the importance of each feature. In Random Forests, feature importance is calculated based on how much each feature decreases the impurity (e.g., Gini impurity[2]) across all trees.
- Sorting Feature Importances:
  - np.argsort(importances)[:: −1] sorts the importances in descending order and returns their indices. The most important feature has the highest value.

[2] Gini impurity is a measure used to assess the frequency at which any element from the set is incorrectly labeled when it is randomly chosen.

- Printing Feature Rankings:
  - The code iterates through the sorted indices and prints out the feature rankings. Each feature is listed with its rank, importance score, and name.
- Selecting Features Based on a Threshold:
  - The threshold for feature selection is set to the median of the importances (np.median(importances)).
  - selected_features_rf = X_train.columns[importances > threshold] selects features whose importance is greater than this threshold.
- Result:
  - The final result is a list of selected features (selected_features_rf) that have importance scores above the median. These features are considered more influential in the Random Forest model for predicting the target variable.

### Feature Selection Using Decision Trees

```
from sklearn.tree import DecisionTreeClassifier
import numpy as np

# Initialize the Decision Tree classifier
dt = DecisionTreeClassifier()

# Fit the model
dt.fit(X_train, y_train)

# Get the feature importances
importances = dt.feature_importances_

# Sort the feature importances in descending order and get the indices
indices = np.argsort(importances)[::-1]

# Print the feature rankings
print("Feature ranking:")
for f in range(X_train.shape[1]):
    print(f"{f + 1}. feature {indices[f]}
    ({importances[indices[f]]}) - {X_train.columns[indices[f]]}")

# Define a threshold for selecting features, e.g., the median of
    the importances
threshold = np.median(importances)

# Select features based on the threshold
selected_features_dt = X_train.columns[importances >= threshold]
print("Selected Features:", selected_features_dt)
```

Decision Trees, like Random Forests, provide a measure of feature importance based on how much each feature decreases the impurity. Here is the breakdown of the code:

- Decision Tree Initialization and Fitting:
  - A DecisionTreeClassifier is initialized using default settings and then is fitted to X_train and y_train.
- Feature Importances:
  - importances = dt.feature_importances_ retrieves the importance of each feature. The importance in a decision tree is calculated based on how much each feature contributes to splitting nodes and reducing impurity (e.g., Gini impurity) in the tree.
- Sorting Feature Importances:
  - np.argsort(importances)[:: –1] sorts the feature importances in descending order and gets their indices.
- Printing Feature Rankings:
  - The for loop iterates over the sorted indices and prints the feature rankings, including their rank, importance score, and name.
- Selecting Features Based on a Threshold:
  - A threshold is defined as the median of the feature importances: threshold = np.median(importances).
  - selected_features_dt = X_train.columns[importances > = threshold] selects features whose importance is equal to or greater than this threshold.
- Result:
  - The final output is a list of selected features (selected_features_dt) based on the threshold criterion.

## Feature Selection Using Gradient Boosting Machines

```
from sklearn.ensemble import GradientBoostingClassifier
import numpy as np

# Initialize the Gradient Boosting Classifier
gbm = GradientBoostingClassifier(n_estimators=100, learning_
rate=1.0,max_depth=1, random_state=42)

# Fit the model
gbm.fit(X_train, y_train)

# Get the feature importances
importances = gbm.feature_importances_
```

```
# Sort the feature importances in descending order and get the indices
indices = np.argsort(importances)[::-1]

# Print the feature rankings
print("Feature ranking:")
for f in range(X_train.shape[1]):
    print(f"{f + 1}. feature {indices[f]}({importances
    [indices[f]]}) - {X_train.columns[indices[f]]}")

# Define a threshold for selecting features
threshold = np.median(importances)

# Select features based on the threshold
selected_features_gbm = X_train.columns[importances >= threshold]
print("Selected Features:", selected_features_gbm)
```

The code follows the same structure as the code we used for Random Forest and Decision trees. The only difference is the model, which in this case is the GradientBoostingClassifier.

### Feature Selection Using XGBoost

```
import xgboost as xgb
from sklearn.feature_selection import SelectFromModel

# Initialize XGBoost classifier
xgb_clf = xgb.XGBClassifier(objective='multi:softprob',
   num_class=5, random_state=42)

# Fit the model
xgb_clf.fit(X_train, y_train_encoded)

# Feature importance based feature selection
model = SelectFromModel(xgb_clf, prefit=True)
X_train_xgb = model.transform(X_train)
X_test_xgb = model.transform(X_test)

# Extract the indices of the selected features
selected_feature_idx = model.get_support(indices=True)

# Get the feature names using the indices
selected_features = X_train.columns[selected_feature_idx]
print("Selected Features with XGBoost: ", selected_features)
```

- Model Initialization: An XGBoost classifier is initialized. The objective parameter is set to objective = 'multi:softprob' (multiclass problem), num_class = 5 (use 'binary:logistic' for binary classification tasks).
- Model Fitting: The model is fitted to the training data.
- Feature Selection: SelectFromModel is used with the fitted XGBoost model to perform feature selection. The features are selected based on the importance scores calculated by XGBoost.
- Data Transformation: Both X_train and X_test are transformed to include only the selected features.
- Extracting and Printing Selected Features: The indices of the selected features are determined and their names are printed.

This approach leverages the internal feature importance scores generated by XGBoost to select features that contribute the most to the prediction task.

#### Note on LightGBM

LightGBM works similarly to XGBoost and can be used in the same way for feature selection. The primary difference is in the underlying algorithms and optimizations, with LightGBM being designed for speed and efficiency, particularly on large datasets. You can replace xgb.XGBClassifier with lightgbm.LGBMClassifier and follow the same steps for feature selection using LightGBM.

## *Hybrid Method*

### Variance Threshold and Feature Selection Using RandomForest

```
from sklearn.ensemble import RandomForestClassifier
from sklearn.feature_selection import SelectFromModel,
VarianceThreshold

# Step 1: Filter Method - Variance Threshold
# Removing features with low variance
selector = VarianceThreshold(threshold=0.2)  # Adjust threshold as
    needed
X_reduced = selector.fit_transform(X_train)

# Step 2: Wrapper Method - Feature selection using RandomForest
rf = RandomForestClassifier(n_estimators=100, random_state=42)
rf.fit(X_reduced, y_train_encoded)

# Selecting features based on importance
sfm = SelectFromModel(rf, threshold='median')  # Adjust threshold as
    needed
```

```
X_transformed = sfm.fit_transform(X_reduced, y_train_encoded)

# Getting the selected feature names
selected_features = X_train.columns[sfm.get_support(indices=True)]
print("Selected Features: ", selected_features)
```

Explanation of the Code:

1. Initial Filtering: Using VarianceThreshold, features with variance below a certain threshold are removed. This reduces the number of features, focusing on those with more significant variation.
2. Feature Importance with RandomForest: A RandomForest classifier is trained on the reduced feature set.
3. Feature Selection Method: SelectFromModel is used with RandomForest to select features based on their importance. A median threshold is set, meaning only features with importance above the median are selected.
4. Selected Features: The names of the selected features are extracted.

Important Considerations:

- Variance Threshold: The threshold in VarianceThreshold should be set based on an understanding of the dataset. Too high a threshold may exclude important features.
- Feature Importance Threshold: In SelectFromModel, the threshold can be adjusted. 'Median' is a common choice, but this can be tuned based on the specific requirements.
- Model Compatibility: The choice of model in the next phase should be compatible with the data and the problem at hand. RandomForest is a versatile choice but can be replaced with other models and methods if needed.
- Evaluation: It is important to evaluate the performance of the model using the selected features to ensure the effectiveness of the hybrid method.

## Summary

In this chapter, we looked at feature selection techniques, a crucial aspect of the data science process. We began by discussing the significance of feature selection in enhancing model performance, reducing complexity, and increasing interpretability. Our discussion covered a range of methods, each tailored to different needs and scenarios in the data science workflow. Key highlights of the chapter included a look at filter methods, known for their speed and effectiveness in preliminary feature reduction. These methods rely on statistical measures and operate independently of the machine learning models. We also examined wrapper methods, which, despite their computational intensity, provide a high accuracy level by evaluating feature subsets using specific models. Embedded methods, which integrate feature selection as part of the model training process, were also a focal point. Additionally, we explored hybrid methods that combine the strengths of filter and wrapper or embedded methods. These methods are useful for complex datasets, offering an efficient and effective approach to feature selection. Throughout the chapter, we demonstrated practical Python code examples. These examples illustrated the application of each method and highlighted how they could be adapted to the specific needs of different datasets and analysis goals.

CHAPTER 7

# Animal Research: Supervised and Unsupervised Learning Algorithms

Machine learning models can perform various tasks including regression, classification, and clustering. In supervised learning the objective is to learn a representation from an input to an output based on input-output pairs. It requires labeled training data where inputs (also known as features) have a corresponding label (also known as target). In this type of machine-learning system, the data that can be fed into the algorithm, with the desired solution, are referred to as *labels*. Supervised machine learning primarily involves two key problem types: classification and regression. During the learning stage, input data are passed to the ML systems that generate the outputs based on a learning rule as illustrated in Figure 7.1. In supervised learning or learning with a teacher, the network is assumed to have an external teacher which has knowledge of the outside environment. In this case, the output of the network is benchmarked with the required value. Then, the synaptic weights of the network are updated using the error signal. Backpropagation (BP) represents one of the most powerful algorithms used to train Multilayer Perceptron networks (MLP).

A supervised learning algorithm operates through two distinct stages: the forward pass and the backward pass. In the former, the input data are passed to the network and the weighted sums are calculated and passed to the subsequent layers until the final output is generated, while in the latter, the error which is the difference between the actual and network outputs is propagated backwards to train and adjust the network weights using quality measures such as the sum of the squared errors.

It should be noted that the network's mathematical complexity increases as the number of inputs, hidden neuron and outputs increase. Therefore, the disadvantage of many supervised learning techniques is that they may be limited by their poor scaling behavior. To overcome this problem, unsupervised learning can be used.

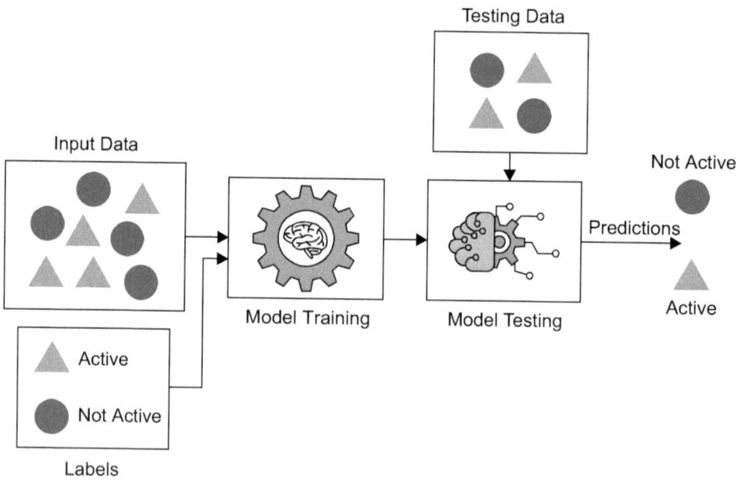


**Figure 7.1: Supervised ML block diagram.**

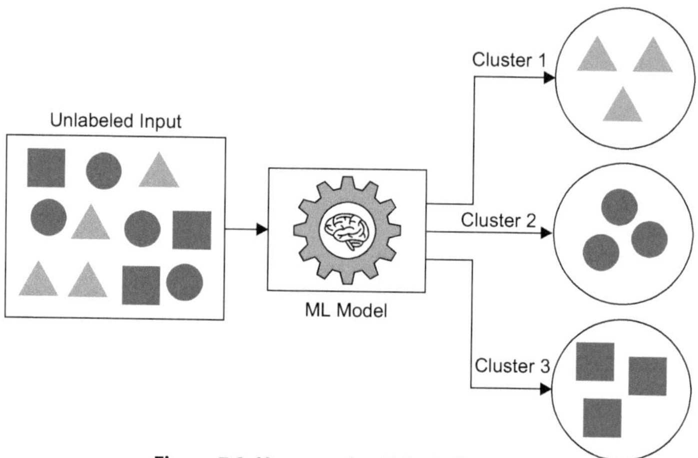


**Figure 7.2: Unsupervised block diagram.**

In contrast to supervised learning, the unsupervised algorithm does not require the use of correct external target values to adjust its parameters. Therefore, the network receives input patterns and produces the output set with no information from the external environment telling the network whether the outputs produced by the network are correct or how they should be. To be able to classify the input patterns, unsupervised learning requires some redundant information in the input data. Unsupervised learning is often used when the data labels are not available, and it is achieved as a part of exploratory data analysis as shown in Figure 7.2. Hence, the validation and evaluation of the unsupervised ML model is difficult to achieve.

In this chapter, we will examine the leading algorithms in both supervised and unsupervised learning for monitoring animal behavior.

## Backpropagation Learning Algorithm

BP is a popular learning algorithm in neural networks. Researchers often refer to it as error-backpropagation or simply backprop. The algorithm can be used for training MLP networks using gradient descent. Various researchers have developed the algorithm independently, for instance the algorithm was researched by (Werbos, 1974), then (David B. Parker, 1985) and independently revived by (Rumelhart et al., 1986).

In the past, researchers noted that single layer neural networks have limitations since there is a limited range of functions that can be represented using these networks. MLP can approximate functions to any required degree of accuracy. After the work of Rumelhart and his colleagues (Rumelhart et al., 1986), interest in MLP increased and backpropagation became central to much of the current work on learning neural networks.

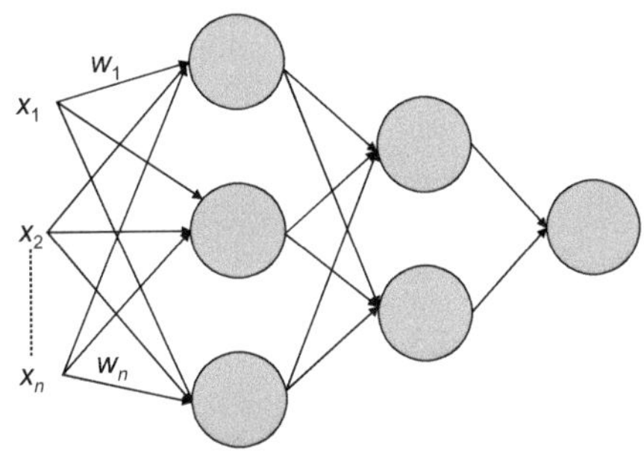


**Figure 7.3: Feedforward neural network.**

Consider a MLP which has two layers, the hidden and the output layers. The index $k$ denotes the input units, the index $i$ denotes the output units, and the index $j$ represents the hidden neurons. Figure 7.3 illustrates the design framework of MLP.

Let the number of inputs, outputs and hidden units to be $M$, $N$, and $S$ respectively. Let $y$ and $x$ represent the $N$-tuple outputs and the $M$-tuple inputs to the network, respectively. The matrix of weights linking the input layer to the hidden layer is denoted by $w_{jk}^1$ with $S \times M$ elements. The weights matrix connecting the hidden and the output layers is represented by $w_{jk}^2$ with $N \times M$ elements. The biases can be presented separately in the MLP or by adding an extra input value of one to each layer of the network.

When $p$ (input pattern) is introduced to the MLP, $j$, which is the hidden neuron, receives a net input $n_j$ defined as:

$$n_j^p = \sum_{k=1}^{M} w_{jk}^1 x_k^p.$$

The output of this unit is:

$$V_j^p = f(n_j^p) = f\left(\sum_{k=1}^{M} w_{jk}^1 x_k^p\right)$$

where, $f$ is the transfer function that must be differentiable.

Since we are considering only one hidden later for simplicity to describe the BP algorithm, the hidden layer's output is then forwarded to the subsequent layer, known as the output layer $i$ in this example. Hence, the input value to the output $i$ can be calculated as:

$$n_i^p = \sum_{j=1}^{S} w_{ij}^2 V_j^p = \sum_{j=1}^{S} w_{ij}^2 f\left(\sum_{k=1}^{M} w_{jk}^1 x_k^p\right)$$

and the output $i$ generates the following value:

$$y_i^p = f(n_i^p) = f\left(\sum_{j=1}^{S} w_{ij}^2 V_j^p\right) = f\left(\sum_{j=1}^{S} w_{ij}^2 f\left(\sum_{k=1}^{M} w_{jk}^1 x_k^p\right)\right).$$

The transfer functions used at the output and the hidden layers respectively can be different. However, for simplicity we will assume that the same transfer function is used in all the layers of the MLP.

The error generated at the output unit $i$ is:

$$e_i^p = d_i^p - y_i^p$$

where, $d_i^p$ represent the target value and $p$ denotes the input pattern.

The total error or the cost function per pattern is defined as:

$$J = \frac{1}{2}\sum_{i=1}^{N} [e_i^p]^2 = \frac{1}{2}\sum_{i=1}^{N} [d_i^p - y_i^p]^2.$$

Substituting the output value $y$, we get,

$$J = \frac{1}{2}\sum_{i=1}^{N} \left[d_i^p - f\left(\sum_{j=1}^{S} w_{ij}^2 f\left(\sum_{k=1}^{M} w_{jk}^1 x_k^p\right)\right)\right]^2.$$

The change in the weight between a hidden and an output unit is defined as:

$$\Delta w_{ij}^2 = -\eta \frac{\partial J}{\partial w_{ij}^2}$$

where, η represents the learning rate, and

$$\begin{aligned}
\frac{\partial J}{\partial w_{ij}^2} &= \frac{\partial}{\partial w_{ij}^2}\left(\frac{1}{2}\sum_{i=1}^{N} [d_i^p - y_i^p]^2\right) \\
&= \frac{\partial}{\partial w_{ij}^2}\left(\frac{1}{2}\sum_{i=1}^{N} \left[d_i^p - f\left(\sum_{j=1}^{S} w_{ij}^2 f\left(\sum_{k=1}^{M} w_{jk}^1 x_k^p\right)\right)\right]^2\right) \\
&= -(d_i^p - y_i^p) f'(n_i^p) V_j^p.
\end{aligned}$$

Therefore, the weight change of the output $i$ is:

$$\Delta w_{ij}^2 = \eta e_i^p f'(n_i^p) V_j^p.$$

Let

$$\delta_i^p = e_i^p f'(n_i^p),$$

then

$$\Delta w_{ij}^2 = \eta \delta_i^p V_j^p.$$

The chain rule can be used to update the weights between the input and the hidden layers as follows:

$$\Delta w_{jk}^1 = -\eta \frac{\partial J}{\partial w_{jk}^1} = -\eta \frac{\partial J}{\partial V_j^p} \frac{\partial V_j^p}{\partial w_{jk}^1}$$

where, we have:

$$\frac{\partial J}{\partial V_j^p} = -\sum_{i=1}^{N} e_i^p f'(n_i^p) w_{ij}^2$$

and

$$\frac{\partial V_j^p}{\partial w_{jk}^1} = f'(n_j^p) x_k^p$$

$$\Delta w_{jk}^1 = \eta f'(n_j^p) x_k^p \sum_{i=1}^{N} \delta_i^p w_{ij}^2$$
$$= \eta \delta_j^p x_k^p.$$

In this case, $\delta_j^p$ is determined as follows:

$$\delta_j^p = f'(n_j^p) \sum_{i=1}^{N} w_{ij}^2 \delta_i^p.$$

Hence, the steps involved in the BP learning algorithm are:

1. Initialize the network weights and biases.
2. Present the inputs patterns to the network.
3. Determine the outputs of each layer until the final network outputs are calculated.
4. Find the δ values for the output layer.
5. Propagate the errors backwards and calculate the delta values.
6. Calculate the updated weights of the network using the following equation:

$$w_{ij}(t+1) = \alpha w_{ij}(t) + \Delta w_{ij}(t).$$

**Backpropagation is the most used learning algorithm for the training MLP network.**

### *Backpropagation Algorithm with Momentum Term Updating*

A primary drawback of the BP as a gradient descent learning algorithm is that a small value of the learning rate denoted by $\eta$ should be selected. As a result, the learning process can be very slow. For large values of $\eta$, the learning process is fast, however it leads to parasitic oscillations that prevent the algorithm from converging to the required response. This problem comes from the fact that, the solution space may have many valleys that have steep sides and shallow slops along the valley floor. This will cause the learning to get trapped in local minima or very flat plateaus (Cichocki & Unbehauen, 1993). To address this issue, a momentum term $\alpha$ can be added to the updating equations as follows:

$$w_{ij}(t+1) = \alpha w_{ij}(t) + \Delta w_{ij}(t).$$

The value of the momentum is usually selected between zero and one and a typical value is 0.9.

The purpose of the momentum term is to give inertia to every weight connection of the network such that it will move in the direction of decreasing the cost function, rather than oscillating widely with every update.

The momentum term can be used for both the online learning and batch learning, although it has first been used in online learning (Hertz et al., 1991).

**The momentum term is added to the backpropagation to improve convergence and training of MLP.**

## Machine Learning Algorithms

Multiple machine learning methods have been employed for animals' activity recognition. In this section, some popular classification algorithms will be described.

### *K-nearest Neighbors*

K-nearest neighbors (KNN) is a widely recognized machine learning algorithm applicable to both classification and regression tasks. In classification tasks, KNN determines the predicted class for a new data point by adopting the most frequent class among its $k$ nearest neighbors. In the case of regression, the mean or weighted mean of the target values of the $k$ closest neighbors is used as the predicted value. KNN algorithm (Géron & Russell, 2019) takes into account only the single nearest neighbor, which is the training data point closest to the predicted value as shown in Figure 7.4. The prediction is the known output for this specific training data point. KNN classifier typically relies on the Euclidean distance calculated between a test sample and a designated training sample.

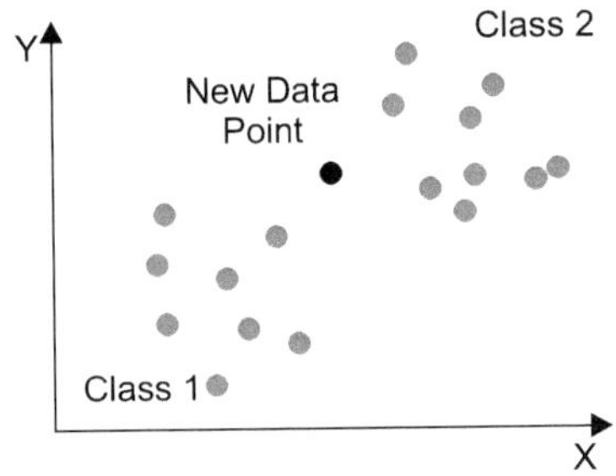


**Figure 7.4: KNN plot.**

## *Logistic Regression*

Logistic regression (LR) is one of the most important analytic tools in the social and natural sciences (Jurafsky & Martin, 2014, 2009). In natural language processing, LR is the baseline supervised machine learning algorithm for classification and has a very close relationship with neural networks. LR is used to classify observations into two classes, such as *animal has a disease* or *animal doesn't have a disease* as shown in Figure 7.5. It can also be used to classify observations into multiple classes, which is known as multinomial logistic regression.

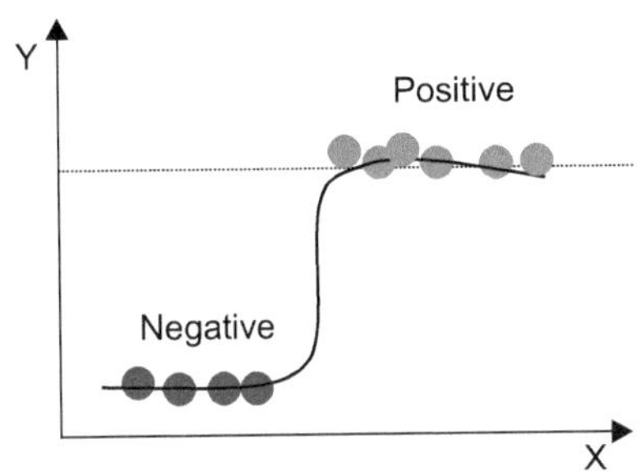


**Figure 7.5: Logistic regression plot.**

## *Support Vector Machines (SVM)*

SVM (Gaye et al., 2021) is a training algorithm for learning classification and regression rules from data. SVM is suitable for linear classification tasks, where data points can be separated by a straight line, a hyperplane (in higher dimensions), or a hyperplane in the transformed feature space (with kernel functions). For non-linear data, SVM can use kernel functions like polynomial, radial basis function (RBF), or sigmoid to project the data into a higher-dimensional space, where it becomes linearly separable. SVM is a traditional machine-learning method based on classification and can handle both linear and non-linear data. SVM's robustness in high-dimensional spaces and its capacity to control the trade-off between margin and classification errors makes it a valuable tool in ML and data analysis. These are some SVM concepts and variables that are important to know:

- Margin is defined as the distance between the hyperplane and the nearest data point from each class. The SVM aims to maximize this margin, as it indicates a more robust separation between classes.
- Support Vectors are the data points that are closest to the hyperplane and have the smallest margin. These are critical in defining the position and orientation of the hyperplane.
- Kernel Trick: SVM can handle data that is not linearly separable by transforming it into a higher-dimensional space through the application of a kernel function (e.g., polynomial, radial basis, or sigmoid). This transformation often makes it possible to identify a hyperplane that distinguishes the classes.
- Regularization: SVM employs a regularization parameter (C) to manage the balance between maximizing the margin and reducing the classification error. A small C value creates a larger margin but may allow some misclassifications, while a larger C value reduces the margin but aims to minimize misclassifications.

Figure 7.6 shows the SVM plot which includes the margin and the supporting vectors.

## *Ensemble Machine Learning*

Ensemble machine learning algorithms are powerful approaches to enhance the effectiveness and robustness of predictive models through the combination of predictions from several base models (Dietterich, 2000). Ensemble learning techniques are based on the principle that when multiple models' outputs are combined, it often leads to more accurate and stable predictions compared to relying on a single model. Some commonly used ensemble learning techniques include (C. D. Sutton, 2005):

- Bagging: It is a technique where several base models are trained independently on bootstrapped subsets of the training data. These subsets are generated by randomly selecting data points with replacements. The final prediction is typically obtained by averaging or voting on the predictions of the base models. Random Forest is a popular ensemble method based on bagging.

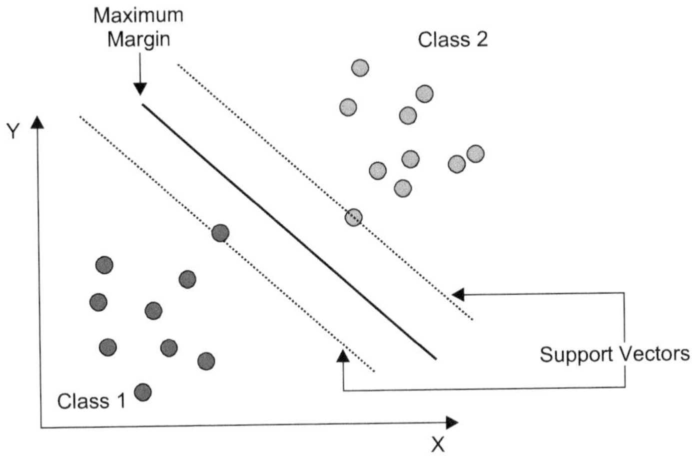


**Figure 7.6: SVM plot.**

- Boosting: Boosting focuses on improving the weaknesses of individual models. It trains a sequence of base models where each successive model gives more weight to the misclassified samples of the previous model. AdaBoost and Gradient Boosting are well-known boosting algorithms.

**The ensemble machine learning algorithms are powerful approaches to enhance the effectiveness and stability of predictive models through the combination of the predictions of various base models.**

## Decision Trees (DT)

DTs (Géron & Russell, 2019) are commonly utilized models for both classification and regression. They learn a hierarchy of conditional statements, leading to a decision. The learning in a DT involves learning the sequence of conditional statements that arrives at the correct answer faster. Within the concept of machine learning setting, these conditions are called tests. To construct the tree, the algorithms perform a search around the possible tests and determine the one which has the most informative details. A simple architecture of the Decision tree is illustrated in Figure 7.7. Due to their interpretability, they are considered valuable for understanding the decision-making process. DT methods are versatile, and applied in various fields for classification, prediction, interpretation, and data manipulation, offering a powerful tool for both beginners and experts in data analysis and machine learning applications. Below are some of the DT key variables.

- Tree Structure: The top node is called the root, and it branches into various nodes below it. Nodes in the tree represent decisions or test conditions based on input features. Terminal nodes at the bottom of the tree are called leaf nodes. These nodes represent the final predicted output.
- Splitting Criteria: DT chooses the top features and conditions for splitting the dataset into smaller groups at each node. The goal is to create splits that result in the most homogeneous subsets in terms of the target variable.
- Entropy and Information Gain: In classification, DTs often utilize metrics such as entropy and information gain to evaluate the impurity or disorder of a dataset. The goal is to reduce entropy and increase information gain when making splits.
- Gini Impurity: Another common metric for classification trees is Gini impurity, which measures the probability of misclassifying a randomly chosen element from the dataset. Decision trees aim to minimize Gini impurity.
- Pruning: DTs can be prone to overfitting, where they become too complex and fit the training data noise. Pruning techniques are used to simplify the tree by eliminating branches that are not significant for the model's performance.
- Decision Rules: Each path from the root to a leaf node represents a decision rule. These rules can be easily interpreted, making DT a useful tool.

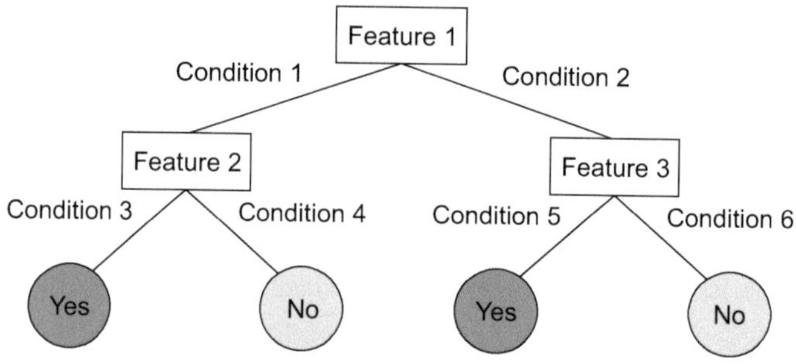


**Figure 7.7: Decision tree plot.**

## Random Forest

Random Forest (RF) consists of a number of decision trees. In this case, every tree exhibits slight variations from the others. The concept of RF revolves around the concept that while each tree can effectively classify data to some extent, it is prone to overfitting certain portions of the dataset. Ensemble learning involves building multiple DTs, each of which may overfit the data differently. By averaging the predictions of these trees, the ensemble can reduce overfitting and improve generalization. This technique is based on statistical learning theory and leverages the law of large numbers to mitigate individual model variance. Methods like bagging in random forests represent this approach, demonstrating its effectiveness in ML by combining the strengths of multiple models while addressing overfitting concerns.

Figure 7.8 shows the structure of RF architecture.

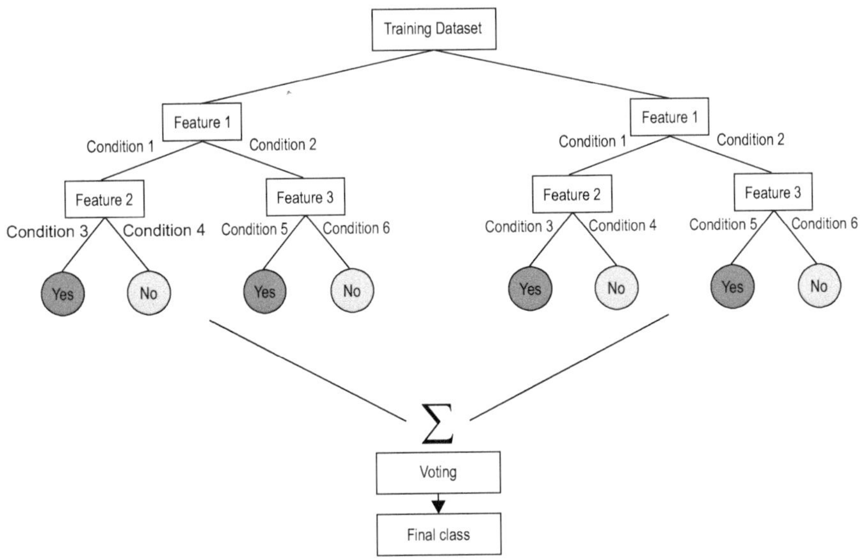


**Figure 7.8: RF architecture.**

## Classification Using Python

In this section, we will train various classification algorithms using X_train, y_train, X_test, y_test datasets that you can find in our GitHub repository in the Chapter_7 folder. The Jupyter notebook is located in the same directory under the name Chapter_7_Classification.ipynb. To provide a clear and focused view on classification, we will not perform additional preprocessing steps on these datasets. Instead, we will use them as they are, to illustrate how different classification algorithms can be applied in Python for practical data science scenarios. This approach allows us to concentrate on the specific implementations of each classification technique.

To begin, we will first upload and inspect the pre-saved datasets (X_train, y_train, X_test, y_test) using the following code:

```
import pandas as pd

def load_and_inspect_data():
    # Paths to the saved datasets
    file_paths = {
        'X_train': 'X_train.csv',
        'y_train': 'y_train.csv',
        'X_test': 'X_test.csv',
        'y_test': 'y_test.csv'
    }

    datasets = {}

    # Load and inspect datasets
    for name, path in file_paths.items():
        data = pd.read_csv(path)
        datasets[name] = data

        # Inspecting the data
        print(f"--- {name} ---")
        print("Head:\n", data.head())
        print("Shape:", data.shape)
        print("NaN Values:", data.isnull().sum().sum(), "\n")

    return datasets

# Load datasets
datasets = load_and_inspect_data()

# Access individual datasets
X_train = datasets['X_train']
y_train = datasets['y_train']
X_test = datasets['X_test']
y_test = datasets['y_test']
```

To check the distribution of the labels in both defined X_train and X_test, we can write a function that plots the label distributions using a bar plot. An example using a bar plot with matplotlib and seaborn is shown below:

```
import seaborn as sns
import matplotlib.pyplot as plt

def plot_label_distribution(y_train, y_test):
```

```
    # Plotting
    fig, axes = plt.subplots(1, 2, figsize=(12, 5), sharey=True)
    fig.suptitle('Label Distribution in Training and Testing Sets')

    # Training set distribution
    sns.countplot(ax=axes[0], x=y_train.iloc[:, 0])
    axes[0].set_title('Training Set')
    axes[0].set_xlabel('Labels')
    axes[0].set_ylabel('Count')

    # Testing set distribution
    sns.countplot(ax=axes[1], x=y_test.iloc[:, 0])
    axes[1].set_title('Testing Set')
    axes[1].set_xlabel('Labels')
    axes[1].set_ylabel('Count')

    plt.tight_layout()
    plt.savefig('datasets_distribution')
    plt.show()

# Call the function
plot_label_distribution(y_train, y_test)
```

For the classification algorithms it is important to perform appropriate preprocessing steps to ensure optimal performance of each model. Below is a reminder of common preprocessing steps and specific considerations for the mentioned algorithms:

Common Preprocessing Steps:

- Scaling/Standardization: Most algorithms perform better when features are on a similar scale. Scaling is important for algorithms like KNN, SVM, and

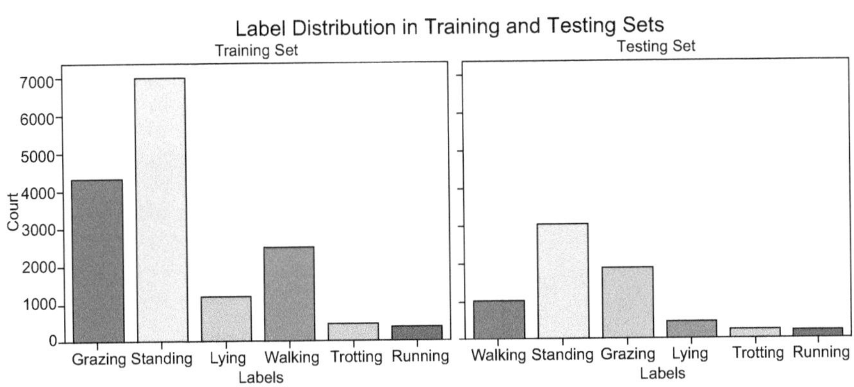


**Figure 7.9: Distribution of labels in training and testing sets.**

MLP, which are sensitive to feature scaling. The StandardScaler from scikit-learn is used for this purpose.

- Label Encoding: Since our labels are categorical, they should be encoded into numerical values. This can be done using LabelEncoder from scikit-learn. This step is essential for all algorithms.
- Handling Missing Values: We will ensure there are no missing values in our datasets.
- Feature Encoding: If we have categorical features, we should encode them using techniques like one-hot encoding, especially for algorithms that do not inherently handle categorical variables well (e.g., SVM, KNN).

In the following code we introduce a comprehensive function that evaluates various classifiers on our training and test sets. This function is useful for ML practitioners who are investigating different ML models to find the best fit for their data.

```
# Importing the libraries
from sklearn.preprocessing import StandardScaler, LabelEncoder
from sklearn.linear_model import LogisticRegression
from sklearn.svm import SVC
from sklearn.neighbors import KNeighborsClassifier
from sklearn.tree import DecisionTreeClassifier
from sklearn.ensemble import RandomForestClassifier,
                             GradientBoostingClassifier,
                             AdaBoostClassifier
from sklearn.naive_bayes import GaussianNB
from sklearn.neural_network import MLPClassifier
from sklearn.discriminant_analysis import
LinearDiscriminantAnalysis,

QuadraticDiscriminantAnalysis
from xgboost import XGBClassifier
from lightgbm import LGBMClassifier
from tqdm import tqdm
import itertools

# Step 1: Convert the y datasets to 1D array
y_train_np = y_train.to_numpy().ravel()
y_test_np = y_test.to_numpy().ravel()

# Step 2: Apply Label Encoding to y_train and y_test
label_encoder = LabelEncoder()
y_train_encoded = label_encoder.fit_transform(y_train_np)
y_test_encoded = label_encoder.transform(y_test_np)
```

```
# Step 3: Scaling the X_train and X_test datasets
scaler = StandardScaler()
X_train_scaled = scaler.fit_transform(X_train)
X_test_scaled = scaler.transform(X_test)

# Step 4: Convert scaled arrays back to DataFrame
X_train_scaled_df = pd.DataFrame(X_train_scaled, columns=X_train.
columns)
X_test_scaled_df = pd.DataFrame(X_test_scaled, columns=X_test.
    columns)

# Step 5: Renaming the datasets
y_train = y_train_encoded
y_test = y_test_encoded
X_train = X_train_scaled_df
X_test = X_test_scaled_df

# Step 6: Create a function to train and test various classifiers

def test_various_classifiers(X_train, y_train, X_test, y_test):
    # Dictionary to hold the classifiers and their parameters
    classifiers = {
        'Logistic RegressionM': {
            'model': LogisticRegression(),
            'params': {'multi_class': ['multinomial'],
                        'solver': ['newton-cg', 'sag', 'saga'],
                       'max_iter': [10000],
                       'random_state' : [42]}
        },
        'Logistic RegressionOvR': {
            'model': LogisticRegression(),
            'params': {'multi_class': ['ovr'], 'solver':
                       ['liblinear'],
                       'random_state' : [42]}
        },
        'SVM Radial': {
            'model': SVC(),
            'params': {'C': [1, 10], 'gamma': ['scale', 'auto'],
                        'random_state': [42]}
        },
        'KNN': {
            'model': KNeighborsClassifier(),
            'params': {'n_neighbors': [3, 5, 7], 'weights':['uniform',
                        'distance']}
        },
        'Decision Trees': {
            'model': DecisionTreeClassifier(),
            'params': {'max_depth': [None, 10, 20], 'random_state'
              : [42]}
```

```
        },
        'Random Forest': {
            'model': RandomForestClassifier(),
            'params': {'n_estimators': [50, 100, 200],
                       'max_features': ['sqrt']}
        },
        'AdaBoost': {
            'model':AdaBoostClassifier(),
            'params':{}
        },
        'XGBoost': {
            'model': XGBClassifier(eval_metric='mlogloss'),
            'params': {}
        },
        'GBM': {
            'model': GradientBoostingClassifier(),
            'params': {}
        },
        'LightGBM': {
            'model': LGBMClassifier(),
            'params': {}
        },
        'Naive Bayes': {
            'model': GaussianNB(),
            'params': {}
        },
        'MLP': {
            'model': MLPClassifier(),
            'params': {'hidden_layer_sizes': [(50,50,50),
                       (50,100,50), (100,)],
                       'activation': ['tanh', 'relu'],
                       'solver': ['sgd', 'adam'],
                       'learning_rate': ['constant','adaptive'],
                       'alpha': [0.0001, 0.001, 0.5],
                       'max_iter': [1000, 2000],
                       'random_state' : [42]}
        },
        'Linear Discriminant Analysis': {
            'model': LinearDiscriminantAnalysis(),
            'params': {}
        },
        'Quadratic Discriminant Analysis': {
            'model': QuadraticDiscriminantAnalysis(),
            'params': {}
        }
        # We can add more classifiers and parameters here
}
```

```
    results = []

    # Iterate through each classifier and parameter combination
    for clf_name, clf_dict in tqdm(classifiers.items(),
                                   desc="Evaluating Classifiers"):
        for param_combination in
        itertools.product(*clf_dict['params'].values()):
            param_dict = dict(zip(clf_dict['params'].keys(),
                                  param_combination))
            clf = clf_dict['model'].set_params(**param_dict)

            # Training and evaluating
            clf.fit(X_train, y_train)
            score = clf.score(X_test, y_test)
            results.append((clf_name, param_dict, score))
            print(f"Classifier: {clf_name}, Params: {param_dict},
                  Score: {score}")

    return results

# Call the function with the datasets
results = test_various_classifiers(X_train, y_train, X_test, y_test)

# Find the highest score and its corresponding model and parameters
best_model, best_params, best_score = max(results, key=lambda x: x[2])

# Print the results
print("Best Model:", best_model)
print("Best Parameters:", best_params)
print("Best Score:", best_score)

# Output
Best Model: XGBoost
Best Parameters: {}
Best Score: 0.9617422012948793
```

Code Breakdown:

- Converting $y$ Datasets to 1D Arrays: The target variables are converted from DataFrames into 1-dimensional numpy arrays to fit the requirements of scikit-learn's models.
- Label Encoding: The target variables are label-encoded to transform categorical labels into a numerical format.
- Feature Scaling: The StandardScaler is applied to scale the features in X_train and X_test.

- DataFrame Conversion: The scaled data is converted back into DataFrames with the original column names. This step retains the interpretability of the features.

Classification Algorithms and their Hyperparameters:

- LR:
  - Suitable for binary and multiclass classification.
  - Hyperparameters: multi_class (approach for multiclass classification), solver (algorithm for optimization), max_iter (maximum iterations).
- SVM Radial:
  - Effective in high-dimensional spaces.
  - Hyperparameters: C (regularization parameter), gamma (kernel coefficient).
- KNN:
  - A simple classifier that predicts based on the majority label of nearest neighbors.
  - Hyperparameters: n_neighbors and weights (weighting function for prediction).
- DT:
  - A tree-like model for decision making.
  - Hyperparameters: max_depth (maximum depth of the tree).
- RF:
  - An ensemble of decision trees, offering more robustness and accuracy.
  - Hyperparameters: n_estimators, and max_features.
- AdaBoost:
  - Boosting algorithm that focuses on correctly classifying previously misclassified instances.
  - Hyperparameters: Can be tuned for the base estimator.
- XGBoost:
  - An efficient and scalable implementation of gradient boosting.
  - Hyperparameters: Evaluated using mlogloss for multiclass classification.
- GBM:
  - Boosting technique that builds models sequentially correcting errors of the previous model.
- LightGBM:
  - A fast, distributed gradient boosting framework.

  - Hyperparameters: force_col_wise optimizes column-wise histogram building.
- Naive Bayes:
  - Assumes independence between predictors, simple and effective for large datasets.
- MLP:
  - A neural network model, suitable for complex datasets.
  - Hyperparameters: hidden_layer_sizes (architecture), activation (activation function), solver (optimizer for weight optimization), learning_rate (rate at which the model learns), alpha (regularization term), and max_iter (maximum iterations).
- Linear Discriminant Analysis (LDA):
  - LDA finds a linear combination of feature variables separating different classes.
  - Typically requires minimal hyperparameter tuning.
- Quadratic Discriminant Analysis (QDA):
  - Similar to LDA but assumes each class has its own covariance matrix.
  - Suitable for datasets where classes have distinct distribution.

The Function: test_various_classifiers

- Purpose: To evaluate multiple classifiers on a dataset, with a focus on different parameter settings for each classifier.
- Process:
  - Iterates through a predefined dictionary of classifiers and their respective hyperparameters.
  - For each classifier, it tests various combinations of parameters using itertools.product. All the classifiers undergo training on the X_train and evaluated on the X_test. The accuracy of each model with its specific parameters is printed out.
  - Finally, the best model and its hyperparameters are also printed.

Note to the Reader:

- The provided code is for illustration purposes, showcasing how various classifiers can be implemented and evaluated using scikit-learn.
- Readers are encouraged to experiment with different parameter settings and tuning methods to understand their impact on model performance better. We will talk about hyperparameter tuning in the next chapter (Chapter 8)

- The selection of appropriate hyperparameters is often dataset-specific and requires experimentation.

Now let's move to some practical examples for regression analysis.

## *Regression Analysis Using Python*

This section will focus on demonstrating how to use regression techniques in Python, specifically in the context of farm animal management. Despite the predominant use of classification in animal activity recognition, this section explores regression in estimating continuous outcomes like weight gain in farm animals, using a dataset, *Farm Animal Growth Dataset*.

The *Farm Animal Growth Dataset* is a synthetic dataset generated solely for the purpose of illustrating how regression analysis can be applied using Python in the context of farm animal management. Please note that this dataset does not represent a real-world agricultural dataset but has been crafted to simulate some characteristics and complexities often encountered in agricultural data.

The dataset consists of features related to animal nutrition, and environmental conditions, with the target variable being animal weight gain.

To create this dataset, we followed a structured process:

- We generated 100 samples.
- Random values were assigned to features such as food intake and temperature, simulating variability.
- The target variable, weight gain, was computed as a function of food intake, temperature, and random noise, reflecting a simplified relationship between these factors and animal growth.

You can find the code by visiting the Chapter_7 folder in our GitHub Repository. The file is titled Chapter_7_Regression.ipynb.

### Linear Regression

Linear regression serves as a foundational approach, aiming to portray the correlation between two variables (exploratory variable $x$ and dependent variable $y$) through the application of a linear equation to fit the gathered data.

Formula:

$$y = w_0 + w_1 x + \varepsilon$$

where,

- $y$ is the dependent variable (predicted variable e.g., weight gain).
- $x$ is the independent variable (predictor e.g., food intake).
- $w_1$ is the slope (regression coefficient).

- $w_0$ is the $y$-intercept of the line. This parameter represents the value of $y$ when $x$ is 0.
- ε is the random error term.

The objective of simple linear regression is to establish a line (often referred to as the optimal fit line) that passes through the data points in such a way as to minimize the total of the squared residuals. Here, a residual represents the discrepancy between the actual value and the value predicted by the linear model. To estimate the parameters within a linear regression framework, the Ordinary Least Squares (OLS) technique is employed. This method aims to reduce the total squared deviations between the observed dependent variable in the dataset and the predictions made by the linear equation.

### *Fitting Line and Residuals*

- Fitting Line: In the context of regression, fitting a line means finding the values of that minimize the residuals for the data points.
- Residuals: The residuals are the vertical distances between the data points and the fitting line. They represent the error in the prediction.

### *Implementation for Linear Regression Using the Normal Equation*

Normal Equation is a mathematical approach to find the best fitting model for linear regression. The following code represents the implementation of Normal Equation using python.

```
from sklearn.preprocessing import add_dummy_feature

# Adding an intercept term to the food intake feature

X_intercept = add_dummy_feature(food_intake)

# Calculating the optimal model parameters using the Normal Equation
optimal_params = np.linalg.inv(X_intercept.T.dot(X_intercept)).
                 dot(X_intercept).dot(weight_gain)
optimal_params

# Output
array([[52.12061003],
       [ 0.49769395]])
```

- add_dummy_feature(food_intake): Adds a column of ones to the food_intake array. This is necessary for incorporating the intercept term in the linear model.

- The optimal model parameters (optimal_params) are computed using the Normal Equation. This equation is a mathematical approach to determine the best-fitting line to the data in a linear regression model. The computation involves:
  - Taking the transpose of X_intercept and multiplying it by X_intercept itself.
  - Calculating the inverse of this product.
  - Multiplying the result by the transpose of X_intercept and then by the weight_gain vector.
- optimal_params will contain the coefficients (including the intercept term) that define the best-fit line for predicting weight_gain from food_intake.

### *Fitting a Linear Regression Model Using Scikit-learn*

The process of fitting a linear regression model using scikit-learn can be simpler compared to the Normal Equation method, especially for larger datasets.

### *Scikit-learn vs. Normal Equation Method*

- Normal Equation Method: Involves mathematical calculations using matrices to find the best-fit parameters for the regression line. It is direct and does not require iterative optimization. But it can be computationally expensive for datasets that are large as it requires matrix inversion, which has a computational complexity of $O(n^3)$.
- Scikit-learn Method: Utilizes efficient numerical optimization methods under the hood, making it more scalable for large datasets. It is designed to handle various types of linear models and provides additional functionalities like regularization. The computational complexity is generally lower compared to the Normal Equation, especially for datasets with a large number of features or samples.

Python implementation of Linear Regression using scikit-learn:

```
import pandas as pd
import numpy as np
from sklearn.linear_model import LinearRegression
import matplotlib.pyplot as plt

# 'data' is the DataFrame containing the synthetic dataset
X = data[['Food_Intake']]  # Independent variable
y = data['Weight_Gain']    # Dependent variable

# Create and fit the OLS model
model = LinearRegression()
model.fit(X, y)
```

```
# Display model parameters
print(f"Intercept: {model.intercept_}")
print(f"Coefficient for Food Intake: {model.coef_[0]}")

# Visualizing the Regression Line
plt.figure(figsize=(6, 4))
plt.plot(X, y, 'g.')  # Actual data points
plt.plot(X, model.predict(X), 'r-', label='Regression Line')  #
Regression line
plt.xlabel('Food Intake')
plt.ylabel('Weight Gain')
plt.grid()
plt.legend()
plt.title('Simple Linear Regression with OLS')
plt.show()

# Output
Intercept: 52.120610025626576
Coefficient for Food Intake: 0.4976939463543877
```

In this code:

- Data Preparation: The dataset is defined as X and y, representing the independent variable (Food Intake) and dependent variable (Weight Gain), respectively.
- Model Creation and Fitting: A linear regression model is created using LinearRegression() and then fitted to the data with model.fit(X, y). This step involves internally calculating the best-fit line.
- Displaying Model Parameters: After fitting the model, the intercept and coefficient for the Food Intake variable are printed. These values represent the $y$-intercept and slope of the regression line, respectively.
- Visualization: The code then plots the original data points (X, y) as green dots and the regression line as a red line. This visualization helps in understanding the fit of the model and the relationship between Food Intake and Weight Gain.
- Plot Formatting: The plot is formatted with labels, grid, legend, and title for better readability and understanding (refer to Figure 7.10).

### Multiple Linear Regression

Unlike linear regression, multiple linear regression is employed when there are two or more predictors.

Formula:

$$y = w_0 + w_1x_1 + w_2x_2 + \cdots + w_nx_n + \varepsilon$$

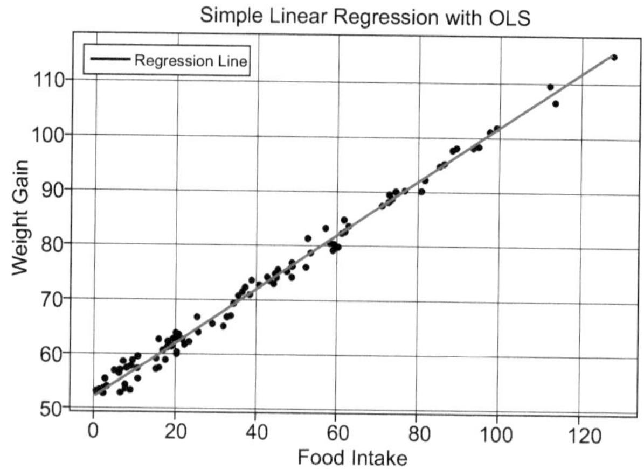


**Figure 7.10: Linear regression model.**

where,

- $y$ is the output variable.
- $x_1, x_2, \ldots, x_n$ are the predictors.
- $w_1, w_2, \ldots, w_n$ are the coefficients that represent the weight of each predictor.
- $w_0$ is the intercept.
- $\varepsilon$ is the random error.

### *Applying Multiple Linear Regression*

Based on our previous examples with the synthetic dataset, now we will predict the Weight_Gain (dependent variable) using both Food_Intake and Temperature (independent variables).

Here is how you can implement multiple linear regression using scikit-learn:

```
from mpl_toolkits.mplot3d import Axes3D

#'data' is the DataFrame containing the synthetic dataset
X = data[['Food_Intake', 'Temperature']]  # Independent variables
y = data['Weight_Gain']  # Dependent variable

# Create and fit the multiple linear regression model
multi_model = LinearRegression()
multi_model.fit(X, y)

# Display the model parameters
print(f"Intercept: {multi_model.intercept_}")
print(f"Coefficients: {multi_model.coef_}")
```

```
# Creating a meshgrid for the values of Food_Intake and Temperature
x_surf, y_surf = np.meshgrid(np.linspace(data['Food_Intake'].min(),
                 data['Food_Intake'].max(), 100),
                 np.linspace(data['Temperature'].min(),
                 data['Temperature'].max(), 100))
onlyX = pd.DataFrame({'Food_Intake': x_surf.ravel(),
                      'Temperature': y_surf.ravel()})
fittedY = multi_model.predict(onlyX)

# Reshape the predicted values to match x_surf and y_surf
fittedY = fittedY.reshape(x_surf.shape)

# Plotting
fig = plt.figure(figsize=(8, 6))
ax = fig.add_subplot(111, projection='3d')

# Scatter plot with the actual data
ax.scatter(data['Food_Intake'], data['Temperature'], data['Weight_
Gain'], c='blue', marker='o', alpha=0.5)

# Plotting the surface (plane of best fit)
    ax.plot_surface(x_surf, y_surf, fittedY, color='r', alpha=0.3)

# Labels and titles
ax.set_xlabel('Food Intake')
ax.set_ylabel('Temperature')
ax.set_zlabel('Weight Gain')
plt.savefig('Multiple Linear Regression.png', dpi=300)
plt.title('Multiple Linear Regression: Weight Gain vs Food Intake &
    Temperature')

# Show the plot
plt.show()

# Output
Intercept: 49.72927664354033
Coefficients: [0.50184912 0.10654829]
```

In this example, LinearRegression is used, but this time with multiple features to predict Weight Gain. The model's coefficients indicate the influence of each feature. We also create a meshgrid using np.meshgrid (Figure 7.11). This meshgrid is used to predict values across the entire grid, providing data for the surface plot. Then, multi_model.predict(onlyX) is used to predict the Weight_Gain over the meshgrid. The ax.scatter plots the actual data points in the 3D space. Additionally, the ax.plot_surface plots the plane of best fit using the predicted values. This plane represents the multiple linear regression model.

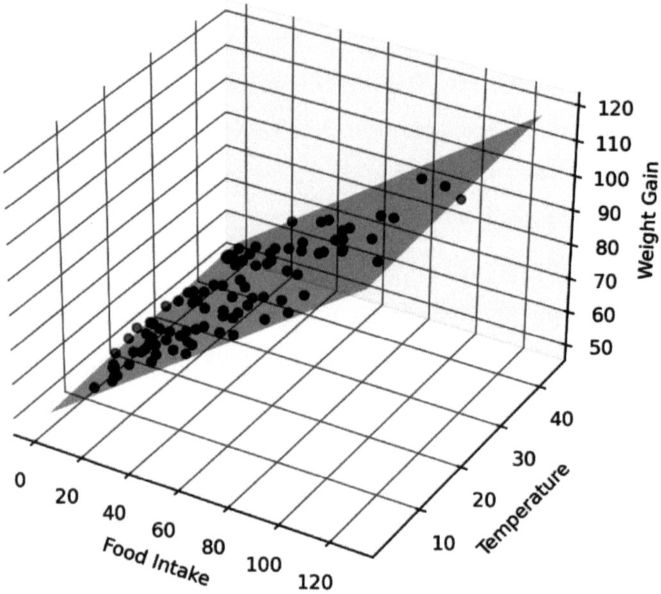


**Figure 7.11: Multiple linear regression fit.**

### Polynomial Regression

In polynomial regression the predictor $x$ and the dependent variable $y$ is modelled as a $n$ degree polynomial.

Advantages:

- It can model a wider range of relationships between variables.
- Provides a better fit for datasets that are not well-described by a linear relationship.

Disadvantages:

- Prone to overfitting, especially with high-degree polynomials.
- The interpretability of the model can be challenging compared to simple linear regression.

Parameters in Polynomial Regression:

- Degree of the Polynomial ($n$): Determines the curving of the fit.
- Regularization: Techniques like Ridge or Lasso can help to avoid overfitting, especially for higher-degree polynomials.

Python code to implement polynomial regression:

```
import numpy as np
import pandas as pd
from sklearn.linear_model import LinearRegression
from sklearn.preprocessing import PolynomialFeatures
import matplotlib.pyplot as plt

# 'data' is the DataFrame with 'Food_Intake' and 'Weight_Gain'
X = data['Food_Intake'].values.reshape(-1, 1)  # Independent
variable
y = data['Weight_Gain']  # Dependent variable

# Degrees of the polynomial to be evaluated
degrees = [4, 15, 40]

# Setting up the plot
plt.figure(figsize=(10, 6))
plt.scatter(X, y, label='Data Points')

# Iterate through various degrees and fit the polynomial regression
model for degree in degrees:
    poly_features = PolynomialFeatures(degree=degree, include_
    bias=False)
    X_poly = poly_features.fit_transform(X)

    poly_reg_model = LinearRegression()
    poly_reg_model.fit(X_poly, y)

    # Predictions for visualization
    X_range = np.linspace(X.min(), X.max(), 100).reshape(-1, 1)
    y_pred = poly_reg_model.predict(poly_features.transform(X_range))

    # Visualizing the Polynomial Regression results
    plt.plot(X_range, y_pred, label=f'{degree}-Degree Polynomial')

# Plot formatting
plt.xlabel('Food Intake')
plt.ylabel('Weight Gain')
plt.legend()
plt.title('Polynomial Regression with Different Degrees')
plt.show()
```

In this example we use PolynomialFeatures from scikit-learn to transform our features into polynomial features. We then fit a LinearRegression model on these transformed features. The degree of the polynomial is set to [4, 15, 40]. Finally, we visualize the polynomial regression fit along with the original data points.

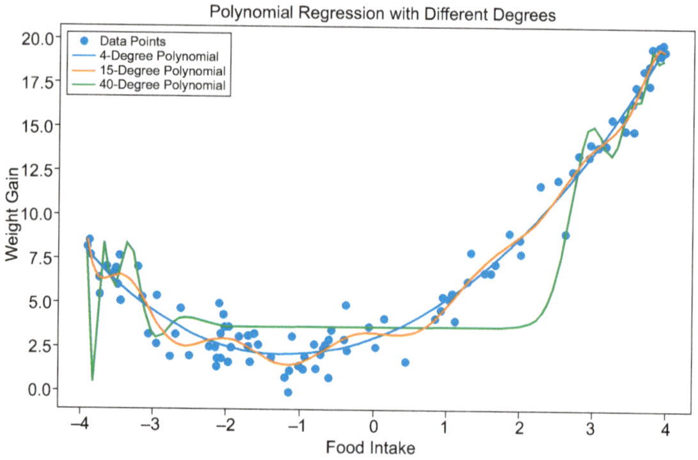


**Figure 7.12: Polynomial regression of various degree values.**

Figure 7.12 illustrates how each polynomial degree fits the data. The low-degree polynomial will appear as a smooth curve, the moderate degree as a more flexible curve, and the high degree might show extreme bends and turns, indicating overfitting. This demonstration highlights the importance of choosing the right degree for polynomial regression.

## Other Regression Algorithms

While polynomial regression is useful for non-linear relationships, scikit-learn offers a variety of other regression algorithms, each suitable for different kinds of data and problems. Below are some common regression techniques available in scikit-learn, with their typical use cases:

### *Common Setup for All Examples*

```
import numpy as np
import pandas as pd
from sklearn.model_selection import train_test_split

X = data.drop('Weight_Gain', axis=1)
y = data['Weight_Gain']

# Splitting the dataset into training and testing sets
X_train, X_test, y_train, y_test = train_test_split(X, y, test_
size=0.2, random_state=42)
```

- Ridge Regression (L2): Useful for handling multicollinearity and overfitting through L2 regularization.

 - Common Hyperparameters: alpha (regularization strength), solver (algorithm to use in the optimization problem).

```
# Ridge Regression

from sklearn.linear_model import Ridge
from sklearn.metrics import mean_squared_error, r2_score

# Fit the model
ridge_reg = Ridge(alpha=1.0)
ridge_reg.fit(X_train, y_train)

# Evaluation
y_pred = ridge_reg.predict(X_test)

# Evaluate the model
mse = mean_squared_error(y_test, y_pred)
r2 = r2_score(y_test, y_pred)

# Print evaluation metrics
print(f"Mean Squared Error (MSE): {mse:.2f}")
print(f"R-squared (R2) Score: {r2:.4f}")

# Output
Mean Squared Error (MSE): 15.48
R-squared (R2) Score: 0.4744
```

- Lasso Regression (L1): Effective for models where some features are irrelevant, using L1 regularization to promote sparsity.
  - Common Hyperparameters: alpha (regularization strength).

```
# Lasso regression

from sklearn.linear_model import Lasso

lasso_reg = Lasso(alpha=0.1)
lasso_reg.fit(X_train, y_train)

# Evaluation
y_pred = lasso_reg.predict(X_test)

# Evaluate the model
mse = mean_squared_error(y_test, y_pred)
r2 = r2_score(y_test, y_pred)
```

```
# Print evaluation metrics
print(f"Mean Squared Error (MSE): {mse:.2f}")
print(f"R-squared (R2) Score: {r2:.4f}")

# Output
Mean Squared Error (MSE): 15.41
R-squared (R2) Score: 0.4765
```

- Elastic Net (ElasticNet): Combines L1 and L2 regularization, useful when various features are correlated with each other.
  - Common Hyperparameters: alpha (regularization strength), l1_ratio (mixing parameters between L1 and L2).

```
# Elastic net

from sklearn.linear_model import ElasticNet

elastic_net = ElasticNet(alpha=1, l1_ratio=0.5)
elastic_net.fit(X_train, y_train)

# Evaluation
y_pred = elastic_net.predict(X_test)

# Evaluate the model
mse = mean_squared_error(y_test, y_pred)
r2 = r2_score(y_test, y_pred)

# Print evaluation metrics
print(f"Mean Squared Error (MSE): {mse:.2f}")
print(f"R-squared (R2) Score: {r2:.4f}")

# Output
Mean Squared Error (MSE): 14.74
R-squared (R2) Score: 0.4992
```

- Support Vector Regression (SVR): Suitable for both linear and non-linear relationships, especially effective in high-dimensional spaces.
  - Common Hyperparameters: C (regularization parameter), kernel (type of kernel used), epsilon (specifies the epsilon where there is no associated penalty).

```
# Support Vector Regression
from sklearn.svm import SVR

svr_reg = SVR(C=1.0, epsilon=0.2)
svr_reg.fit(X_train, y_train)

# Evaluation
y_pred = svr_reg.predict(X_test)

# Evaluate the model
mse = mean_squared_error(y_test, y_pred)
r2 = r2_score(y_test, y_pred)

# Print evaluation metrics
print(f"Mean Squared Error (MSE): {mse:.2f}")
print(f"R-squared (R2) Score: {r2:.4f}")

# Output
Mean Squared Error (MSE): 2.45
R-squared (R2) Score: 0.9166
```

- Decision Tree Regression (DecisionTreeRegressor): Can model complex datasets without requiring feature scaling or much data preprocessing.
  - Common Hyperparameters: max_depth, min_samples_split.

```
from sklearn.tree import DecisionTreeRegressor

tree_reg = DecisionTreeRegressor(max_depth=3)
tree_reg.fit(X_train, y_train)

# Evaluation
y_pred = tree_reg.predict(X_test)

# Evaluate the model
mse = mean_squared_error(y_test, y_pred)
r2 = r2_score(y_test, y_pred)

# Print evaluation metrics
print(f"Mean Squared Error (MSE): {mse:.2f}")
print(f"R-squared (R2) Score: {r2:.4f}")

# Output
Mean Squared Error (MSE): 2.63
R-squared (R2) Score: 0.9107
```

- Random Forest Regression (RandomForestRegressor): A robust method that uses an ensemble of decision trees, good for high accuracy without overfitting.
  - Common Hyperparameters: n_estimators, max_depth.

```
from sklearn.ensemble import RandomForestRegressor

forest_reg = RandomForestRegressor(n_estimators=100)
forest_reg.fit(X_train, y_train)

# Evaluation
y_pred = forest_reg.predict(X_test)

# Evaluate the model
mse = mean_squared_error(y_test, y_pred)
r2 = r2_score(y_test, y_pred)

# Print evaluation metrics
print(f"Mean Squared Error (MSE): {mse:.2f}")
print(f"R-squared (R2) Score: {r2:.4f}")

# Output
Mean Squared Error (MSE): 1.95
R-squared (R2) Score: 0.9339
```

- Gradient Boosting Regression (GradientBoostingRegressor): Builds an additive model in a forward stage-wise fashion; it is great for handling a variety of data types.
  - Common Hyperparameters: n_estimators (number of boosting stages), learning_rate (rate at which the contribution of each tree shrinks).

```
# Gradient Boosting regression
from sklearn.ensemble import GradientBoostingRegressor

gbrt = GradientBoostingRegressor(n_estimators=100, learning_
rate=0.1)
gbrt.fit(X_train, y_train)

# Evaluation
y_pred = gbrt.predict(X_test)

# Evaluate the model
mse = mean_squared_error(y_test, y_pred)
r2 = r2_score(y_test, y_pred)
```

```
# Print evaluation metrics
print(f"Mean Squared Error (MSE): {mse:.2f}")
print(f"R-squared (R2) Score: {r2:.4f}")

# Output
Mean Squared Error (MSE): 1.75
R-squared (R2) Score: 0.9404
```

This book provides an introduction to various regression models available in scikit-learn, giving a broad overview of options for different scenarios. For a more details about implementation and explanation of each regression technique, the interested reader can refer to the documentation of scikit-learn.

## Unsupervised Learning

### *Unsupervised Competitive Learning and Self-organizing Feature Maps*

Unsupervised learning requires no target values, this means that the network uses the information available from the input patterns to classify its outputs. Therefore, the input patterns must have some redundant information for the network to extract the features, regularities, correlations, or categories embedded in the input data.

In a competitive learning algorithm only one output or only one output per group is activated. The activated output unit is called the winner-take-all unit or the grandmother cell (Hertz et al., 1991). A network that uses such a learning rule aims at categorizing or clustering the input data into separate groups. An input pattern recognized as belonging to a particular group should fire the same output unit as the other members of the same group.

Self-organising feature maps (SOFMs) are a class of neural networks that use the unsupervised competitive learning rule. The network needs only one output for each group involved, this means that for $N$ groups, the network requires $N$ output units. Furthermore, the output units are placed in a line or plane geometrical structure. Hence, the position of the winning units gives some information about its neighboring units. The locations of cells represent a particular domain of their input patterns.

SOFMs have been used as an alternative to the traditional structures of neural networks. An analysis of these networks has been carried out from the technical rather than the biological viewpoint, however the learning results look natural which indicates that the adaptive processes may be like those encountered in the brain. SOFMs have various applications like BP networks; they can be applied to pattern recognition, robotics, and process control.

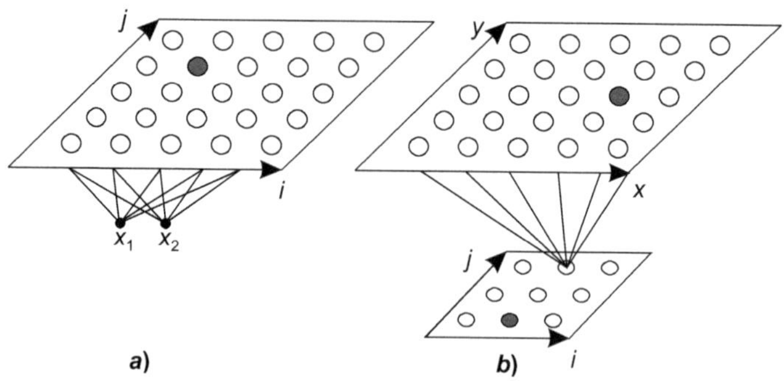


**Figure 7.13: Two common types of feature mapping architectures. (a) The conventional type of feature map with its continuous value of inputs. (b) The second type of self-organizing feature map which has more biological importance such as the mapping from the ear to the auditory cortex.**

There are two common types to the structure of SOFMs as shown in Figure 7.13. For both cases, the outputs are organized in a two-dimensional plane, while the inputs are presented to the networks as continuous values for the first type and the inputs themselves are arranged in a two-dimensional plane in the second type.

For the first type of SOFM (refer to Figure 7.13(a)), the inputs $x_1$ and $x_2$ are connected to each output unit via the weights and are presented directly to the network. The organization of the output does not have to be in the square plane and the inputs can have higher dimensions. The second type (illustrated in Figure 7.13(b)) has more biological importance, however it is less studied than the first type. In this case, the inputs are organized in a two-dimensional plane that defines the input space. In the simplest form, one of the input patterns defined in the input space is turned on. The aim of the training process is to find a suitable mapping from the input to the output spaces. The importance of this type of feature map comes from the fact that such mapping frequently occurs in the brain such as in the connections of the sensing organs including the eye, ear and skin to the cortex and the connections of the different areas of the cortex.

## *Kohonen Self-organizing Feature Map Learning Algorithm*

Consider a SOFM as shown in Figure 7.13(a) with $M$ inputs. The inputs are written in the vector form as $X = [x_1\ x_2\ \ldots.\ x_M]$ and for the $j$th output unit there exists a $M$ dimensional weight vector with $w_j = [w_{j1}\ w_{j2}\ \ldots.\ w_{jM}]$ that the input vector is connected to. The output plane can be rectangular, hexadecimal or even have an irregular structure (Kohonen, 1989). For high-dimensional input vectors $X$, SOFM is a non-linear mapping of the probability density function $P(X)$ to the two-dimensional output space.

The input vector, $X$, is compared to the weights of each element of the output units. A simple way of comparison is using the Euclidean distance, where we have:

$$\|X - W_j\| = \sqrt{\sum_{i=1}^{M} (x_i - w_{ji})^2}.$$

The best-matching node is defined as the output node in which the Euclidean distance between the input pattern and the weight vector is minimum, i.e.,

$$\|X - W_c\| = \min_j\{\|X - W_j\|\}$$

where, $c$ represents the location of the node with the best-matching.

The learning process starts by defining initial values to the weights of the SOFM. Then, the input patterns are presented to the network once at a time. For each input data, the best-matching output unit is determined, and the weights of this unit and its neighbors are also updated according to the following equation:

$$W_j(t+1) = W_j(t) + h_{cj}(t)(X(t) - W_j(t)).$$

If we let $Nc$ to be the number of neighborhood cells around the best-matching unit $c$, then the network updates the value of the weights of the best matching cell and its neighbors around $Nc$. The initial number of the neighborhood units, $Nc$, is selected to have large value and reduced monotonically as time proceeds. At the start of the training, the large value of $Nc$ will produce a rough global ordering in the weights of the network. When shrinking the value of $Nc$, this will cause an improvement in the spatial resolution of the map.

The value of $h_{cj}(t)$ can be selected to be equal to $\alpha(t)$, which is a scalar value called the adaptation rate. A general representation of the value of $h_{cj}(t)$ in terms of the Gaussian function is:

$$h_{cj}(t) = \alpha(t) \cdot exp\left(-\frac{\|r_c - r_j\|^2}{2\sigma^2(t)}\right)$$

where, $\alpha(t)$ represents the learning rate, while $\sigma(t)$ defines the width of the kernel function. Both parameters $\alpha(t)$ and $\sigma(t)$ are monotonically decreasing functions of time. The values of $r_c$ and $r_{cj}$ are the coordinates of cells $c$ and $i$, respectively.

The initial value of $\alpha(t)$ can be selected to be near to one and then its value is decreased monotonically with time. The function that controls the value of $\sigma(t)$ can be selected as linear, exponential, or inversely proportional to $t$. An alternative to the value of $\alpha(t)$ is $0.9(1 - t/1000)$.

The number of steps used to train the network has great influence on the final accuracy of the mapping. The learning process is a stochastic, meaning that the accuracy of the mapping is based on the steps in the last convergence phase. A proper selection for the number of steps can be around 500 times the number of units in the network.

## Clustering Using Python

Clustering is an unsupervised learning technique that analyzes data to find natural groupings. It is performed without prior knowledge or labels, making it a form of *learning by observation* rather than *learning by examples*. Clustering algorithms process the input features of the data points and group similar items based on a chosen similarity measure, such as Euclidean distance in the case of numerical data.

### Applications in Farm Animal Activity Recognition

In the context of farm animal activity recognition, clustering can be employed in various ways:

- Behavioral Analysis: By clustering activity data is collected from the sensors or observation logs, one can identify common behaviors among animals. For instance, clustering can reveal patterns such as grazing, resting, or roaming behaviors.
- Health Monitoring: Clustering can help in early detection of health issues. Animals that suddenly change their movement patterns or show different behavior from their usual cluster might be exhibiting signs of illness or distress.
- Breed Characterization: Different breeds may exhibit unique behavior patterns. Clustering can help in characterizing these patterns, contributing to genetic and behavioral research.
- Optimizing Resource Allocation: Clustering can be used to group animals based on their dietary needs or milk production, helping to manage feeding schedules and optimize resource distribution.

### Applications in Farm Animal Management

For farm management, clustering offers significant benefits:

- Grouping by Productivity: Identifying clusters of high-yielding versus low-yielding animals can inform breeding decisions and resource allocation.
- Spatial Distribution: Clustering can analyze the spatial distribution of animals within a farm to optimize space utilization and enhance animal welfare.
- Temporal Behavior Analysis: Clustering can reveal how animal activities change over time, which can be crucial for planning daily operations and long-term strategies.

- Anomaly Detection: Unusual patterns identified by clustering can flag potential issues in animal behavior, prompting further investigation.

### *Understanding make_blobs for Clustering Examples*

Before going into the clustering examples, let's discuss make_blobs, a function that is helpful for creating synthetic datasets, which is useful for illustrating clustering algorithms.

### *What is make_blobs?*

make_blobs is a utility from Scikit-learn's datasets module. It generates isotropic Gaussian blobs for clustering, meaning it creates clusters of data points centered around specified locations, with a certain degree of variance and number of centers.

### *Creating a Synthetic Dataset with make_blobs*

In the following Python script we create a synthetic dataset using make_blobs that will be used to demonstrate our clustering examples.

```
from sklearn.datasets import make_blobs
import matplotlib.pyplot as plt

# Generate synthetic data with make_blobs
X, y = make_blobs(
    n_samples=1000,
    centers=5,
    n_features=2,
    cluster_std=[0.7, 1.0, 0.4, 0.7, 1.0],  # Different standard
deviations for each cluster
    random_state=42
)

# Visualize the generated clusters
plt.scatter(X[:, 0], X[:, 1], c=y, edgecolor='k', s=20, alpha=0.4)
plt.title('Synthetic Clusters with Different Standard Deviations')
plt.xlabel('Feature 1')
plt.ylabel('Feature 2')
plt.savefig('clusters.png', dpi=300)
plt.show()
```

Parameters used in this function:

- n_samples = 1000: This specifies that the generated dataset will contain 1000 data points.
- centers = 5: This indicates that the synthetic dataset should contain 5 distinct clusters.

- n_features = 2: This sets the number of features (or dimensions) for each sample to 2. This means every data point will have two coordinates, making it easy to visualize on a 2-dimensional plot.
- cluster_std = [0.7, 1.0, 0.4, 0.7, 1.0]: This is a list specifying the standard deviation of the clusters. The standard deviation controls how much the data points in each cluster are spread out around the cluster center.
- random_state = 42: This sets the seed for the random number generator that make_blobs use to create the dataset.

The result of this function call is two variables, *X* and *y*:

- *X* will be a NumPy array of shape (1000, 2), containing the coordinates for the 1000 generated samples across 2 features.
- *y* will be a NumPy array of shape (1000,), containing the cluster labels for each sample, ranging from 0 to 4 (since there are 5 centers).

### Gaussian Mixture Models (GMMs)

GMM is a probabilistic model for representing normally distributed subpopulations within an overall population. In the context of clustering, GMMs offer a method for identifying the latent Gaussian distributions from which the data points are generated. GMMs assume that the data is created from a combination of various Gaussian distributions with unknown parameters. They belong to the category of soft clustering algorithms that allow for a data point to belong to multiple

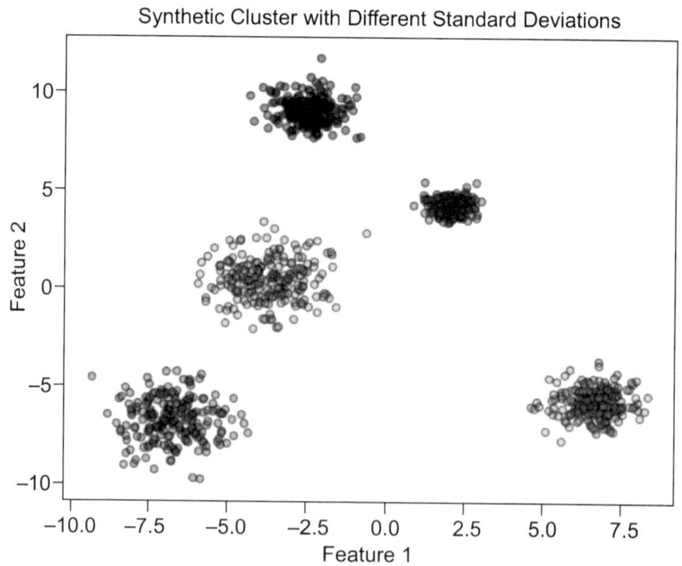


**Figure 7.14: Visualization of the synthetic dataset.**

clusters with different levels of membership. This is in contrast to hard clustering algorithms, where each data point is strictly assigned to one cluster.

Algorithm:

- Initialize the number of Gaussian distributions.
- Expectation-Maximization (EM): Iteratively assign points to clusters (Expectation) and then update the clusters' parameters (Maximization).

Formula:

The probability of observing a data point $x$ is given by the sum over all possible clusters (Gaussian components):

$$P(x) = \sum_{i=1}^{k} \pi_i \cdot N(x|\mu_i, \Sigma_i)$$

where,

- $k$ is the number of clusters.
- $\pi_i$ is the weighting factor of the $i$th cluster.
- $N(x|\mu_i, \Sigma_i)$ is the Gaussian distribution with mean $\mu_i$ and covariance $\Sigma_i$.

Here we apply a GMM to the synthetic dataset we created with make_blobs. We will use Scikit-learn's GaussianMixture class.

```
from sklearn.mixture import GaussianMixture
from sklearn.datasets import make_blobs
import matplotlib.pyplot as plt

# Fit Gaussian Mixture Model
gmm = GaussianMixture(n_components=5, random_state=42)
gmm.fit(X)

# Predict cluster labels
labels = gmm.predict(X)

# Visualize the clusters
plt.scatter(X[:, 0], X[:, 1], c=labels, cmap='viridis',
   edgecolor='k', s=20, alpha=.3)
plt.title('Gaussian Mixture Clustering')
plt.xlabel('Feature 1')
plt.ylabel('Feature 2')
plt.show()
```

We are using the GaussianMixture class to fit the model to our data. We specify n_components = 5, corresponding to the number of clusters we know our data to have. The GaussianMixture object uses the Expectation-Maximization algorithm

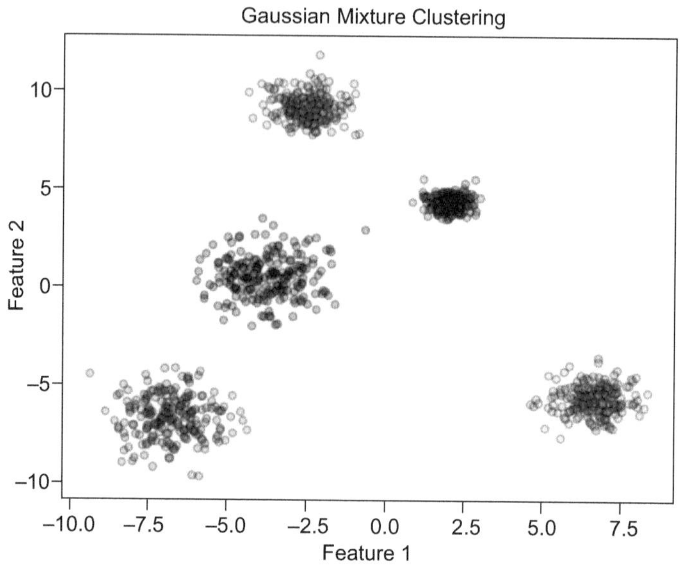


**Figure 7.15: Effective cluster identification using Gaussian Mixture model: A close match with predefined groups in the synthetic dataset.**

to find the parameters of the Gaussians that best fit our data. After fitting the model, we use it to predict labels for our dataset, which we then plot to visualize the resulting clusters.

The application of GMM on our synthetic dataset has yielded good results. The clusters identified by the GMM align closely with the five groups we initially defined using the make_blobs function.

## Hierarchical Clustering

Hierarchical clustering is an approach to cluster analysis that aims to create a hierarchy of clusters. The results of hierarchical clustering are usually presented in a dendrogram, which illustrates the arrangement of the clusters produced by the analysis. In the context of farm animal activity recognition, hierarchical clustering can be particularly useful for understanding behaviors that operate at different scales or granularities.

### *Agglomerative Hierarchical Clustering*

This represents the most frequently utilized form of hierarchical clustering, employed to categorize objects using a bottom-up methodology:

- Child and Parent Clusters: At the start, each data point is considered an individual cluster (child). As the algorithm proceeds, clusters are merged based on their distance, until all points converge into a single cluster (parent).

- Dendrograms: The process of merging can be visualized using a dendrogram, which is a tree-like diagram that records the sequences of merges or splits.
- Application: In farm management, agglomerative clustering could be used to discern hierarchical relationships in animal behaviors. For example, animals might be grouped based on minute-by-minute activities, and these groups could then be clustered based on daily patterns.

Formulas and Metrics:

The choice of distance metrics and linkage criteria has a significant impact on the clusters formed by hierarchical clustering:

- Single Linkage: min(dist $(a, b)$), where $a$ and $b$ are objects in different clusters.
- Complete Linkage: max(dist $(a, b)$).
- Average Linkage: avg(dist $(a, b)$).
- Ward's Method: Reduces the total of squared differences within all clusters. This method focuses on minimizing variance and shares similarities with the objective of K-Means.

Python Example:

```
# Agglomerative Clustering

from sklearn.datasets import make_blobs
from sklearn.cluster import AgglomerativeClustering
import scipy.cluster.hierarchy as sch
import matplotlib.pyplot as plt

# Fit the Agglomerative Clustering
agg_clustering = AgglomerativeClustering(n_clusters=5)
agg_labels = agg_clustering.fit_predict(X)

# Plot the dendrogram
plt.title('Hierarchical Clustering Dendrogram')
dendrogram = sch.dendrogram(sch.linkage(X, method='ward'))
plt.show()

# Plot the clusters
plt.scatter(X[:, 0], X[:, 1], c=agg_labels, cmap='viridis', s=20,
    alpha=.3)
plt.xlabel('Feature 1')
plt.ylabel('Feature 2')
plt.title('Agglomerative Clustering Results')
plt.savefig('Agglomerative.png', dpi=300)
plt.show()
```

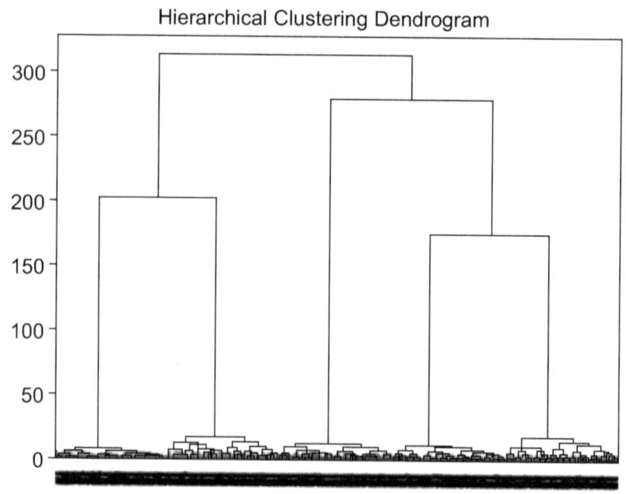


**Figure 7.16: Hierarchical clustering dendrogram.**

Code Explanation:

- Agglomerative Clustering: AgglomerativeClustering from scikit-learn is used with 5 clusters. The model is fitted to X, and a cluster label is given to each data point.
- Dendrogram Visualization (Figure 7.16): The dendrogram is plotted using sch.dendrogram with Ward's method as the linkage criterion. This visualization helps in understanding the cluster formations and their hierarchical relationships.
- Cluster Plotting: The data points are plotted with colours indicating their cluster membership, showing how they are grouped into clusters.

The flexibility and specific constraints of metric choices in scikit-learn are key factors when exploring distance metrics for linkage computation in agglomerative clustering:

- Metric Options: The metric used to compute the linkage in agglomerative clustering can vary. Acceptable metrics include euclidean, l1, l2, manhattan, cosine, or precomputed. This range of metrics allows the algorithm to be tailored to different types of data and different notions of *distance* or *similarity*.
- Constraint for Ward Linkage: If the linkage method chosen is *ward*, the only accepted metric is *euclidean*. This is because Ward's method inherently relies on the minimization of variance within clusters, which is directly tied to the Euclidean distance.
- Precomputed Distance Matrix: When using a *precomputed* metric, the input for the fit method needs to be a distance matrix rather than the raw data points.

This is useful when the distance computation is complex or non-standard and is pre-calculated before the clustering process.

- The default value is *euclidean.*

This flexibility in the choice of metric allows agglomerative clustering to be effectively applied to a wide variety of datasets, each possibly requiring a different notion of distance to best reflect the inherent structure of the data. The euclidean distance remains the most common and default choice, suitable for many standard clustering tasks.

The provided dendrogram (Figure 7.16) illustrates the hierarchical clustering of the dataset. The vertical axis quantifies the dissimilarity between clusters, with higher values indicating less similarity. Additionally, Figure 7.17 shows the results of an agglomerative clustering algorithm, revealing five distinct groups within the data based on two features.

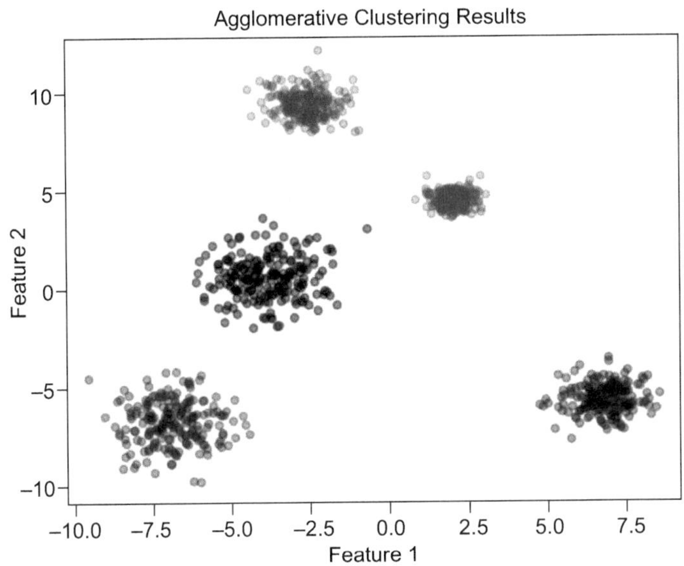


**Figure 7.17: Agglomerative clustering results.**

## Partitional Clustering

Unlike hierarchical clustering, partitional clustering divides the data set into disjoint clusters without any explicit structure that would relate clusters to each other.

Characteristics:

- Disjoint Clusters: Each object belongs to one and only one cluster.
- No Hierarchical Relationship: There in no concept of child or parent clusters; all clusters are on the same level.

### *K-Means*

K-Means clustering is straightforward and a widely utilized algorithm for grouping data. Its objective is to divide a collection of observations into a predefined number of clusters, minimizing the within-cluster variances.

How K-Means Works:

1. Initialization: K-Means starts by randomly selecting '*k*' centroids, where '*k*' represents the total count of clusters you choose. These centroids are the initial guesses for the locations of the cluster centers.
2. Assignment Step: Every data point is allocated to the closest centroid, determined by the squared Euclidean distance. This forms '*k*' clusters.
3. Update Step: The centroids of the clusters are recalculated. This is typically done by calculating the average of all data points assigned to that cluster's centroid.
4. Iterative Optimization: Steps 2 and 3 continue until there is minimal to no movement in the centroids, indicating that convergence has been achieved.

Formula:

The objective function of K-Means is defined as:

$$J = \sum_{i=1}^{k} \sum_{x \in S_i} \|x - \mu_i\|^2$$

where,

- $J$ is the cost function to be minimized.
- $k$ is the number of clusters.
- $S_i$ is the set of data points in the $i$th cluster.
- $x$ is a data point in cluster $S_i$.
- $\mu_i$ is the centroid of the $i$th cluster.

Python Example for K-Means:

```
# K-Means Clustering

from sklearn.datasets import make_blobs
from sklearn.cluster import KMeans
import matplotlib.pyplot as plt

# Apply K-Means clustering
k_means = KMeans(n_clusters=5, random_state=42, n_init='auto')
k_means.fit(X)
```

```
# Visualize the clusters
plt.scatter(X[:, 0], X[:, 1], c=k_means.labels_, cmap='viridis',
s=20, alpha=.3)
plt.scatter(k_means.cluster_centers_[:, 0], k_means.cluster_centers_
    [:, 1], s=20, c='red', marker='x', label='Centroids')
plt.xlabel('Feature 1')
plt.ylabel('Feature 2')
plt.legend()
plt.title('K-Means Clustering Results')
plt.savefig('kmeans.png', dpi=300)
plt.show()
```

Figure 7.18 illustrates the results of the K-Means clustering analysis above, showing five distinct groups across two features with their respective centroids marked by black crosses.

### *The Elbow Method*

The Elbow Method is a heuristic used in cluster analysis to determine the optimal number of clusters ($k$) in a dataset. It is commonly used in conjunction with the K-Means clustering algorithm, although it can be applied to other clustering methods as well.

How the Elbow Method Works:

- Variance Within Clusters: The basic idea behind the Elbow Method is to run the clustering algorithm across a range of values of $k$ (number of clusters) and

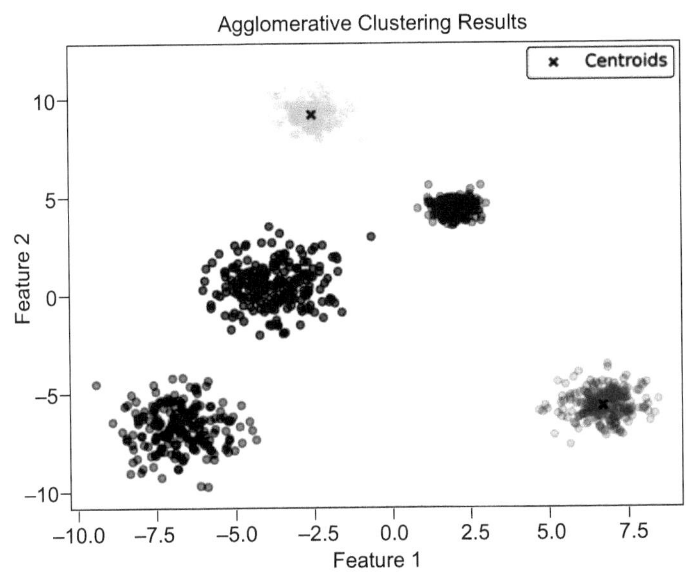


**Figure 7.18: K-Means clustering plot.**

for each value, calculate the sum of squared distances from each point to its assigned center (within-cluster sum of squares(WCSS)).

- Plotting the Results: The values of WCSS are then plotted against the number of clusters. As the number of clusters increases, the WCSS will typically decrease (since the points are closer to their respective centers).
- Identifying the *Elbow*: The key is to find the point where the rate of decrease changes, which typically represents a situation where adding more clusters does not provide better data modelling. This point is often referred to as the *elbow*, similar to the angle in the human arm.

The Benefits of Using the Elbow Method:

- Trade-off Between Simplicity and Accuracy: This method helps in restoring equilibrium between the utmost data compression through a single cluster and the highest accuracy by allocating each data point to its distinct cluster.
- Intuitive and Easy to Implement: It provides a simple way to visually assess the optimal number of clusters.

Limitations of the Elbow Method:

- Subjectivity: At times, the *elbow* point may not be distinctly visible or pronounced, leading to subjective interpretations
- Not Applicable to All Datasets: Some datasets might not demonstrate a clear elbow, or the elbow method may not be appropriate for datasets with structures that do not align well with spherical clusters, as assumed by K-Means.

Example of Elbow Method in Python:

```
# The Elbow Method

wcss = []
for i in range(1, 11):  # experimenting on different numbers of
clusters
    kmeans = KMeans(n_clusters=i, init='k-means++', random_state=42)
    kmeans.fit(X)
    wcss.append(kmeans.inertia_)  # inertia_ is the WCSS for each model

# Plotting the Elbow Method graph
plt.plot(range(1, 11), wcss)
plt.title('Elbow Method')
plt.xlabel('Number of clusters')
plt.ylabel('WCSS')
plt.show()
```

In the above script, the WCSS (inertia) is calculated for different values of $k$ and plotted. The *elbow* of the plot is where you would typically choose the optimal number of clusters to use for your K-Means clustering.

The plot generated by the Elbow Method provides a visual way to determine the optimal number of clusters for K-Means. In this graph, the $x$-axis represents the number of clusters tested, while the $y$-axis shows the WCSS. The WCSS is a measure of variance within each cluster; lower values indicate that the data points are closer to their respective cluster centroids.

In Figure 7.19, we can notice that as the number of clusters increases, the WCSS initially decreases rapidly. This decrease slows down after reaching a certain number of clusters, beyond which adding more clusters does not significantly improve the compactness of the clusters. This point, where the rate of decrease changes and the plot starts to level off, is the *elbow*.

### *The Silhouette Score*

The silhouette score is used to measure the quality of clusters created by a clustering algorithm. It provides a succinct graphical representation of how well each object lies within its cluster, which is a critical aspect when you are working with unsupervised learning algorithms where the ground truth labels are not known.

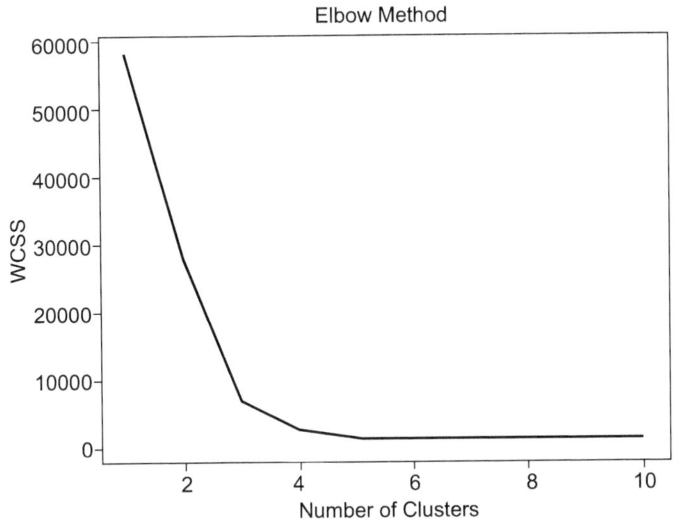


**Figure 7.19: Elbow method for optimal K-Means cluster count.**

When to Use the Silhouette Score:

- Assessing Cluster Cohesion and Separation: The silhouette score is useful for determining how close each point in one cluster is to points in the neighbouring clusters. This measure has both a cohesion and a separation component: It considers how similar the points within a cluster are (cohesion) and how distinct a cluster is from other clusters (separation).
- Comparing Different Clustering Algorithms: When you have applied different clustering algorithms to your dataset, or the same algorithm with different parameters (like the number of clusters), the silhouette score can be a valuable metric to compare the performance of these models. It helps in identifying which model or parameter setting does the best job in clustering the data effectively.
- Determining the Optimal Number of Clusters: In many clustering algorithms, you need to specify the number of clusters beforehand. The silhouette score can be used to determine the most appropriate number of clusters by running the algorithm with different cluster counts and choosing the one with the highest silhouette score.
- Visualizing Cluster Stability: Beyond just providing a numeric score, the silhouette analysis can also be visualized to understand the stability of the clusters formed. Each data point's silhouette score gives an indication of how well it fits within its assigned cluster, and a plot of these scores can highlight any potential issues with the clustering (like too many or too few clusters or overlapping clusters).

The Silhouette Score for each point is calculated using the following formula:

$$S(i) = \frac{b(i) - a(i)}{max(a(i), b(i))}$$

where,

- $a(i)$ represents the mean distance of the $i$th point from the other points within the same cluster, assessing the cluster's tightness (measures cohesion).
- $b(i)$ denotes the least mean distance of the $i$th point to points in another cluster, determined by finding the lowest among clusters, indicating how well-separated it is from other clusters (measures separation).

The silhouette score, which can vary between –1 and +1, serves as a measure of how well an object fits into its own cluster compared to neighbouring clusters. A higher score suggests a good fit within its own cluster and a poor fit with adjacent clusters. A clustering setup is considered suitable if most objects score highly. Conversely, a prevalence of low or negative scores may indicate an inappropriate number of clusters.

Python Example:

```
# Silhouette Score
from sklearn.metrics import silhouette_score

# Trying different numbers of clusters
range_n_clusters = [2, 3, 4, 5, 6, 7, 8, 9, 10]
silhouette_avg_scores = []

for n_clusters in range_n_clusters:
    # Apply K-Means
    kmeans = KMeans(n_clusters=n_clusters, random_state=42)
    cluster_labels = kmeans.fit_predict(X)

    # Calculate silhouette score
    silhouette_avg = silhouette_score(X, cluster_labels)
    silhouette_avg_scores.append(silhouette_avg)
    print("For n_clusters =", n_clusters, "the average silhouette_
          score is :", silhouette_avg)

# Plotting the silhouette scores
plt.plot(range_n_clusters, silhouette_avg_scores)
plt.title('Silhouette Scores for Various Numbers of Clusters')
plt.xlabel('Number of Clusters')
plt.ylabel('Silhouette Score')
plt.show()

# Output
For n_clusters = 2 the average silhouette_score is : 0.6104719594142752
For n_clusters = 3 the average silhouette_score is : 0.7281412339755425
For n_clusters = 4 the average silhouette_score is : 0.7749732690936251
For n_clusters = 5 the average silhouette_score is : 0.7286939542403489
For n_clusters = 6 the average silhouette_score is : 0.6616726780183116
For n_clusters = 7 the average silhouette_score is : 0.6420350016075017
For n_clusters = 8 the average silhouette_score is : 0.6438986259781332
For n_clusters = 9 the average silhouette_score is : 0.5477904966277551
For n_clusters = 10 the average silhouette_score is : 0.45689258731284543
```

Code Explanation:

- The K-Means clustering algorithm is applied to this dataset with a varying number of clusters (n_clusters).
- After clustering, the silhouette score for each number of clusters is calculated using silhouette_score from sklearn.metrics.
- Finally, we plot the silhouette scores against the number of clusters to visually determine the best number of clusters.

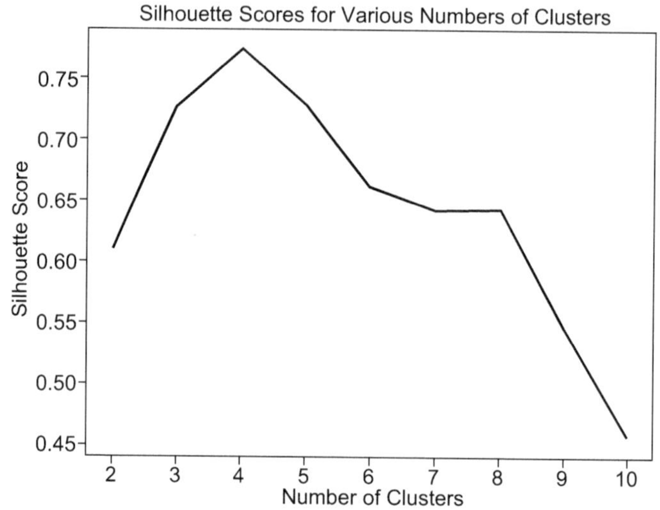


**Figure 7.20: Silhouette scores for K-Means clustering with different numbers of clusters.**

Figure 7.20 shows silhouette scores for different numbers of clusters, ranging from 2 to 10. The key observations are:

- The silhouette score is highest when the number of clusters is 4, with an average score of approximately 0.775. This indicates a very good structure, as the score is close to 1, suggesting that the clusters are well separated and distinct.
- As the number of clusters increases beyond 4, the silhouette score starts to decline. This suggests that adding more clusters does not contribute to better-defined or more distinct clusters. Particularly, there is a notable decrease in the score when moving from 4 to 7 clusters and beyond.
- With 2 or 3 clusters, the scores are lower than with 4 clusters, indicating that the data points are not as appropriately grouped into distinct clusters as they are with 4 clusters.
- For higher numbers of clusters (from 8 to 10), the silhouette scores decrease further, suggesting that such a high number of clusters may lead to overfitting and clusters that are not very meaningful.

Note to the Reader: Impact of Initial Centroid Selection on Model Accuracy in K-Means Clustering.

When applying the K-Means clustering algorithm, it is important to note that the selection of initial centroids can significantly influence the model's accuracy. The initial placement of these centroids essentially sets the starting condition for the iterative process of K-Means, which seeks to optimize the positions of centroids to minimize within-cluster variances. Different initialization methods can lead to different clustering outcomes, and in some cases, affect the convergence speed of the algorithm.

Initialization Methods in Scikit-learn's K-Means:

Scikit-learn offers several options for initializing centroids in K-Means, each with its own advantages:

- K-Means++ (Default): This method improves upon simple random initialization by spacing out the initial centroids. It does so by selecting centroids that are likely to be distant from each other, reducing the chances of poor cluster formation and speeding up convergence. The process involves multiple sampling steps where the most suitable centroid is chosen from several candidates.
- Random: In this straightforward approach, a set number of data points are randomly chosen from the dataset to serve as the initial centroids. While simple, this method can sometimes lead to less optimal clustering, particularly if the randomly chosen centroids are not well distributed.
- Custom Array: If you have prior knowledge or a specific strategy, you can directly pass an array of shape (n_clusters, n_features) that specifies the exact initial positions for the centroids.
- Custom Function: For even greater control, a callable function can be provided. This function should accept the data (X), the number of clusters (n_clusters), and a random state as arguments and return the initial centroid coordinates. This approach allows for a custom initialization strategy tailored to specific data characteristics or requirements.

Selecting the right initialization method depends on the nature of your data and the specific requirements of your clustering task. While the K-Means++ method is generally a good starting point due to its balanced approach, exploring other methods can be beneficial in certain scenarios, such as when dealing with datasets that have unusual distributions or when prior knowledge about the data can be leveraged for more informed centroid placement.

### *Density-Based Spatial Clustering of Applications with Noise (DBSCAN)*

DBSCAN is another popular clustering algorithm that is mainly known for its ability to discover clusters of arbitrary shapes and sizes. Unlike K-Means, DBSCAN eliminates the need for pre-specifying the number of clusters.

How DBSCAN Works:

- DBSCAN clusters points that are densely grouped together, identifying those that are isolated in areas of low density as outliers. This characteristic makes it suitable for data with noise or irregularly shaped clusters.

- Core, Border, and Noise Points: DBSCAN categorizes points into core points, border points, and noise. A core point has a minimum number of points (minPts) within a given radius (ε), a border point is within the radius of a core point but with fewer neighbors than minPts, and a noise point is neither a core nor a border.
- Forming Clusters: Clusters are formed by connecting core points that are within a distance ε of each other and including any border points that are within this radius of core points.

Parameters:

The DBSCAN python algorithm has two primary parameters:

- eps (ε): This is the maximum distance permitted for one sample to be considered within the neighborhood of another.
- minPts: This is the number of points necessary to establish a dense region (minimum points to form a cluster).

Python Example:

We will use DBSCAN on a synthetic dataset to illustrate its clustering capabilities, especially in identifying noise and handling arbitrary cluster shapes.

```
# DBSCAN

import matplotlib.pyplot as plt
from sklearn.datasets import make_blobs
from sklearn.cluster import DBSCAN

# Generate synthetic data with non-spherical shapes
X, _ = make_moons(n_samples=1000, noise=0.05, random_state=42)

# Apply DBSCAN
dbscan = DBSCAN(eps=0.2, min_samples=5)
dbscan.fit(X)
labels = dbscan.labels_

# Create subplots
fig, axs = plt.subplots(1, 2, figsize=(12, 5))
# Plot original dataset
axs[0].scatter(X[:, 0], X[:, 1], s=20, alpha=.3)
axs[0].set_title('Original Dataset')
axs[0].set_xlabel('Feature 1')
axs[0].set_ylabel('Feature 2')
# Plot DBSCAN clustering result
axs[1].scatter(X[:, 0], X[:, 1], c=labels, s=20, alpha=.5)
```

```
axs[1].set_title('DBSCAN Clustering Result')
axs[1].set_xlabel('Feature 1')
axs[1].set_ylabel('Feature 2')
# Layout adjustments
plt.tight_layout()

# Save the figure
plt.savefig('dbscan_clustering_results.png', dpi=300)
plt.show()
```

In this example, make_moons is used to create a dataset with two half circles. DBSCAN is then applied to this dataset. The algorithm is capable of identifying the two moon-shaped clusters while marking any outliers as noise (refer to Figure 7.21).

Applying DBSCAN:

- DBSCAN is imported from Scikit-learn and applied to the dataset with eps = 0.2 and min_samples = 5.
- dbscan.fit(X) fits the DBSCAN model to the data X.
- labels contains the cluster labels for each data point. Noise points are labeled as −1.

The world of supervised and unsupervised algorithms extends beyond what we have covered. We encourage readers who are interested in expanding their knowledge to explore further resources and literature. The field of machine learning is rapidly evolving, and staying updated with the latest developments and research can provide deeper insights and more sophisticated tools for data analysis.

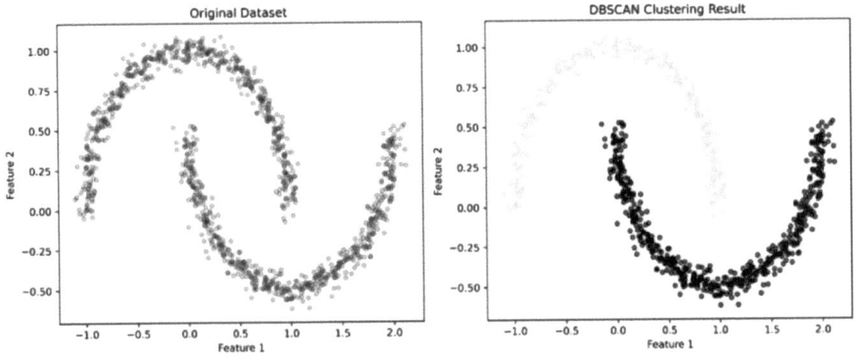


**Figure 7.21: Comparative visualization of original dataset and DBSCAN clustering.**

## Machine Learning Applications in Farm Animal Activity Recognition

In this section, we introduce some studies that have utilized ML for the detection of animal activities, showcasing approaches and findings in this research area. Barwick et al., 2018 investigated the capability of tri-axial accelerometers, mounted on the collar, leg, and ear, to discriminate between sound and lame gait movements in sheep. Through the application of Quadratic Discriminant Analysis on data segmented into 10-second behaviour epochs, Barwick's study pinpointed challenges in accurately classifying lame grazing events and underscored the performance of ear-mounted accelerometers. The final classification model achieved notable prediction accuracies, especially with the leg accelerometer, showing the complexities of accurate gait classification and the importance of sensor placement in monitoring animal health.

Continuing the exploration of lameness detection, Kaler et al. embarked on a study focusing on the automated detection of lameness in sheep (Kaler et al., 2020). By employing accelerometers and gyroscopes attached to ear sensors, they collected data from 23 datasets encompassing both lame and non-lame sheep. Their methodology involved developing algorithms such as random forest, neural networks, support vector machines, AdaBoost and k-nearest neighbors, capable of distinguishing lameness across three activities: walking, standing, and lying. With an accuracy rate exceeding 80% in activity-specific classifications, their study demonstrated that features extracted from accelerometer and gyroscope signals could effectively differentiate between lame and non-lame sheep. The random forest algorithm emerged as the most effective for lameness classification, highlighting the potential for developing automated systems for early lameness detection.

Further extending the application of machine learning in animal behavior recognition, Tran et al. presented an Internet of Things (IoT)-based design for cow behavior recognition (Tran et al., 2021). This study leveraged the synergy between leg-mounted and collar-mounted accelerometers to enrich the dataset for machine learning analysis. By synchronizing data from both sensors, their system was able to distinguish between similar behaviors such as feeding and standing. Utilizing the Random Forest algorithm, the authors claimed that good performance can be achieved in identifying four key cow behaviors: walking, feeding, lying, and standing.

Random forests was also evaluated by Kleanthous et al. (Kleanthous et al., 2019). The authors combined accelerometer and gyroscope features enabling the random forest algorithm to achieve an accuracy of 96.43% and a kappa value of 95.02%. The study also found that relying solely on accelerometer data slightly reduced accuracy and kappa value by 0.40% and 0.56%, respectively, suggesting that gyroscopes, while beneficial, might not be essential for high levels of accuracy.

To classify the behaviors of walking, grazing, and resting in cattle, data from the Global Positioning System (GPS) was analyzed using linear discriminant analysis, which identified 71% of the behaviors (Schlecht et al., 2004). Similarly, another research effort implemented collars equipped with both accelerometer and GPS sensors on cattle to differentiate behaviors such as foraging, ruminating, resting, and other active states. By applying a method based on threshold decision trees, this study claimed to achieve an accurate classification of 90.5% of the collected data points (González et al., 2015). Furthermore, classification tree analysis and K-Means clustering were employed on GPS data for the behavioral categorization of cattle (Ungar et al., 2005; Schwager et al., 2007).

In the domain of equine monitoring, activities of horses were examined using sensors for acceleration, gyroscope, and magnetometry, with data processing conducted through an embedded multilayer perceptron algorithm (Gutierrez-Galan et al., 2018). This approach resulted in an 81% accuracy rate in recognizing horse behaviors in real-world conditions. Additionally, accelerometer data was utilized in conjunction with threshold-based statistical methods to monitor standing and feeding behaviors in cows (Arcidiacono et al., 2017).

Another research used a Boruta feature selection technique, in conjunction with several machine learning algorithms, including multilayer perceptron, random forests, extreme gradient boosting, and K-nearest Neighbors. Among these, the random forests algorithm stood out, delivering results with an accuracy of 96.47% and a kappa value of 95.41% (Kleanthous et al., 2018).

The recognition of animal activities holds considerable significance for the agricultural community, animal behaviorists, and conservationists, as it serves as a crucial indicator of an animal's health and nutritional intake, especially when observations are made throughout their daily cycles. Leveraging machine learning techniques offers a sophisticated means to discern the activities of livestock, facilitating the differentiation of complex behavioral patterns that are challenging and laborious to identify through human observation alone.

Together, these studies underscore the transformative potential of machine learning and IoT technologies in advancing the field of animal behavior monitoring. By leveraging data science approaches, researchers are enhancing our understanding of animal welfare as well as paving the way for more sustainable and efficient agricultural practices.

## Summary

In this chapter, we took a look at supervised and unsupervised learning algorithms and their applications within animal research. We started with a foundational overview of supervised machine learning models, highlighting the significance of various algorithms. We then presented Python examples tailored for both classification and regression tasks. Moving into the domain of unsupervised learning, we discussed Unsupervised Competitive Learning and Self-organizing Feature Maps, further enriching the reader's understanding with Python examples that illustrate the utilization of some commonly used clustering techniques. The chapter concludes with a focus on the application of these machine learning techniques in monitoring farm animal activities, incorporating insights from relevant literature to underscore the practical implications of these methodologies in animal research.

CHAPTER 8

# Evaluation, Model Selection and Hyperparameter Tuning

In ML, the construction of a model represents an important phase of the overall analytical process. Post-development, the model must undergo thorough evaluation to determine its performance. This stage is key, as it determines the model's ability to generalize and accurately interpret new instances.

To evaluate the performance of the model effectively, we use various metrics. While we introduced the concept of evaluation metrics in Chapter 3, within the context of our practical, step-by-step machine learning project in Python, our discussion there was not complete. In Chapter 3, we focused on applying these metrics to estimate model performance preliminarily, without looking into the theoretical foundations or the span of metrics available. In this chapter, we aim to fill that gap by providing a comprehensive overview of evaluation metrics.

Evaluation metrics differ based on the type of machine learning task at hand, classification or regression. For classification tasks, metrics such as accuracy, precision, recall, F1-score, and the confusion matrix provide insights into the model's ability to correctly classify the different behaviors of animals. In regression tasks, metrics like Mean Squared Error (MSE), Root Mean Squared Error (RMSE), $R^2$, $R^2$ Adjusted, and Mean Absolute Error (MAE) help in assessing the model's accuracy in predicting continuous outcomes.

However, evaluating a model is not the end goal. The subsequent steps involve model selection and hyperparameter tuning. Once the model is selected, then the next step involves adjusting the model parameters that are not directly learned from the data. Techniques such as grid search, and random search, are employed to find the optimal set of hyperparameters that enhance the model's ability to make accurate predictions.

Throughout this chapter, we will explore these aspects, using Python as our tool. The application of these procedures in animal behavior research is important, as it ensures that the models we develop are accurate, reliable and robust.

## Evaluation Metrics for Classification

As we said earlier, Model evaluation is an integral part of the machine learning workflow. It provides a measure of how well our model will likely perform on unseen data. Several metrics are available to evaluate models, with the choice of metric depending on the task at hand.

The dataset that we will use for the classification evaluation metrics comprises of accelerometer data, aimed at identifying various behaviors in sheep. This dataset represents a comprehensive record of animal movements, captured and transformed into a format suitable for machine learning analysis. The Python code related to model evaluation is available for access and use in our dedicated GitHub repository in Chapter_8 folder.

### *Confusion Matrix*

The Confusion Matrix is a fundamental tool in evaluating the performance of classification models. It is especially valuable when dealing with multiple classes, but its utility is also evident in binary classification scenarios. Essentially, a Confusion Matrix is a tabular representation that illustrates the accuracy of a model by comparing the actual and predicted classifications.

#### Structure of the Confusion Matrix

In a binary classification ($N = 2$), the matrix is a $2 \times 2$ table. However, this structure expands to a $N \times N$ matrix for problems with more than two classes ($N > 2$), where N represents the number of distinct classes.

The matrix comprises four distinct elements:

- **True Positives (TP):** Instances correctly predicted as positive.
- **True Negatives (TN):** Instances correctly predicted as negative.
- **False Positives (FP)**, also known as **Type I Error:** Instances incorrectly predicted as positive.
- **False Negatives (FN)**, also known as **Type II Error:** Instances incorrectly predicted as negative.

While the binary confusion matrix is straightforward, the multiclass scenario involves a larger matrix, where each row and column correspond to a specific class. The principles remain the same, but the matrix provides a detailed view of the model's performance across multiple classes.

#### Python Example: Generating a Confusion Matrix

To generate and visualize a confusion matrix in Python, you can use the following code:

```
# Confusion matrix
from sklearn.metrics import confusion_matrix
import seaborn as sns
import matplotlib.pyplot as plt

# generating the confusion matrix
cm = confusion_matrix(y_test, y_pred)

# Plotting the confusion matrix with class names
plt.figure(figsize=(10, 7))
sns.heatmap(cm, annot=True, fmt='d', cmap='Blues',
            xticklabels=label_encoder.classes_,
            yticklabels=label_encoder.classes_)
plt.xlabel('Predicted')
plt.ylabel('Actual')
plt.title('Confusion Matrix')
plt.savefig('cm.png', dpi=300)
plt.show()
```

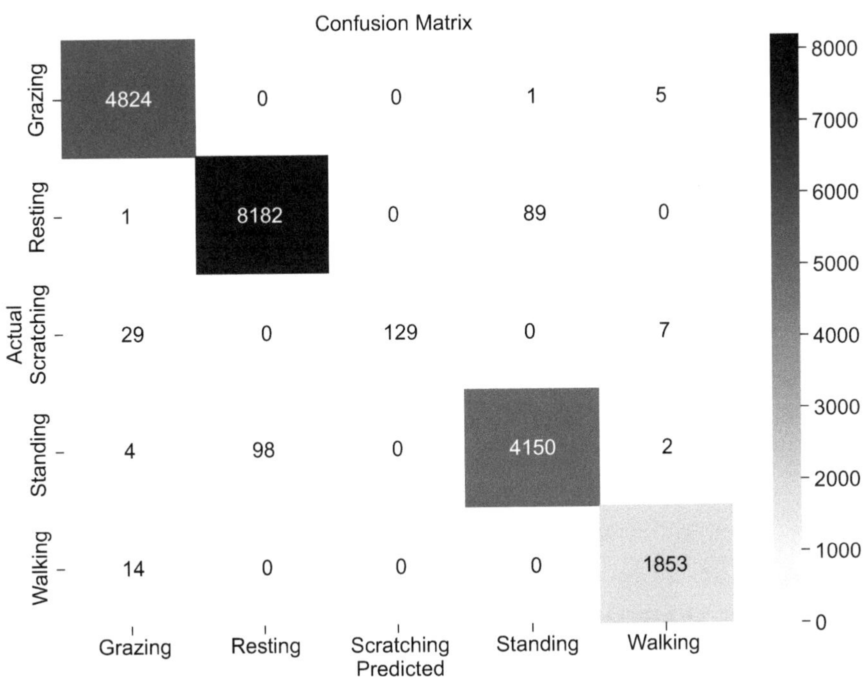


**Figure 8.1: Confusion matrix visualization for multiclass sheep behavior classification.**

This code uses seaborn and matplotlib for visualization, providing a heatmap representation of the confusion matrix. The confusion matrix is generated by typing confusion_matrix(y_test, y_pred). This matrix offers a visual and quantitative insight into the performance of our classification model, making it easier to identify the areas where the model excels or needs improvement. Figure 8.1 shows the generated confusion matrix.

In the generated confusion matrix, each cell represents the count of predictions for the actual labels (rows) versus the predicted labels (columns).

Here is how to interpret the TP, TN, FP, FN values from this matrix:

- TP: Diagonal cells where the predicted label matches the actual label (e.g., 'grazing' predicted as 'grazing').
- TN: For a specific class, these are all the cells that correctly predict the negative class. In a multiclass matrix, this would be calculated for each class by considering all the correct predictions that are not the current class.
- FP: Off-diagonal cells in a row where the predicted label is the class in question and the actual label is not (e.g., actual 'grazing' predicted as 'walking').
- FN: Off-diagonal cells in a column where the actual label is the class in question and the predicted label is not (e.g., actual 'resting' predicted as 'grazing').

Please note that for multiclass confusion matrices, the concept of TN is less straightforward than in binary classifications because it involves all the true negatives for each class across the matrix. The visual representation in the heatmap above makes it easier to identify the TP and FN directly. However, calculating the TN and FP for each class involves a bit more consideration of the other classes.

- Main Diagonal (True Predictions):
  - 'Grazing': 4824 instances were correctly predicted as grazing.
  - 'Resting': 8182 instances were correctly predicted as resting.
  - 'Scratching': 129 instances were correctly predicted as scratching.
  - 'Standing': 4150 instances were correctly predicted as standing.
  - 'Walking': 1853 instances were correctly predicted as walking.
- Off-Diagonal (Incorrect Predictions):
  - 'Grazing' was once incorrectly predicted as 'walking' and 5 times as 'standing'.
  - 'Resting' was once incorrectly predicted as 'grazing' and 89 times as 'standing'.
  - 'Scratching' had 29 instances incorrectly predicted as 'grazing' and 7 as 'standing'.

  - 'Standing' had 4 instances incorrectly predicted as 'grazing' and 2 as 'walking'.
  - 'Walking' had 14 instances incorrectly predicted as 'grazing'.
- Type I Errors (False Positives):
  - For instance, 'walking' as predicted when actually 'grazing' (14 times).
- Type II Errors (False Negatives):
  - For example, 'grazing' was the actual behavior but 'standing' or 'walking' was predicted (6 times in total).
- Class Imbalance:
  - Some behaviors like 'scratching' have fewer instances, which might indicate a class imbalance in the dataset.
- Performance Insights:
  - The model is good at predicting 'grazing', 'resting', and 'standing', as indicated by the high numbers on the diagonal.
  - The model struggles with 'scratching', which may be due to fewer training samples or similarities with other behaviors that make it harder to distinguish.
  - 'Walking' has a substantial number of false negatives (instances where the model predicted a different behavior), which could suggest that the features describing 'walking' are not distinct enough for the model or that there's some confusion with other behaviors.

This confusion matrix is a powerful tool for understanding the overall accuracy of the model, and its specific strengths and weaknesses in predicting each class. It provides a detailed breakdown of predictions and can help in identifying where the model needs improvement.

### *Accuracy*

This is one of the most straightforward metrics. It calculates the proportion of correct predictions in the overall predictions. In essence, it measures the overall correctness of the model by answering the question: "*Of all the predictions made, how many were correct?*"

$$Accuracy = \frac{TP+TN}{TP+TN+FP+FN}.$$

Accuracy is particularly useful when the class distribution is even, meaning each class has a roughly equal number of instances. However, its usefulness weakens when dealing with imbalanced datasets where one class significantly outnumbers the other(s), as it can give a misleading impression of the model's performance.

### Limitations of Accuracy

While accuracy is an excellent measure for giving a quick overview of model performance, it has its limitations:

- Imbalanced Classes: In datasets where some classes are underrepresented, accuracy can be skewed, as the model may bias towards the majority class.
- Misleading Interpretation: A model with high accuracy might still be performing poorly in one or more classes, which is why it is essential to look at other metrics like precision and recall for a more comprehensive evaluation.

### Python Example: Calculating Accuracy

Let's calculate accuracy using Python with the help of scikit-learn:

```
# Accuracy

from sklearn.metrics import accuracy_score

# Assuming y_test and y_pred are available
# y_test: actual truth labels
# y_pred: predicted labels by the classification model

accuracy = accuracy_score(y_test, y_pred)
print(f"Accuracy: {accuracy:.2f}")

# Output
Accuracy: 0.987
```

In this example, accuracy_score is a function that computes the accuracy metric, comparing the true and predicted labels. The y_test array holds the actual labels, while y_pred contains the predictions made by the classifier. The printed result is the accuracy of the model formatted to two decimal places.

## *Precision*

Precision, also referred to as the positive predictive value, is a metric that estimates the accuracy of the positive predictions made by a classification model. It answers the question: "*Of all the instances classified as positive, how many are actually positive*"? Precision is a critical measure when the costs of False Positives are high.

The formula to calculate precision for a binary or multiclass classification task is:

$$Recall = \frac{TP}{TP + FN}.$$

### Precision in Multiclass Classification

In a multiclass setting, precision must be calculated for each class separately and then averaged, which can be done in various ways:

- **Macro-average Precision:** Compute the precision for each class and then take the average.
- **Weighted-average Precision:** Calculate the precision for each class, but when taking the average, give each class's precision a weight equal to the class's presence in the dataset.

### Limitations of Precision

Precision alone does not tell the complete story of a model's performance. It does not take into account the false negatives, or the positive instances that the model incorrectly classified as negative. Therefore, it is often used in conjunction with recall (also known as sensitivity) to provide a more comprehensive evaluation of a classifier's performance.

### Python Example: Calculating Precision

We calculate precision in Python with scikit-learn's precision_score function:

```
# Precision
from sklearn.metrics import precision_score

# Assuming y_test and y_pred are available
# y_test: actual truth labels
# y_pred: predicted labels by the classification model

# For binary classification
#precision_binary = precision_score(y_test, y_pred,
   pos_label='positive_class_name')
#print(f"Binary Classification Precision: {precision_binary:.2f}")

# For multi-class classification with macro-average
precision_macro = precision_score(y_test, y_pred, average='macro')
print(f"Multi-Class Precision (Macro-average): {precision_
macro:.3f}")

# For multi-class classification with weighted-average
precision_weighted = precision_score(y_test, y_pred,
average='weighted')
print(f"Multi-Class Precision (Weighted-average): {precision_
   weighted:.3f}")

# Output
Multiclass Precision (Macro-average): 0.990
Multiclass Precision (Weighted-average): 0.987
```

In the above example, replace 'positive_class_name' with the actual name or label of the positive class in your dataset. The average parameter defines the averaging method used for multiclass classification. If not specified, the default is binary classification, which requires the pos_label parameter to be set for imbalanced or non-binary classification tasks. The printed results show the precision of the model, formatted to three decimal places.

## *Recall*

Recall, also known as sensitivity, addresses the question: "*Of all the actual positives, how many were identified correctly*"? Recall is especially important in situations where missing a positive instance is costly, such as in animal medical diagnosis.

The formula for recall is:

$$Recall = \frac{TP}{TP+FN}.$$

### Recall in Multiclass Classification

For multiclass classification problems, recall can be computed for each class individually and can be summarized using:

- Macro-average Recall: Calculate the recall for each class independently and then take the average.
- Weighted-average Recall: Calculate the recall for each class and then take the average, weighting the recall of each class by the number of true instances for that class in the dataset.

### Limitations of Recall

While recall is an essential metric, it does not consider False Positives. A model with a high recall rate might also have a high number of False Positives, which would not be ideal in every situation. As highlighted before, recall is often used in conjunction with precision and other metrics to get a balanced view of the model's performance.

### Python Example: Calculating Recall

Here is how you can calculate recall using Python with the recall_score function:

```
# Recall
from sklearn.metrics import recall_score

# Assuming y_test and y_pred are available
# y_test: actual truth labels
# y_pred: predicted labels by the classification model
```

```
# For binary classification
#recall_binary = recall_score(y_test, y_pred, pos_label='positive_
    class_name')
#print(f"Binary Classification Recall: {recall_binary:.2f}")

# For multiclass classification with macro-average
recall_macro = recall_score(y_test, y_pred, average='macro')
print(f"Multiclass Recall (Macro-average): {recall_macro:.3f}")

# For multiclass classification with weighted-average
recall_weighted = recall_score(y_test, y_pred, average='weighted')
print(f"Multiclass Recall (Weighted-average): {recall_
weighted:.3f}")

Multiclass Recall (Macro-average): 0.948
Multiclass Recall (Weighted-average): 0.987
```

The average parameter is used to specify how the recall should be calculated in a multiclass scenario. The printed results show the recall metric for the model, formatted to three decimal places.

## *F1-score*

The F1-score is a metric that combines precision and recall into a single value, providing a balance between the two. It is particularly useful when you need to find a balance between precision and recall, especially in scenarios with an uneven class distribution or when the cost of false positives and false negatives differs.

The F1-score is the harmonic mean of precision and recall, giving both metrics equal weight.

The formula for the F1-score is:

$$\text{F1-}score = 2 \times \frac{Precision \times Recall}{Precision + Recall}.$$

This formula ensures that the F1-score takes both false positives and false negatives into account. Consequently, a high F1-score indicates that the model has a robust balance of precision and recall.

### F1-score in Multiclass Classification

Just like precision and recall, the F1-score can be calculated for each class in a multiclass classification problem. The overall F1-score can be computed using:

- Macro-average F1-score: Calculating the F1-score independently for each class and then taking the average, giving equal weights to each class.

- Weighted-average F1-Score: Computing the F1-score for each class and then taking the average, giving weights to each class according to their presence in the dataset.

#### Limitations of the F1-score

The F1-score is a valuable metric when you are seeking a balance between precision and recall, especially if there is a class imbalance. However, by averaging the two, it may mask the model's weaknesses in either precision or recall. Sometimes a model may have excellent recall but poor precision, or vice versa, and the F1-score might suggest a performance that is not fully representative of the underlying issues.

#### Python Example: Calculating the F1-score

We calculate the F1-score using Python with scikit-learn's f1_score function:

```
# F1-score
from sklearn.metrics import f1_score

# For binary classification
# f1_binary = f1_score(y_test, y_pred, pos_label='positive_class_
name')
# print(f"Binary Classification F1-Score: {f1_binary:.2f}")

# For multiclass classification with macro-average
f1_macro = f1_score(y_test, y_pred, average='macro')
print(f"Multiclass F1-Score (Macro-average): {f1_macro:.3f}")

# For multiclass classification with weighted-average
f1_weighted = f1_score(y_test, y_pred, average='weighted')
print(f"Multi-Class F1-Score (Weighted-average): {f1_weighted:.3f}")

# Output
Multiclass F1-Score (Macro-average): 0.966
Multiclass F1-Score (Weighted-average): 0.987
```

### *Classification Report for Evaluation Metrics*

After discussing key classification evaluation metrics such as accuracy, precision, recall, and F1-score, it is important to note that these metrics can be conveniently viewed together using the classification_report. This tool groups these metrics into a single report, offering a thorough overview of a model's performance across all classes.

The classification_report function generates a report that includes:

- Accuracy
- Precision
- Recall
- F1-score
- Support: Indicates how many occurrences of each class exist in the dataset. It is useful for identifying class imbalances.

Each of these metrics is presented for individual classes and as weighted averages, making the report particularly useful for multiclass classification problems.

**Python Code**

```
# Classification report
from sklearn.metrics import classification_report

# Generate and print the classification report
report = classification_report(y_test, y_pred)
print(report)

# Output
              precision    recall  f1-score   support

     grazing       0.99      1.00      0.99      4830
     resting       0.99      0.99      0.99      8272
  scratching       1.00      0.78      0.88       165
    standing       0.98      0.98      0.98      4254
     walking       0.99      0.99      0.99      1867

    accuracy                           0.99     19388
   macro avg       0.99      0.95      0.97     19388
weighted avg       0.99      0.99      0.99     19388
```

The classification_report function is used to generate a summary of key metrics, which is printed out.

### *Area Under Receiver Operating Characteristic Curve (AUC-ROC)*

AUC-ROC is a performance measure for classification problems at various threshold settings. The ROC curve visually represents how well a binary classification system performs by showing its diagnostic capability as the discrimination threshold changes.

The ROC curve is created by plotting the TPR against the FPR at several settings of the threshold. The TPR is plotted on the $y$-axis, while the FPR is plotted on the $x$-axis.

The formulas for these are:

$$TPR = \frac{TP + FN}{TP}$$

$$FPR = \frac{FP + TN}{FP}.$$

The AUC-ROC metric indicates the model's proficiency in differentiating between classes. A higher AUC signifies a superior model, while an AUC of 0.5 implies an absence of discriminatory power, which is equivalent to random guessing.

The ROC AUC is more straightforward and traditionally used for binary classification problems. In binary classification, the ROC AUC provides a clear and interpretable measure of a model's ability to distinguish between the two classes.

But, the answer to the question, 'why it is more suitable for binary classification' is provided below:

- Clear Interpretation: In binary classification, the ROC curve plots the TPR against the FPR at various threshold settings. This gives a clear and intuitive understanding of the trade-off between correctly identifying positive cases and incorrectly identifying negative cases as positive.
- Threshold-Independent: The ROC AUC provides a summary of model performance across all possible classification thresholds, making it a robust measure that is not dependent on a particular threshold value.
- Balanced View of Performance: It considers both classes (positive and negative) equally, which is particularly useful when the classes are balanced or when both types of classification errors (false positives and false negatives) are equally important.

### AUC-ROC in Multiclass Classification

For multiclass classification, the AUC-ROC metric is calculated for each class against all others combined, and then the average AUC is taken. There are two common approaches to this:

- One-vs-Rest (OvR): Compute the AUC for each class by treating it as the positive class and the rest of the classes as the negative class, and then average the AUCs.
- One-vs-One (OvO): Compute the AUC for each pair of classes, then average these AUCs.

### Limitations of AUC-ROC

While AUC-ROC is a powerful metric, it has its limitations:

- Class Imbalance: AUC-ROC may be overly optimistic with imbalanced datasets, as it averages over all possible thresholds.
- Not Sensitive to Threshold Selection: The ROC curve does not reflect the impact of threshold selection on the number of false positives.

### Python Example: Calculating AUC-ROC (Binary Classification)

In the following example, we will demonstrate how to calculate AUC scores and plot the ROC curve using binary classification, which is more suitable for this analysis as previously mentioned. We will use a dataset named binary_data.csv in Folder_8, which consists of two classes represented by the labels *active* and *inactive*. These classes will be treated as our binary outcomes, with *inactive* designated as the positive class for the purpose of ROC analysis.

```
# AUC-ROC

from sklearn.metrics import roc_curve
from sklearn.metrics import roc_auc_score

# We load the binary_data.csv
df = pd.read_csv("binary_data.csv")

# Split the dataset into features and target variable
X_b = df.drop('label', axis=1)
y_b = df['label']

# We split into train and test sets
X_train_b, X_test_b, y_train_b, y_test_b = train_test_split(X_b, y_b,
                                                            test_size=0.7,
                                                            stratify=y,
                                                            random_state=2)

# Fit a Random Forest model
rf_model = RandomForestClassifier(random_state=42)
rf_model.fit(X_train_b, y_train_b)

# Predict probabilities
rf_probs = rf_model.predict_proba(X_test_b)

# Retain probabilities for the class of interest (positive class
    'inactive')
rf_probs = rf_probs[:, 1]
```

```
# Calculate AUC scores for a classifier with no predictive power
    (baseline) and the trained Random Forest
baseline_auc = roc_auc_score(y_test_b, [0 for _ in range(len(y_
    test_b))])
rf_auc = roc_auc_score(y_test_b, rf_probs)

# Display the AUC scores for the baseline and Random Forest
    classifiers
print('Baseline: ROC AUC=%.4f' % (baseline_auc))
print('Random Forest: ROC AUC=%.4f' % (rf_auc))

# Output
Baseline: ROC AUC=0.5000
Random Forest: ROC AUC=1.0000
```

Explanation of the Code:

- We load and split our dataset into features (X_b) and the target variable (y_b). The features will be used to train our model, and the target variable is what we aim to predict. Then we divide our data into training and testing sets to evaluate the model's performance on unseen data. The stratified parameter ensures that both sets have a similar distribution of classes.
- We train Random Forest and predict the probabilities of the class ('inactive') for the test set.
- AUC scores for both the *baseline* prediction and the Random Forest model are calculated to compare their performances.
- Results Interpretation: The printed results show that the *baseline* classifier has an AUC of 0.5000, which is what we expect from random guessing. The Random Forest model has an AUC of 1.0000, indicating that it was able to discriminate between 'active' and 'inactive' classes with 100% accuracy.

Now we calculate and plot the ROC curves:

```
# Calculate ROC curves for the baseline and Random Forest
classifiers
baseline_fpr, baseline_fpr_tpr, _ = roc_curve(y_test_b, [0 for _ in
                                    range(len(y_test_b))], pos_
label='inactive')
rf_fpr, rf_tpr, _ = roc_curve(y_test_b, rf_probs, pos_
label='inactive')

# Plot the ROC curve for the model
```

```
plt.figure(figsize=(8, 6))
plt.plot(baseline_fpr, baseline_fpr_tpr, linestyle='--',
label='Baseline',
    color='gray')
plt.plot(rf_fpr, rf_tpr, marker='.', lw = 2, label='Random Forest',
    color='purple')
plt.xlabel('False Positive Rate')
plt.ylabel('True Positive Rate')
plt.title('Receiver Operating Characteristic (ROC) - Binary
Classification')
plt.legend()
plt.show()
```

- The ROC curve is calculated for both the *baseline* classifier and the Random Forest model using the roc_curve function.
- Plot ROC Curves (Figure 8.2): The ROC curves are plotted with matplotlib.pyplot. The *baseline* classifier's curve is a dashed line, while the Random Forest's curve is marked with points. The *x*-axis represents the false positive rate, and the *y*-axis represents the true positive rate.

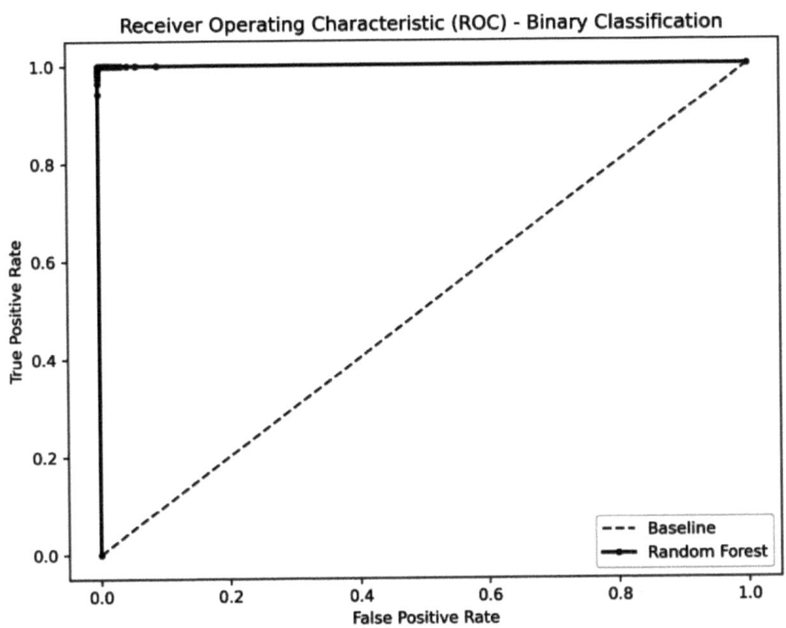


**Figure 8.2: ROC curve comparison between baseline classifier and Random Forest model.**

## *Log Loss*

Log Loss is another metric in classification models that measures the accuracy by considering the predicted probabilities. It focuses on the reliability of the predictions by penalizing false classifications based on the predicted probabilities.

**Definition**: Log Loss is the negative average of the logarithm of corrected predicted probabilities for each instance in the dataset. It emphasizes the probability estimates of the true class for each observation.

Formula:

$$Log\ Loss = -\frac{1}{N}\sum_{i=1}^{N}\left(y_i \cdot \log\left(p_i\right) + \left(1 - y_i\right) \cdot \log\left(1 - p_i\right)\right)$$

where, $N$ is the number of samples, $y_i$ is the actual label, and $p_i$ is the predicted probability for the $i$th sample.

### Python Example for Log Loss

```
# Log Loss
from sklearn.metrics import log_loss

# Assuming y_test contains the actual labels and y_pred_proba
contains the
  predicted probabilities
logloss_value = log_loss(y_test, y_pred_probam)
print(f"Log Loss: {logloss_value:.4f}")

# Output
Log Loss: 0.0668
```

Notably, a lower Log Loss value indicates a better-performing model in terms of classification accuracy and reliability. It is important to understand, however, that there is no universally *good* benchmark for Log Loss. Its interpretation is context-dependent, varying according to the specific application or use-case. In the context of our analysis, the obtained Log Loss stands at 0.0668, suggesting a high level of model accuracy, though its suitability should be evaluated relative to the specific requirements and expectations of the application at hand.

## *Kolmogorov-Smirnov (K-S)*

The Kolmogorov-Smirnov (K-S) is a valuable tool in assessing the predictive power of classification models, particularly in binary classification scenarios. The K-S statistic measures the degree of separation between the distributions of positive and negative cases. It evaluates how well the model distinguishes between these two classes.

Key Concepts of the K-S Chart:

- Degree of Separation: The K-S statistic is a measure of how distinctly the model separates the positive and negative cases. It calculates the maximum difference between the cumulative distribution functions of positive and negative cases.
- Range of K-S Values: The K-S statistic ranges from 0 to 100:
  - K-S = 100: This is an ideal scenario where the model perfectly separates positive and negative cases, with no overlap.
  - K-S = 0: This indicates that the model is no better than random guessing at separating the two classes.
  - Typical K-S Values: In practice, the K-S value for most models will fall between 0 and 100, with higher values indicating better model performance.
- Use in Model Evaluation: A higher K-S statistic suggests that the model has a stronger ability to differentiate between positive and negative cases, making it a useful metric for evaluating classification models.

## Python Example: Calculating the K-S Statistic

```
# Kolmogorov-Smirnov

import numpy as np

# Converting labels to binary format
y_binary = np.where(y_b == 'inactive', 1, 0)

# Split the dataset into training and testing sets
X_traink, X_testk, y_traink, y_testk = train_test_split(X_b, y_
binary, test_size=0.2, random_state=42)

# Fit a Random Forest model
rf_model = RandomForestClassifier(random_state=42)
rf_model.fit(X_traink, y_traink)

# Predict probabilities for the test set
y_pred_proba = rf_model.predict_proba(X_testk)[:, 1]

# Sorting the Predictions and calculating the Cumulative
    Distributions
sorted_indices = np.argsort(y_pred_proba)[::-1]
sorted_y_test = y_testk[sorted_indices]
sorted_y_pred_proba = y_pred_proba[sorted_indices]
```

```
# Calculating the cumulative sum of positives and negatives
cumulative_positives = np.cumsum(sorted_y_test)
cumulative_negatives = np.cumsum(1 - sorted_y_test)

# Normalizing the cumulative sums to get cumulative distributions
cumulative_positives = cumulative_positives / max(cumulative_
    positives)
cumulative_negatives = cumulative_negatives / max(cumulative_
negatives)

# Finding the maximum difference (K-S statistic)
ks_values = cumulative_positives - cumulative_negatives
max_ks_value = np.max(ks_values)
max_ks_index = np.argmax(ks_values)
max_ks_threshold = sorted_y_pred_proba[max_ks_index]

# Plotting the Cumulative Good and Bad
plt.figure(figsize=(10, 6))
plt.plot(sorted_y_pred_proba, cumulative_positives,
label='Cumulative Positive', color='blue')
plt.plot(sorted_y_pred_proba, cumulative_negatives,
label='Cumulative Negative', color='red')
plt.axvline(x=max_ks_threshold, color='green', linestyle='--',
label=f'Max KS at
    threshold {max_ks_threshold:.2f}')
plt.title('Cumulative Distribution of Positives and Negatives')
plt.xlabel('Predicted Probability')
plt.ylabel('Cumulative Distribution')
plt.legend()
plt.show()

print(f'Maximum K-S Value: {max_ks_value:.4f} at threshold
    {max_ks_threshold:.4f}')

# Output
Maximum K-S Value: 0.9996 at threshold 0.5800
```

The provided Python code executes a sequence of steps to compute the K-S statistic for a binary classification task using the RF algorithm.

- The code converts the string labels in our dataset to binary format, with 'inactive' as the positive class (represented by 1) and the other class as the negative class (represented by 0).
- Splits the dataset into training and test subsets.
- Trains RF classifier on the training subset.
- Predicts probabilities for the test subset.

- Calculates the ROC, which provides the TPR and FPR at various threshold levels.
- Calculates the K-S statistic as the maximum difference between the TPR and FPR, which represents the best threshold for distinguishing between the two classes.
- Finally, it plots the cumulative distributions of the classes to visualize the model's discriminatory power.

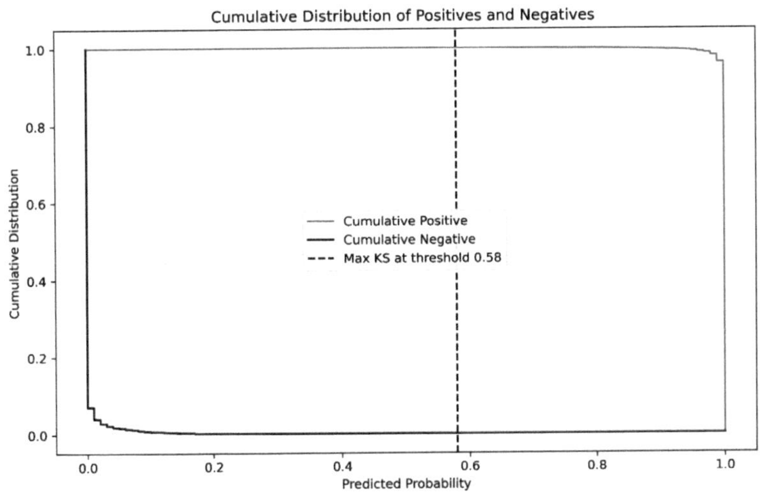


**Figure 8.3: Kolmogorov-Smirnov chart demonstrating model's ability to separate positive and negative classes.**

The Kolmogorov-Smirnov chart (Figure 8.3) reveals a near-perfect model performance with a maximum K-S value of 0.9996 at a threshold of 0.58, indicating a high degree of separation between the positive and negative classes.

With the discussion of Kolmogorov-Smirnov, we end the section on evaluation metrics for classification models. These metrics provide a comprehensive framework to assess the performance of classification algorithms. They offer valuable insights into aspects such as prediction accuracy, model reliability, probability calibration, and the equality of distribution in predicted outcomes.

## Evaluation Metrics for Regression

We now move to evaluation metrics for regression models. In this section, we will look into various metrics used to quantify the performance of regression models, understanding how these metrics can guide us in assessing the accuracy and efficacy of our predictive models in continuous outcome scenarios.

For this purpose, we will generate a dataset using the make_regression() function. This approach allows us to tailor the dataset's characteristics to effectively illustrate various regression evaluation metrics.

## *Generating the Synthetic Dataset*

A dataset is generated using the make_regression() function. The make_regression() function enables us to create a dataset for a regression problem with configurable parameters, such as the number of samples, number of features, noise level, and more.

### Python Code to Generate the Dataset

```
from sklearn.datasets import make_regression
import pandas as pd

# Configurable parameters for dataset generation
n_samples = 1000  # Number of samples
n_features = 2   # Number of features
noise_level = 4  # Noise in the dataset

# Generating the dataset
X, y = make_regression(n_samples=n_samples, n_features=n_features,
    noise=noise_level, random_state=42)

# Converting to a Pandas DataFrame for ease of use
df_X = pd.DataFrame(X, columns=[f'feature_{i}' for i in range(n_
    features)])
df_y = pd.DataFrame(y, columns=['target'])

# Combining features and target into a single DataFrame
synthetic_dataset = pd.concat([df_X, df_y], axis=1)

from sklearn.linear_model import LinearRegression
from sklearn.model_selection import train_test_split

# Split the dataset
X_train, X_test, y_train, y_test = train_test_split(df_X, df_y,
    test_size=0.2, random_state=42)

# Train a Linear Regression model
model = LinearRegression()
model.fit(X_train, y_train)

# Make predictions
y_pred = model.predict(X_test)
```

Key Points:

- n_samples and n_features determine the size and dimensionality of the dataset.
- noise_level adds a specified amount of noise to the output, simulating real-world data imperfections.
- Splitting the dataset using train_test_split().
- Applying linear regression to get the y_pred results.

This synthetic dataset provides a controlled environment to understand how different metrics respond to various regression outcomes and model behaviors. For our examples, we first apply the linear regression model and predict the results.

In regression analysis, *accuracy* is not defined the same way as it is for classification tasks. Unlike classification, where accuracy refers to the percentage of correctly predicted instances, regression tasks involve predicting continuous outcomes, making the concept of exact *accuracy* less straightforward. Instead, we use metrics like MAE, MSE, and RMSE to evaluate the model's performance.

## Mean Absolute Error (MAE)

MAE is a widely used metric that measures the mean magnitude of errors in a collection of predictive variables, ignoring their direction. It is calculated as the average of the absolute differences between the output values and the observed true values.

The formula for MAE is:

$$MAE = \frac{1}{N}\sum_{i=1}^{N} |y_i - \hat{y}_i|$$

where, $N$ is the number of samples, $y_i$ is the actual value of the sample, and $\hat{y}_i$ is the predicted value.

Interpretation:

- Scale-dependent: MAE is scale-dependent and should be used to compare models on the same dataset. Its value indicates how close the predictions are to the actual values, on average.
- Robust to Outliers: MAE is not sensitive to outliers. Large errors have a linearly proportional impact, making MAE a robust metric in the presence of outliers.

```
# Mean Absolute Error

from sklearn.metrics import mean_absolute_error

# Calculate MAE
mae = mean_absolute_error(y_test, y_pred)
print(f"Mean Absolute Error (MAE): {mae:.4f}")

# Output
Mean Absolute Error (MAE): 1.6051
```

The MAE gives a straightforward interpretation:

- MAE Value Interpretation: A lower MAE value indicates a better model performance, meaning the predictions are closer to the actual values. However, MAE is scale-dependent and should be interpreted in the context of the data's scale. A MAE of 0 means perfect predictions with no error, which is rare in real-world scenarios.
- Contextual Meaning: The value of MAE must be assessed relative to the scale of our target variable. For example, an MAE of 5 may be very small if the target variable ranges in the thousands but significant if the target ranges around 10.

## MSE and RMSE

### MSE

MSE is a metric in regression analysis that measures the average of the squares of the errors or deviations. Essentially, it quantifies the difference between the predicted values and the actual values.

Formula:

$$MSE = \frac{1}{N} \sum_{i=1}^{N} (y_i - \hat{y}_i)^2.$$

Interpretation:

- Sensitivity to Outliers: Unlike MAE, MSE is more sensitive to outliers because the errors are squared, thereby magnifying the effect of large errors.
- Scale: MSE is in squared units of the target variable, which can sometimes make interpretation less intuitive.

### RMSE

RMSE is the square root of the MSE. It is a measure of the average error and is widely used in regression analysis. Compared to MAE, RMSE tends to give higher weights to larger errors, punishing models with larger deviations more.

Formula:

$$RMSE = \sqrt{MSE} = \sqrt{\frac{1}{N}\sum_{i=1}^{N}(y_i - \hat{y}_i)^2}.$$

Key Points about RMSE:

- Emphasizing Large Deviations: The RMSE places a higher emphasis on larger errors by squaring the deviations before averaging them. This aspect ensures that larger errors are more prominent in the final metric.
- Robustness against Error Sign: The squaring of errors in RMSE ensures that both positive and negative deviations contribute positively to the overall error metric. This characteristic prevents the possibility of error cancellation that might occur in metrics considering absolute errors.
- Reliability with More Samples: With an increasing number of samples, the error distribution represented by RMSE becomes more reliable and indicative of the model's performance.
- Sensitivity to Outliers: RMSE is notably sensitive to outliers. Large deviations have a squared effect on RMSE, making it crucial to address outliers in your dataset before employing this metric.

### Python Example Using Synthetic Data to Calculate MSE and RMSE

```
# MSE and RMSE
from sklearn.metrics import mean_squared_error
import numpy as np

# Calculate MSE and RMSE
mse = mean_squared_error(y_test, y_pred)
rmse = np.sqrt(mse)

print(f"Mean Squared Error (MSE): {mse:.4f}")
print(f"Root Mean Squared Error (RMSE): {rmse:.4f}")

# Output
Mean Squared Error (MSE): 4.1272
Root Mean Squared Error (RMSE): 2.0316
```

In this example, we calculate both MSE and RMSE on our synthetic dataset. These metrics provide different perspectives on the model's error magnitude, with RMSE often being more representative due to its scale alignment with the target variable.

### *Root Mean Square Log Error (RMSLE)*

RMSLE is a variation of the MSE used in regression models, particularly when the target variable involves exponential growth or when the data encompasses a wide range of values. RMSLE is useful in scenarios where you want to penalize underestimations more than overestimations.

The formula for RMSLE is:

$$RMSLE = \sqrt{\frac{1}{N}\sum_{i=1}^{N}\left(\log\left(y_i+1\right)-\log\left(\hat{y}_i+1\right)\right)^2}$$

where, a logarithmic transformation is applied to both the actual and predicted values.

Key Characteristics:

- Handling of Asymmetric Data: RMSLE is particularly useful when the data is asymmetric or when the target variable can have exponential growth, such as population counts.
- Penalizing Underestimation More than Overestimation: The logarithmic transformation implies that underestimations (predicting values lower than the actual) are penalized more heavily than overestimations. This is useful in many business scenarios where underestimation costs are higher.
- Stabilizing Varying Ranges: The log transformation stabilizes the varying ranges of values, bringing them onto a comparable scale and reducing the effect of large outliers.

#### Python Example on How to Calculate RMSLE

```
# RMSLE
from sklearn.metrics import mean_squared_log_error
from sklearn.ensemble import RandomForestRegressor

# Split the dataset
X_train, X_test, y_train, y_test = train_test_split(df_X, df_y,
test_size=0.2, random_state=42)

y_train = np.abs(y_train).to_numpy().ravel()
y_test = np.abs(y_test).to_numpy().ravel()

# Train a Random Forest Regressor
model = RandomForestRegressor(random_state=42)
model.fit(X_train, y_train)
```

```
# Make predictions
y_pred = model.predict(X_test)

# Calculate RMSLE
rmsle = np.sqrt(mean_squared_log_error(y_test, y_pred))
print(f"Root Mean Squared Logarithmic Error (RMSLE): {rmsle:.4f}")

# Output
Root Mean Squared Logarithmic Error (RMSLE): 0.2256
```

In this example, RMSLE is calculated for a Random Forest regression model. It is crucial to ensure that there are no negative values in the target variable, as the logarithm of negative numbers is undefined. RMSLE provides an understanding of the model's performance, especially in datasets where managing the scale of errors is critical.

## R-Squared ($R^2$) and Adjusted $R^2$

### R-Squared

$R^2$ (Coefficient of Determination), is a measure in regression analysis indicating the fraction of variance in the dependent variable that can be predicted by the independent variables. It provides an indication of the *goodness of fit* of a model.

Formula:

$$R^2 = 1 - \frac{Sum\ of\ Squares\ of\ Residuals\ (SSR)}{Total\ Sum\ of\ Squares\ (SST)}$$

where,

- $SSR = \sum_{i=1}^{N} (y_i - \hat{y}_i)^2$.
- $SST = \sum_{i=1}^{N} (y_i - \bar{y})^2$, where $\bar{y}$ is the mean of the data.

The interpretation is as under:

- $R^2$ values range from 0 to 1. A $R^2$ value of 0 means that the model explain the 0% of the variability in the response data around the mean, while a value of 1 implies that it explains 100% of the variability around it.
- It is often used to compare the fit of different regression models. A higher $R^2$ value indicates a model that better fits the data.

### Adjusted R-Squared

- While $R^2$ is a useful metric, it has its limitations, particularly it tends to increase as more predictors are added to a model, regardless of their usefulness. Adjusted R-Squared addresses this issue by penalizing the addition of irrelevant predictors. It is particularly useful when comparing models with a different number of predictor variables.

Formula:

$$Adjusted\ R^2 = 1 - (1 - R^2) \times \frac{n-1}{n-k-1}$$

where,

- $n$ is the number of observations.
- $k$ is the number of predictor variables.

```
# R-squared and Adjusted R-squared

from sklearn.metrics import r2_score
from sklearn.linear_model import LinearRegression

# Assuming df_X and df_y are features and target of our synthetic
    dataset

# Split the dataset
X_train, X_test, y_train, y_test = train_test_split(df_X, df_y,
    test_size=0.2, random_state=42)

# Train a Linear Regression model
model = LinearRegression()
model.fit(X_train, y_train)

# Make predictions
y_pred = model.predict(X_test)

# Calculate R-Squared
r_squared = r2_score(y_test, y_pred)
print(f"R-Squared: {r_squared:.4f}")

# Calculate Adjusted R-Squared
n = len(y_test)
k = X_test.shape[1]

adjusted_r_squared = 1 - (1 - r_squared) * (n - 1) / (n - k - 1)
print(f"Adjusted R-Squared: {adjusted_r_squared:.4f}")

# Output
R-Squared: 0.9974
Adjusted R-Squared: 0.9973
```

In this example, both R-Squared and Adjusted R-Squared are calculated for a linear regression model. These metrics provide insights into how well the model is capturing the variance in the data, with the adjusted version offering a more nuanced view that accounts for the number of predictors used.

The results obtained for R-Squared and Adjusted R-Squared are 0.9974 and 0.9973, respectively. These values are exceptionally high, indicating that our regression model fits the data very well. These results are indicative of a strong predictive performance by our regression model. However, it is always beneficial to complement these metrics with other forms of validation, such as cross-validation, to confirm the model's effectiveness across different subsets of the data.

## Model Selection and Model Performance Assesment

Now that we have established a solid understanding of various evaluation metrics for both classification and regression models, we are well-equipped to look into the process of model selection. This stage is important in the development of ML projects, as the choice of model significantly influences the effectiveness and efficiency of the solution.

### The Process of Model Selection

Model selection extends beyond picking the algorithm with the best performance metrics. It involves a comprehensive assessment that balances accuracy, computational efficiency, and suitability to the problem domain.

- Understanding Data and Objectives: The first step is to thoroughly understand the nature of the data and the specific goals of the ML task. This understanding guides the initial selection of potential models.
- Applying Evaluation Metrics: Use the evaluation metrics such as accuracy, precision, recall, F1-score, MSE, and RMSE to assess the performance of various models.
- Considering Model Complexity: It is important to consider the complexity of the model. A more complex model might give a slightly better performance but could be prone to overfitting and may require more computational resources.
- Cross-Validation for Generalization: Employ cross-validation techniques to ensure that the model performs consistently across different data subsets and is not just tailored to the peculiarities of the training data.
- Practical Constraints: Consider any practical constraint when developing the ML project, such as computation time, resource availability, and ease of model interpretation.

- Domain-specific Requirements: Ensure that the model aligns with domain-specific requirements and constraints. The model should not only be statistically sound but also contextually relevant.

## *Assessing Performance with the Holdout Method*

In this section we will discuss methods for assessing model performance. A common approach is the Holdout Method, often used as a preliminary step in evaluating the generalizability of a model.

### The Holdout Method

The Holdout Method involves splitting the dataset into distinct subsets: typically, a training set, a validation set, and a test set. The model is trained on the training set, tuned using the validation set, and finally evaluated on the test set.

#### *Limitations of the Holdout Method*

While straightforward and often effective, the Holdout Method has limitations, particularly in its judgment of a model's generalization capabilities.

- Data Dependency: The performance estimation can be highly dependent on how the data is split. Different splits might lead to different results.
- Limited Data Utilization: In cases where data is scarce, setting aside a portion for validation and testing can lead to training on a significantly reduced dataset.
- Variability: The Holdout Method might not capture the model's performance variability across different subsets of the dataset.

### Cross-Validation

Given the limitations of the Holdout Method, Cross-Validation appears as a more robust technique for model evaluation, regarding generalization.

#### *Cross-Validation Explained*

Cross-Validation is a resampling procedure used to evaluate machine learning models on a limited data sample. The data is divided into '$K$' folds, and the model is trained and validated '$K$' times, each time using a different fold as the test set while training on the remaining folds. This process helps in mitigating the issues associated with the Holdout Method.

- Comprehensive Data Utilization: Each data point gets to be in a test set exactly once, and in a training set '$K$–1' times. This approach is beneficial when dealing with limited datasets.
- Reduced Bias: Cross-Validation reduces the bias associated with the random sampling of the Holdout Method.

- Performance Stability: It offers a better assessment of the model's stability and performance across different data samples.

### *Types of Cross-Validation*

- **K-Fold Cross-Validation**: This is the most common type of cross-validation. The data is divided into '*K*' folds, and the model is trained and tested '*K*' times, using a different fold as the test set each time.
- **Leave-One-Out Cross-Validation (LOOCV)**: In this type of cross-validation, '*K*' equals the number of data points in the dataset. This means that for each iteration, the model is trained on all data points except one, which is used as the test set. While LOOCV is exhaustive and eliminates bias, it is computationally expensive.

### *Choosing the Value of 'K'*

- General Use of '*K*': A common choice for '*K*' is either 5 or 10. These values are considered a good balance between obtaining a reliable estimate and maintaining computational efficiency.
- Determining '*K*': The choice of '*K*' can depend on the size and specifics of the dataset. A larger '*K*' reduces bias but increases variance and computational cost. Conversely, a smaller '*K*' increases bias but decreases variance and computational cost.

### *Stratified K-Fold Cross-Validation*

Stratified K-Fold Cross-Validation is an enhanced version of K-Fold Cross-Validation used in machine learning to ensure that each fold of the dataset is a good representative of the whole.

### *Key Characteristics of Stratified K-Fold*

- Maintaining Class Proportions: In a stratified K-fold approach, the dataset is divided in such a way that each fold has approximately the same percentage of samples of each target class as the complete set. This is crucial in handling imbalanced datasets where one or more classes are underrepresented.
- Reducing Bias and Variance: By preserving the class distribution in each fold, Stratified K-Fold reduces the possibility of introducing bias and variance in the model evaluation, leading to a more accurate estimate of the model's performance.
- Improved Model Assessment: It provides a more reliable assessment of the model's performance, especially in cases where the target variable's classes are imbalanced.

## *Python Example: Implementing Stratified K-Fold Cross-Validation*

In this section, we will look into the practical application of stratified cross-validation. The accompanied file is named Chapter_8_Model_Selection_and_hyperparamerter_tuning.ipynb. This example utilizes the *sheep_data.csv* dataset, which can also be found within the same folder. The notebook already includes necessary preprocessing steps such as data loading, label encoding, data inspection, and splitting. Given that these foundational procedures have been covered extensively in previous chapters, we will not revisit them in depth here. Our focus in this section is the application of stratified cross-validation.

**Python Example: Implementing Stratified K-Fold Cross-Validation**

```
from sklearn.ensemble import RandomForestClassifier,
GradientBoostingClassifier
from sklearn.tree import DecisionTreeClassifier
from sklearn.svm import SVC
from sklearn.neural_network import MLPClassifier
from sklearn.model_selection import cross_val_score, KFold
from lightgbm import LGBMClassifier
from xgboost import XGBClassifier
import numpy as np
from tqdm import tqdm

# Define the models
models = {
    "Random Forest": RandomForestClassifier(random_state=42),
    "LightGBM": LGBMClassifier(random_state=42),
    "Decision Tree": DecisionTreeClassifier(random_state=42),
    "XGBoost": XGBClassifier(random_state=42, use_label_
    encoder=False, eval_metric='logloss'),
    "SVM Radial": SVC(kernel='rbf', random_state=42),
    "MLP": MLPClassifier(random_state=42)
}

# Define the Stratified K-Fold Cross-Validator
stratified_kfold = StratifiedKFold(n_splits=5, shuffle=True, random_
state=42)

# Perform Stratified K-Fold and store the results
model_scores = {}

for name, model in models.items():
    scores = cross_val_score(model, X, y, cv= stratified_kfold, n_
    jobs=-1)
    model_scores[name] = np.mean(scores)
    print(f"{name}: Average Accuracy = {np.mean(scores):.4f}")
```

```
# Selecting the best model based on the highest average accuracy
best_model = max(model_scores, key=model_scores.get)
print(f"\nBest Model: {best_model} with Average Accuracy =
    {model_scores[best_model]:.4f}")
# Output
Random Forest: Average Accuracy = 0.9878
LightGBM: Average Accuracy = 0.9891
Decision Tree: Average Accuracy = 0.9663
XGBoost: Average Accuracy = 0.9918
SVM Radial: Average Accuracy = 0.7568
MLP: Average Accuracy = 0.8979

Best Model: XGBoost with Average Accuracy = 0.9918
```

The above code initializes a dictionary named models, where key-value pairs represent different algorithms such as Random Forests, LightGBM, Decision Trees, XGBoost, SVM Radial, and MLP. Each key is a string denoting the name of the model, and the corresponding value is the model object itself from scikit-learn or other libraries like XGBoost and LightGBM.

A StratifiedKFold object is created with 5 splits. shuffle = True shuffles the data before splitting it into folds, adding randomness which helps in reducing bias.

In the loop, each model is evaluated using Stratified K-Fold Cross-Validation. The cross_val_score computes the model's accuracy for each fold of the cross-validation process and returns a list of scores. The np.mean(scores) calculates the average accuracy across all folds for each model, and the results are stored in the model_scores dictionary. The n_jobs = –1 parameter enables the function to use all available CPU cores for parallel computation, speeding up the evaluation process.

After evaluating all models, the one with the highest average accuracy is determined using the max function with model_scores.get as the key function. This model is identified as the best model for the given task.

The stratified cross-validation results are as follows:

- Random Forest: Average Accuracy = 0.9878
- LightGBM: Average Accuracy = 0.9891
- Decision Tree: Average Accuracy = 0.9663
- XGBoost: Average Accuracy = 0.9918
- SVM Radial: Average Accuracy = 0.7568
- MLP: Average Accuracy = 0.8979.

Based on these results, the model that stands out with the highest average accuracy is XGBoost, with an accuracy of 0.9918. This performance indicates that XGBoost is highly effective at classifying the different behaviors in the sheep dataset. Given the high accuracy of the XGBoost model in our cross-validation process, the consequent steps would involve fine-tuning this model, followed by a thorough evaluation on a separate test set to confirm its generalizability and effectiveness in practical scenarios.

## Techniques for Improving Model Performance

After selecting the most suitable model for our task, which in our case is XGBoost, the next step is to enhance its performance. One of the key strategies in this process is hyperparameter tuning, which involves adjusting the model's parameters to optimize its performance.

In the context of machine learning models, particularly when discussing hyperparameter tuning and model configuration, we must first differentiate between two types of parameters: model parameters and hyperparameters.

Model parameters are the configuration variables internal to the model and are learned from the data during the training process. They are intrinsic to the model's ability to make predictions and are adjusted automatically to fit the model to the training data.

Characteristics:

- Learned Automatically: Model parameters are learned directly from the training data.
- Model Specific: They differ from one model to another. For example, the weights in a neural network or the coefficients in a regression model.

Hyperparameters, on the other hand, are external to the model and are not learned from the data. Instead, they are set prior to the training process and govern the overall behavior of the model.

Characteristics:

- Set Manually: Hyperparameters are set by the practitioner before training.
- Control Learning Process: They influence the structure of the model and how it learns, but they are not involved in making predictions.
- Model Tuning: Hyperparameters are tuned for optimal performance, often using methods like grid search or random search.

### *Grid Search for Hyperparameter Tuning*

Grid Search is a method used to perform hyperparameter optimization, where a grid of hyperparameter combinations is exhaustively searched to find the best possible configuration for a model. It is called *Grid Search* because it searches across a manually specified subset of the hyperparameter space of the chosen model.

Key Features:

- Exhaustive Search: Tests every combination of hyperparameters in the grid.
- Precision: Can identify the optimal parameters accurately if they are included in the grid.
- Resource-Intensive: This method can be computationally expensive, especially for large datasets and complex models.

**Python Example: Implementing Grid Search with XGBoost**

Given that we have chosen XGBoost as our model, we can utilize Grid Search to fine-tune its hyperparameters including learning rate, the number of trees, depth of trees, and others.

```
from sklearn.model_selection import GridSearchCV
from xgboost import XGBClassifier

# Define the parameter grid
parameters_grid = {
    'n_estimators': [100, 200, 300],
    'learning_rate': [0.01, 0.1, 0.2],
    'max_depth': [3, 4, 5]
}

# Initialize the XGBoost classifier
xgb_model = XGBClassifier(random_state=42, use_label_encoder=False,
    eval_metric='logloss')

# Create the GridSearchCV object
grid_search = GridSearchCV(estimator=xgb_model, param_grid=
    parameters_grid, cv=5, n_jobs=-1, verbose=2)

# Fit to the data
grid_search.fit(X_train, y_train)

# Print the best parameters and the best score
print("Best Parameters:", grid_search.best_params_)
print("Best Score:", grid_search.best_score_)

# Output
Fitting 5 folds for each of 27 candidates, totalling 135 fits
Best Parameters: {'learning_rate': 0.2, 'max_depth': 5, 'n_
estimators': 300}
Best Score: 0.9902958046871252
```

Explanation of the Code:

- Parameter Grid: A dictionary parameters_grid is defined where the keys are hyperparameters of the XGBoost model, and values are lists of settings to be tested.
- XGBoost Classifier Initialization: An instance of XGBClassifier is created with some basic configurations.
- GridSearchCV Object: This object is initialized with the XGBoost classifier, the parameter grid, and the number of folds (cv = 5) for cross-validation, and verbose = 2 provides detailed logging information.
- Fitting the Model: The GridSearchCV object is then fitted to the training data, which performs the grid search across the specified hyperparameter grid.
- Results: After fitting, the best combination of parameters and the corresponding score are displayed, indicating the most optimal settings found for the XGBoost model.

The best hyperparameters identified in the grid search are ({'learning_rate': 0.2, 'max_depth': 5, 'n_estimators': 300}).

Now we will use the best parameters to finally train and evaluate our model:

```
from sklearn.metrics import classification_report
from sklearn.metrics import confusion_matrix
import seaborn as sns
import matplotlib.pyplot as plt

# Best model from Grid Search
best_model = grid_search.best_estimator_
best_model.fit(X_train, y_train)

# Making predictions on the test set
y_pred = best_model.predict(X_test)

# Generate the classification report
class_report = classification_report(y_test, y_pred,
            target_names=label_encoder.classes_)

# generating the confusion matrix
cm = confusion_matrix(y_test, y_pred)

# Plotting the confusion matrix with class names
plt.figure(figsize=(10, 7))
sns.heatmap(cm, annot=True, fmt='d', cmap='Blues',
            xticklabels=label_encoder.classes_,
            yticklabels=label_encoder.classes_)
```

```
plt.xlabel('Predicted')
plt.ylabel('Actual')
plt.title('Confusion Matrix')
plt.savefig('cm.png', dpi=300)
plt.show()

# Display the results
print("Classification Report:\n", class_report)

# Output
Classification Report:
              precision    recall  f1-score   support
     grazing       1.00      1.00      1.00      4830
     resting       0.99      0.99      0.99      8272
  scratching       0.99      0.90      0.94       165
    standing       0.99      0.99      0.99      4254
     walking       1.00      0.99      1.00      1867

    accuracy                           0.99     19388
   macro avg       0.99      0.98      0.98     19388
weighted avg       0.99      0.99      0.99     19388
```

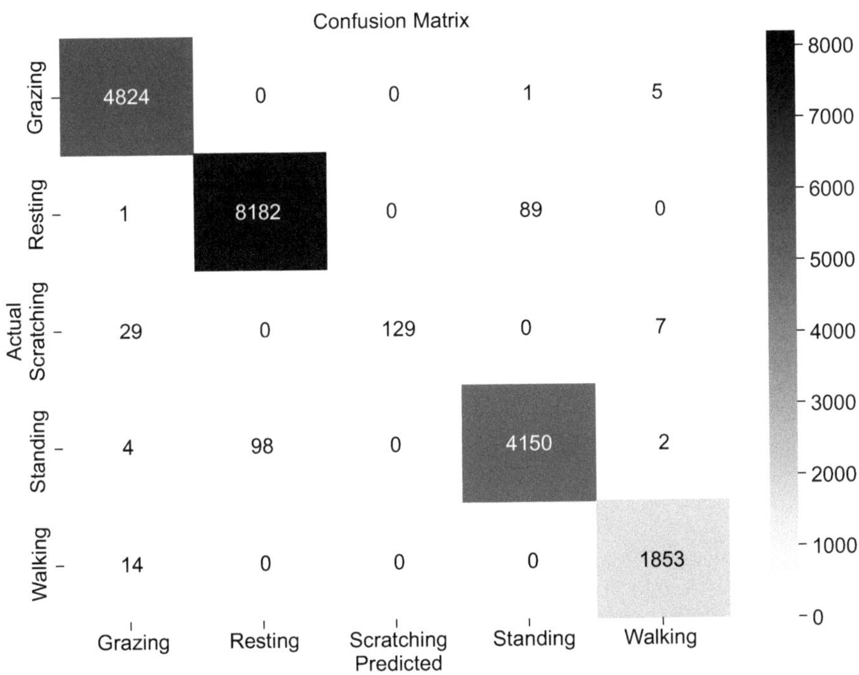


**Figure 8.4: Confusion matrix using XGBoost on the test set.**

The code evaluates the performance of the XGBoost model optimized via Grid Search, on a test dataset. It involves generating a classification report and a confusion matrix (Figure 8.4) to understand the model's effectiveness in classification.

Breakdown of the code and its purpose:

- Retrieve and Fit Best Model: best_model = grid_search.best_estimator_ retrieves the XGBoost model with the best hyperparameters identified during Grid Search. best_model.fit(X_train, y_train) fits this model to the training data.
- Prediction: y_pred = best_model.predict(X_test) uses the fitted model to make predictions on the test dataset.
- Classification Report: This part of the code generates a classification report, which includes key metrics like precision, recall, and F1-score for each class in the test dataset.
- target_names = label_encoder.classes_ translates the encoded class labels back to their original names for better interpretability in the report.
- Confusion Matrix: confusion_matrix(y_test, y_pred) creates a confusion matrix
- Visualization: The confusion matrix is visualized as a heatmap using seaborn and matplotlib.
- Class Labels: The matrix includes class names on both axes for clarity, facilitating a quick understanding of how predictions compare against actual values.
- Final Output: Finally, the classification report is printed, providing a detailed account of the model's performance metrics for each class.

## *Randomized Search for Hyperparameter Tuning*

After discussing Grid Search for hyperparameter tuning, we look into another effective technique: Randomized Search. This method is useful when working with complex models where the hyperparameter space is extensive.

Randomized Search, as the name suggests, involves randomly selecting combinations of hyperparameters to find the best solution for the model. Unlike Grid Search, which exhaustively tries all possible combinations, Randomized Search samples a given number of combinations by selecting random values for each hyperparameter at each iteration.

Advantages of Randomized Search:

- Efficiency: It is typically faster than Grid Search, especially when dealing with a large hyperparameter space or when each evaluation is time-consuming.

- Reduced Overfitting Risk: By randomly sampling the hyperparameter space, it can sometimes find better solutions than Grid Search, as it is not constrained to a fixed set of values.
- Flexibility: It allows you to control the number of iterations, offering a balance between computational cost and optimization.

### Python Example: Implementing Randomized Search with Random Forest

In this example we will apply Randomized Search to tune a Random Forest classifier. We will randomly sample various combinations of hyperparameters and evaluate their performance.

```
# RandomizedSearchCV

from sklearn.model_selection import RandomizedSearchCV
from sklearn.ensemble import RandomForestClassifier
from scipy.stats import randint as sp_randint

# Define the parameter distribution
param_dist = {
    'n_estimators': sp_randint(100, 500),
    'max_depth': sp_randint(3, 10),
    'max_features': ['auto', 'sqrt', 'log2']
}

# Initialize the Random Forest classifier
rf_model = RandomForestClassifier(random_state=42)

# Create the RandomizedSearchCV object
random_search = RandomizedSearchCV(estimator=rf_model,
                                   param_distributions=param_dist,
                                   n_iter=100, cv=5, random_
                                   state=42, n_jobs=-1)

# Fit to the data
random_search.fit(X_train, y_train)

# Print the best parameters and the best score
print("Best Parameters:", random_search.best_params_)
print("Best Score:", random_search.best_score_)

# Output
Best Parameters: {'max_depth': 9, 'max_features': 'auto', 'n_
estimators': 370}
Best Score: 0.9508820574195646
```

Explanation of the Code:

- Parameter Distribution: param_dist defines the distribution of parameters we want to sample. sp_randint is used to generate discrete random values for parameters like n_estimators and max_depth.
- Random Forest Classifier: An instance of RandomForestClassifier is initialized.
- RandomizedSearchCV Object: This object is similar to GridSearchCV but uses param_distributions instead of param_grid. n_iter = 100 controls how many different combinations to try.
- Model Fitting: The model is fitted with the training data, and Randomized Search iteratively samples different hyperparameter combinations.
- Results: After fitting, the best combination of parameters and the corresponding scores are displayed.

## *Halving for Hyperparameter Tuning*

*Halving* is an approach in the domain of hyperparameter optimization, particularly implemented within the context of *Successive Halving* and its more advanced variant, *Hyperband.* These methods are designed to efficiently allocate resources to the most promising hyperparameter combinations, thereby speeding up the search process while minimizing computational expense.

### Successive Halving

Successive Halving operates on the principle of iteratively allocating resources to a subset of configurations based on their performance.

Process:

- Initial Screening: Begins with a large number of randomly selected hyperparameter configurations and a minimal resource budget.
- Performance Evaluation: Each configuration is evaluated using the allocated budget.
- Pruning: The configurations are ranked by performance, and only the top half (or another predefined fraction) is retained for the next round.
- Resource Doubling: The budget for each remaining configuration is doubled, and steps 2-3 are repeated until a satisfactory configuration is found or the resource limit is reached.

### Hyperband

Hyperband improves upon Successive Halving by dynamically adjusting the number of configurations and the allocated resources through multiple brackets, each with a different starting budget. It effectively creates a balance between exploring many configurations with a small budget and exploiting a few configurations with a large budget.

Advantages:

- Efficiency: Both methods significantly reduce computational costs by early-stopping fewer promising configurations.
- Flexibility: They are applicable to a wide range of machine learning models and tasks.
- Scalability: Suitable for parallel and distributed computing environments.

### Python Example: Implementing Halving in Hyperparameter Tuning

Scikit-learn introduced practical implementations of these methods: HalvingGridSearchCV and HalvingRandomSearchCV, which are analogous to the traditional GridSearchCV and RandomSearchCV but incorporate the principles of Successive Halving.

```
# HalvingGridSearchCV

from sklearn.experimental import enable_halving_search_cv
from sklearn.model_selection import HalvingGridSearchCV

# Define the model and parameter space
model = RandomForestClassifier(random_state=42)
param_grid = {
    'n_estimators': [100, 200, 300],
    'max_depth': [5, 10, 15]
}
# Initialize and run the Halving Grid Search
halving_grid_search = HalvingGridSearchCV(model, param_grid, cv=5,
    factor=2, random_state=42) halving_grid_search.fit(X_train,
    y_train)

# Best parameters and score
print("Best Parameters:", halving_grid_search.best_params_)
print("Best Score:", halving_grid_search.best_score_)

# Evaluate on the test set
best_model = halving_grid_search.best_estimator_
test_accuracy = best_model.score(X_test, y_test)
print(f"Test Set Accuracy: {test_accuracy:.4f}")

# Output
Best Parameters: {'max_depth': 15, 'n_estimators': 300}
Best Score: 0.9768948804996072
Test Set Accuracy: 0.9802
```

Initialization and Execution of Halving Grid Search:

- HalvingGridSearchCV: This class implements the *Successive Halving* algorithm for hyperparameter tuning.
- model: This is the machine learning model we are tuning, which in this case is a RandomForestClassifier. The model is passed as an argument to HalvingGridSearchCV.
- param_grid: This is a dictionary that defines the hyperparameter space for the search. Each key in the dictionary is a hyperparameter name, and the corresponding value is a list of values to try for that hyperparameter. The HalvingGridSearchCV will evaluate combinations of these hyperparameters.
- cv = 5: This parameter specifies the number of folds in a K-fold cross-validation. The data is split into 5 parts for each hyperparameter combination.
- factor = 2: The *factor* parameter in HalvingGridSearchCV determines the rate at which the number of configurations is reduced in each iteration of the halving process. A factor of 2 means that half of the configurations are discarded in each iteration. This parameter is crucial in controlling the balance between exploration (trying out many different hyperparameter configurations) and exploitation (focusing on the most promising configurations).

## Bayesian Optimization with Optuna for Hyperparameter Tuning

Another notable method for hyperparameter tuning is Bayesian Optimization, particularly as implemented in the Optuna framework. While we have discussed various approaches like grid search and halving methods earlier in this Chapter, Bayesian Optimization stands out for its efficiency and effectiveness in navigating large hyperparameter spaces.

### Overview of Bayesian Optimization in Optuna

Bayesian Optimization is a probabilistic model-based approach for finding the minimum of a function. In the context of hyperparameter tuning, this function is usually the model's performance metric (like accuracy or loss) as a function of its hyperparameters.

Optuna uses Bayesian Optimization to intelligently propose the next set of hyperparameters to evaluate, based on the past trials. This approach contrasts with grid or random search, where hyperparameters are selected without considering past evaluations.

### Why Optuna?

- Efficient Search: Optuna can find better hyperparameters with fewer trials compared to grid or random search.

- Practicality: It is useful when dealing with high-dimensional hyperparameter spaces or when evaluations of the objective function are expensive.
- Flexibility: Optuna allows the definition of complex search spaces and integrates easily with various machine learning frameworks.

Bayesian Optimization using Optuna represents a sophisticated approach to finding the best hyperparameters for ML models. It illustrates the connection of statistical methods and practical machine learning, offering a pathway to efficiently achieving optimized model performance. For a hands-on example and deeper understanding, please refer back to the earlier chapter where we discussed and demonstrated this approach in detail.

### Reference to Previous Example in the Book

For readers interested in seeing Bayesian Optimization in action using Optuna, we recommend revisiting Chapter 3, subsection 'Hyperparameter Tuning and Model Evaluation'. In that section, we have provided a detailed Python example illustrating how Optuna can be used for tuning hyperparameters of a machine learning model.

## Summary

In this chapter, we have presented model evaluation, model selection, and hyperparameter tuning in machine learning. We started with evaluation metrics, describing accuracy, precision, recall, F1-score and other metrics for classification, and MSE, RMSE, and MAE for regression, to provide a multifaceted understanding of model performance. We then shifted our focus to model selection, emphasizing the importance of cross-validation methods like K-Fold and Stratified K-Fold in ensuring model robustness and reliability across diverse data sets.

In the last part of the chapter, we focused on hyperparameter tuning, comparing traditional approaches like Grid Search and Randomized Search with advanced techniques such as Halving methods and Bayesian Optimization via Optuna. These methods offer significant improvements in efficiency and effectiveness, particularly in handling complex models with large parameter spaces. Practical Python examples were integrated to demonstrate these concepts in action, equipping readers with the skills to evaluate, select, and fine-tune machine learning models efficiently and effectively.

CHAPTER 9

# Deep Learning Algorithms for Animal Activity Recognition

## From Traditional Programming to Machine Learning and Deep Learning

In the field of computing, the evolution from traditional programming to machine learning (ML) and deep learning (DL) marks a significant shift in how we approach problem-solving. This transition mirrors the complexity and diversity of the problems we encounter, particularly in fields like animal activity recognition in our case.

### *Traditional Programming: The Rule-based Approach*

Traditional programming stands on the foundational principle of explicit rule definition. In this paradigm, programmers encode a finite set of rules that a machine follows to produce outputs from given inputs. This approach is highly effective for problems with clear, well-defined rules and predictable outcomes. For instance, a traditional program can efficiently handle tasks like arithmetic operations or database queries where the logic is straightforward, and the rules are unambiguous.

The primary strength of traditional programming lies in its simplicity and predictability. However, its reliance on explicitly defined rules becomes a limitation when dealing with complex problems where rules are not easily detectible or too many to encode manually. This is where machine learning comes to play.

### *Transition to Machine Learning: Learning from Data*

Machine learning represents a paradigm shift from rule-based logic to data-driven learning. Unlike traditional programming, ML does not require explicit programming of all possible rules. Instead, ML algorithms learn these rules by identifying patterns in data. This capability is crucial for tackling problems with inherent complexity and subtle details that are not easily captured through explicit programming.

In ML, the focus shifts from programming specific rules to designing algorithms that can learn these rules from data. As we discussed in the previous chapters, ML algorithms, from simple linear regressions to complex ensemble methods, provide the flexibility and adaptability to learn from data, making them well-suited for a wide range of applications.

## The Arrival of Deep Learning: A Subset of Machine Learning

Deep learning is a specialized subset of machine learning characterized by its use of neural networks. These networks, inspired by the structure and function of the human brain (this is what most people say!), consist of layers of interconnected nodes or *neurons*. Each layer captures different aspects of the data, allowing the model to learn complex and abstract patterns. This layered structure enables deep learning models to handle high-dimensional data like images, sound, and text more effectively than traditional ML models.

Deep learning excels in tasks where the data is rich, and the patterns are highly intricate. Its ability to learn hierarchical representations makes them able to automatically adapt and process data to solve complex tasks. In the context of farm animal activity recognition, deep learning can be instrumental in analyzing visual and auditory data, where the patterns might be too subtle or complex for traditional ML models.

The journey from traditional programming to machine learning, and subsequently to deep learning, is a narrative of increasing complexity and sophistication in problem-solving. Traditional programming, with its rule-based simplicity, is ideal for straightforward tasks. Machine learning offers a more flexible approach, learning from data to handle complex patterns. Deep learning, with its advanced neural network architecture, pushes the boundaries further, tackling problems of even greater complexity, especially those involving large-scale and high-dimensional data.

As we look at the specifics of deep learning, especially in the context of animal activity recognition, we will explore their capabilities and also their limitations, helping readers to make informed choices about the appropriate technology for their specific problems.

## Distinguishing Machine Learning from Deep Learning

While deep learning is a subset of machine learning, there are key differences between the two, as these distinctions influence their applications, capabilities, and limitations.

### Model Complexity and Architecture

- Machine Learning Models: Traditional ML models incorporate a wide range of algorithms, from simple linear regression to complex ensemble methods like random forests and gradient boosting machines. These models often require feature engineering, where domain knowledge is used to create input features that are fed into the model. The complexity of ML models varies, but they are generally less complex than deep learning models.

- Deep Learning Models: DL models, primarily based on neural networks, are characterized by their depth, which refers to the number of layers in the network. These layers enable DL models to learn increasingly abstract features from raw input data, eliminating the need for manual feature engineering. DL models are fundamentally more complex, capable of handling vast and intricate architectures, making them suitable for more complex tasks.

### Data Requirements

- ML Data Requirements: Traditional ML models can often achieve high performance with relatively smaller datasets. They are skilled at capturing relationships in data where the underlying patterns can be observed with fewer examples. This characteristic makes ML models useful in scenarios where data collection is challenging or expensive.
- DL Data Requirements: DL models succeed on large datasets. Their ability to learn and generalize improves significantly with the amount of data fed into them. This data-hungry nature of DL allows them to excel in complex tasks like image and speech recognition, but it also poses challenges in scenarios where data is scarce or expensive to acquire.

### Interpretability and Transparency

- Interpreting ML Models: One of the strengths of many machine learning models, especially simpler ones, is their interpretability. Models like decision trees or linear regression offer insights into how decisions are made, which is crucial in fields where understanding model reasoning is important. Even in more complex ML models, techniques like feature importance scores can provide a level of transparency.
- Interpreting DL Models: Due to their complexity and the way they automatically extract and process features, they often act as *black box*. This term refers to the challenge in understanding how these models arrive at a specific decision. The complex interplay of neurons and layers makes it difficult to determine the exact rationale behind a model's output. This lack of interpretability can be a significant drawback in applications where understanding the decision-making process is as important as the accuracy of the decision itself.

### Computational Resources

- Resources for ML: In general, ML models are less computationally intensive compared to DL models. They can often be trained on standard computing resources, making them more accessible for a wider range of users and applications.

- Resources for DL: Particularly large DL models require significant computational power. They often need the use of specialized hardware like GPUs (Graphics Processing Units) or TPUs (Tensor Processing Units). This requirement for high computational resources makes DL models more expensive and less accessible for casual users or small-scale applications.

### *Choosing Between Machine Learning and Deep Learning*

When faced with a problem requiring a computational solution, one must carefully consider whether to employ ML or DL. This decision centers on various factors including the nature of the problem, the availability of data, and the required interpretability of the model.

#### Rule of Simplicity in Problem Solving

The first consideration should always be the rule of simplicity: If a problem can be solved effectively with a simple, rule-based system, then there is no need to complicate the solution with ML or DL. These advanced techniques are not a one-size-fits-all solution and should be reserved for problems where their specific advantages can be fully leveraged.

#### When Machine Learning is the Right Choice

Machine learning is particularly effective in scenarios where patterns in the data can be discerned with a relatively small dataset. If the problem domain allows for feature engineering — where domain knowledge can be used to create informative features — ML models can be very efficient.

#### Scenarios Favouring Deep Learning

Deep learning comes into its own with large datasets and problems involving high-dimensional data. It has the ability to extract features automatically from raw data and it is proficient in handling complex patterns. However, this comes at the cost of needing substantial computational resources and often results in models that are less interpretable. Deep learning is ideal when the scale and complexity of the data surpass the capabilities of traditional ML methods.

## Deep Learning Foundations: Neural Networks and their Variants

At the heart of deep learning lies the concept of neural networks, which are the foundational building blocks of this advanced form of ML. Neural networks, form the basis from which more complex deep learning architectures evolve.

In this chapter, our primary focus will be on understanding the key types of neural networks and their roles in deep learning. It is important to note, however, that the application of deep learning for wearable sensor data, especially in the context of animal activity recognition, is not as extensively explored as in other domains like computer vision. The deep learning techniques most commonly applied to wearable sensor data include:

- **Multilayer Perceptron Neural Networks (MLPs):** Often referred to as fully connected or dense networks, these represent the simplest form of neural networks and consist of layers where each neuron is connected to all neurons in the preceding and subsequent layers. Their fundamental structure is crucial for understanding the basic workings of neural network architecture.
- **Convolutional Neural Networks (CNNs):** Although widely recognized for their role in processing image data, CNNs can also be adapted for analyzing wearable sensor data, particularly when there is a need to capture spatial hierarchies or patterns in the data.
- **Recurrent Neural Networks (RNNs), Long Short-Term Memory (LSTM), and Gated Recurrent Units (GRUs)**: These networks are designed to handle sequential data, making them suitable for wearable sensor data that often has a time-series nature. RNNs and LSTMs can capture temporal dynamics, which is valuable for analyzing sequences of sensor readings over time. GRUs, like LSTMs, are pivotal for their efficiency in sequence modelling and their suitability for time-series data. Although GRUs share similarities with LSTMs, their simpler structure can offer computational advantages in certain applications.

While this chapter will focus on these specific techniques due to their relevance to wearable sensor data and their applications in animal activity recognition, it is important to acknowledge that the field of deep learning is vast and diverse. There are numerous other techniques and models within deep learning, such as Deep Transfer Learning, Autoencoders, Generative Adversarial Networks (GANs), and more advanced models for Natural Language Processing (NLP). However, these are beyond the scope of this book. Readers interested in these advanced topics are encouraged to seek specialized resources for further study.

Our objective is to provide a comprehensive understanding of the neural network architectures most pertinent to our focus on wearable sensor data for animal activity recognition, guiding readers in applying these techniques effectively within this specific context.

### Practical Examples in PyTorch

As we explore these concepts, we will provide practical examples using PyTorch, a deep learning framework known for its flexibility and ease of use. PyTorch offers an intuitive interface for building and training neural networks, making it an excellent choice for demonstrating DL concepts. All practical examples accompanied with the datasets for this chapter can be found on https://github.com/nkcAna/WSDpython in Chapter_9 folder.

However, readers should be aware that this book is not a comprehensive tutorial on deep learning or PyTorch. Our focus is to apply deep learning principles specifically in the context of farm animal activity recognition using Python. For readers interested in a more detailed exploration of PyTorch, we recommend referring to the official PyTorch website and its extensive documentation and tutorials (https://pytorch.org/tutorials/).

By the end of this chapter, our goal is for readers to have a clear understanding of the fundamental neural network architectures and their applications, all within the framework of PyTorch. This knowledge will serve as a critical step for applying deep learning techniques in the field of animal activity recognition and beyond.

## Neural Networks: The Foundation of Deep Learning

In this section we will start by explaining the fundamental building block of DL: the neural network. So, we will start by introducing the core components and concepts.

### *The Neuron: Basic Unit of a Neural Network*

The *neuron*, or *artificial neuron*, is the basic computational unit in a neural network. Each neuron in a neural network performs a simple but essential computation that contributes to the network's overall task, such as classification or regression.

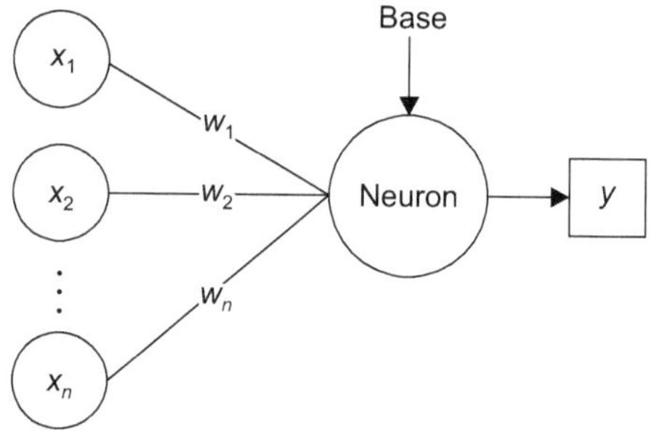


**Figure 9.1: The basic structure of a neuron.**

A neuron receives one or more inputs, processes these inputs, and generates an output. The inputs to a neuron can be features from a dataset or outputs from preceding neurons in the network. Each input $x_i$ (independent variable) is associated with a weight $w_i$, signifying the strength or importance of the inputs. The neuron also includes a bias term , which adjusts the output ($y$) independently of the inputs. The output ($y$) can be continuous, binary or categorical variable.

> **The values of independent variables in a neural network should be either standardized or normalized. This practice ensures that all variables fall within a similar range, enabling the neural network to efficiently process them.**

### Neuron Computation

The output of a neuron is calculated by first computing the weighted sum of its inputs:

$$Weighted\ Sum = \sum_{i=1}^{n} w_i x_i + b$$

where, $n$ is the number of inputs, $w_i$ is the weight associated with the $i$th input ($x_i$), and $b$ is the bias.

After calculating the weighted sum, the neuron applies an activation function ($\phi$) to this sum. A nonlinear activation function introduces nonlinearity into the model, enabling the network to learn complex patterns.

> **Note that the input variables in a dataset are not isolated entities; they represent a single observation. For example, observations like temperature, age of the animal, food intake, are distinct attributes, however, they describe one single observation (the target). When these processed values traverse through the primary neuron, they transform into output values on the other side of the network.**

## *Non-linear Activation Functions: Bringing Non-linearity into the Picture*

Activation functions are critical in neural networks as they introduce non-linearity into the system. This non-linearity allows neural networks to learn and model complex relationships in the data. Without activation functions, a neural network would essentially become a linear regression model, incapable of handling the details of real-world data. Activation functions allow neural networks to make sense of complex and non-linear data. These functions determine whether a neuron should be activated or not, based on the weighted sum of its inputs.

### What Does an Activation Function Do?

An activation function takes the weighted sum of all inputs to a neuron, applies a specific mathematical function, and outputs the result. This output then serves as input to the next layer in the network. The choice of activation function affects the network's ability to converge and the speed of convergence.

#### *Common Activation Functions*

- **Sigmoid Function (Logistic Function)**

  Formula:

  $$\sigma(x) = \frac{1}{1+e^{-x}}.$$

  The Sigmoid function outputs values between 0 and 1, making it particularly useful for models where the output is interpreted as a probability. However, it is vulnerable to the vanishing gradient problem, which can slow down training.

- **Hyperbolic Tangent (tanh) Function**

  The Hyperbolic Tangent function, commonly abbreviated as "tanh", is a widely used activation function in neural networks. It is mathematically defined as:

  $$tanh(x) = \frac{e^x - e^{-x}}{e^x + e^{-x}}.$$

  Here is a breakdown of its characteristics and usage:

  - Range: The tanh function outputs values in the range of –1 to 1. This is one of its primary differences from the sigmoid function, which outputs values between 0 and 1.
  - Zero-centered: Since its output ranges from –1 to 1, the tanh function is zero-centered. This means that its outputs have a mean close to 0, which can help with the convergence of the neural network during training, as it avoids bias in the gradients.
  - Non-linearity: Just like the sigmoid function, tanh is also non-linear. This allows neural networks using tanh to learn complex data patterns and solve classification problems that are not linearly separable.
  - Usage: tanh is commonly used in hidden layers of a neural network. Its zero-centered nature makes it more efficient than the sigmoid function.
  - Vanishing Gradient Problem: Similar to the sigmoid function, tanh suffers from the vanishing gradient problem. When the inputs are very high or very low, the gradient of the tanh function becomes very small. This small gradient can slow down the training process significantly, as it causes very small updates to the weights during backpropagation.

The tanh function is particularly useful in situations when having a zero-centered output is advantageous. However, in many modern neural network architectures, especially deeper networks, other activation functions are often preferred due to their ability to mitigate the vanishing gradient problem.

- **Rectified Linear Unit (ReLU) Function**

  The Rectified Linear Unit (ReLU) function is one of the most widely used activation function in the field of neural networks, particularly in deep learning models. Its simplicity and effectiveness in various network architectures have made it a default choice in many implementations.

  The ReLU function is defined as:

  $$ReLU(x) = max(0, x).$$

  This means that the ReLU function outputs the input itself if it is positive; otherwise, it outputs zero. The key characteristics of the ReLU function include:

  - Non-linearity: Although ReLU is a linear function for all positive inputs, its ability to output zero for all negative inputs introduces non-linearity into the model.
  - Simplicity: One of the primary advantages of ReLU is its computational simplicity. It does not involve any complex operations (like exponentials), making it more efficient than functions like sigmoid or tanh, especially during the training phase.
  - Avoiding the Vanishing Gradient Problem: ReLU helps in mitigating the vanishing gradient problem, common in deep networks with sigmoid or tanh activation functions. Since the gradient for positive inputs is always 1, gradients do not diminish as quickly during backpropagation, allowing for deeper networks.

  *Applications and Usage*

  ReLU is commonly used in hidden layers of neural networks. It is particularly effective in convolutional neural networks (CNNs) and deep learning models where computational efficiency is vital.

  *Limitations*

  Despite its advantages, ReLU has some limitations:

  - Dying ReLU Problem: If a neuron's output is always negative, ReLU will output zero, and the neuron stops adjusting during training, essentially "dying".
  - Non-zero Centered Output: ReLU's output is not zero-centered, which can sometimes lead to issues during optimization.

- **Leaky Rectified Linear Unit (Leaky ReLU) Function**

  The Leaky Rectified Linear Unit (Leaky ReLU) is a variant of the standard

ReLU activation function, designed to address one of its primary limitations: the *dying ReLU* problem. It modifies the ReLU function to allow a small gradient when the unit is not active.

The Leaky ReLU function is defined as:

$$LeakyReLU(x) = \max(ax, x) = \begin{cases} x, & \text{if } x > 0 \\ ax, & \text{if } x \leq 0. \end{cases}$$

*Key Characteristics*

- Non-zero Gradient for Negative Inputs: Unlike ReLU, which outputs zero for all negative inputs, Leaky ReLU allows a small, non-zero output. This ensures that neurons continue to learn and adapt during the training process, even if they are activated infrequently.
- Reduced Risk of Neurons Dying: By maintaining a small gradient for negative inputs, Leaky ReLU reduces the risk of neurons becoming inactive during training. This makes it particularly useful in deeper networks where the dying ReLU problem is more pronounced.
- Computational Efficiency: Similar to ReLU, leaky ReLU is computationally efficient. The additional operation required to implement the "leak" is minimal and does not significantly impact the overall computational cost.

*Usage and Applications*

It is useful in training models where the data is not well normalized, and the inputs to neurons in the network have a significant negative component.

*Limitations*

While Leaky ReLU addresses some issues of ReLU, it also has limitations:

- Tuning the $a$ Parameter: The effectiveness of Leaky ReLU can depend on the choice of $a$. This parameter may need tuning specific to the application and dataset.
- Not a Universal Solution: Leaky ReLU does not always outperform ReLU and needs to be tested empirically for each specific application.

Here is a small constant (typically around 0.01) that gives the function a slight slope for negative values, ensuring that the gradient is never entirely zero. This modification addresses the dying ReLU problem by allowing the flow of gradients even for negative input values.

- **Exponential Linear Unit (ELU)**

  The ELU activation function is another variant that aims to improve upon the ReLU function. It was introduced to address some of the limitations of ReLU and Leaky ReLU.

Formula:

$$ELU(x) = \begin{cases} a\,(e^x - 1), \text{ if } x < 0 \\ x \qquad\quad , \text{ if } x \geq 0. \end{cases}$$

For positive values of $x$, ELU behaves just like ReLU, but for negative values, it has a different behavior. Instead of having a constant value (like in Leaky ReLU), ELU has an exponential curve, which allows it to push the mean activations closer to zero. This zero-centering property helps speed up learning by bringing the average output of the neurons closer to zero, similar to the tanh function.

A key advantage of ELU over ReLU is its non-zero gradient for negative values, which helps alleviate the dying ReLU problem. This characteristic ensures that all neurons in the network continue to learn and adjust during training.

The parameter $a$ in the ELU formula is a constant that defines the value to which an ELU saturates for negative net inputs. It is usually set to 1 but can be tuned based on specific requirements of the neural network.

However, ELU can be computationally more expensive than ReLU and its other variants due to the exponential function, particularly during the backward pass in training.

- **Softmax Activation Function**

  The Softmax activation function stands out in activation functions, especially in the context of classification problems. It is typically used in the output layer of a neural network to transform raw output scores, often called logits, into probabilities, which are easier to interpret.

  The Softmax function is defined as:

$$Softmax(z_i) = \frac{e^{zi}}{\sum_{j=1}^{K} e^{z_j}}$$

  where,

  - $z_i$ is the raw score (logit) for the $i$th class.
  - $K$ is the total number of classes.

  The Softmax function exponentiates each logit and then normalizes these values by dividing each by the sum of all the exponentiated logits. This ensures that the output values (probabilities) are between 0 and 1 and sum up to 1.

  *Key Characteristics*

  - Conversion to Probabilities: The primary role of the Softmax function is to convert logits into probabilities, which are more interpretable and useful for classification tasks.
  - Multi-Class Classification: Softmax is ideal for multi-class classification

problems where each class is mutually exclusive. The function provides a probability for each class.

– Sensitivity to Large Logits: Due to the exponential function, the Softmax function is highly sensitive to large logits, making it more prone to issues with outliers or extreme values.

The Softmax function is commonly used in the output layer of neural networks for multiclass classification problems, such as image classification, where the network must identify an object as belonging to one of several possible categories.

Figure 9.2 illustrates the various activation functions commonly used in neural networks, showcasing various behaviors and outputs. The plotted functions include:

- Sigmoid Function: It is characterized by its S-shaped curve, mapping inputs to a range between 0 and 1.
- Hyperbolic Tangent Function: It is similar to the sigmoid function but outputs values from –1 to 1, making it zero-centered.
- ReLU Function: It outputs the input value if it is positive, else outputs zero, known for its computational efficiency.
- Leaky ReLU Function: This is a variant of the ReLU function that allows a small slope for negative values to prevent neurons from dying.
- ELU Function: This provides an exponential output for negative inputs, aiming to bring mean activations closer to zero.
- Softmax Function: This is typically used in the output layer for multi-class classification, converting logits to probabilities.

Each plot provides a visual representation of how these functions transform their inputs, highlighting their unique properties. These activation functions are integral to neural network designs, influencing their ability to learn and complex model patterns in data.

## *How Neural Networks Work and Learn*

A neural network is composed of layers of neurons, including an input layer, one or more hidden layers, and an output layer. Each neuron in a layer is connected to neurons in the previous and subsequent layers. These connections are weighted and play a pivotal role in the network's learning process.

- **Input Layer**: This receives the initial data for processing.
- **Hidden Layers**: They perform computations using weighted inputs and send the results to the next layer.

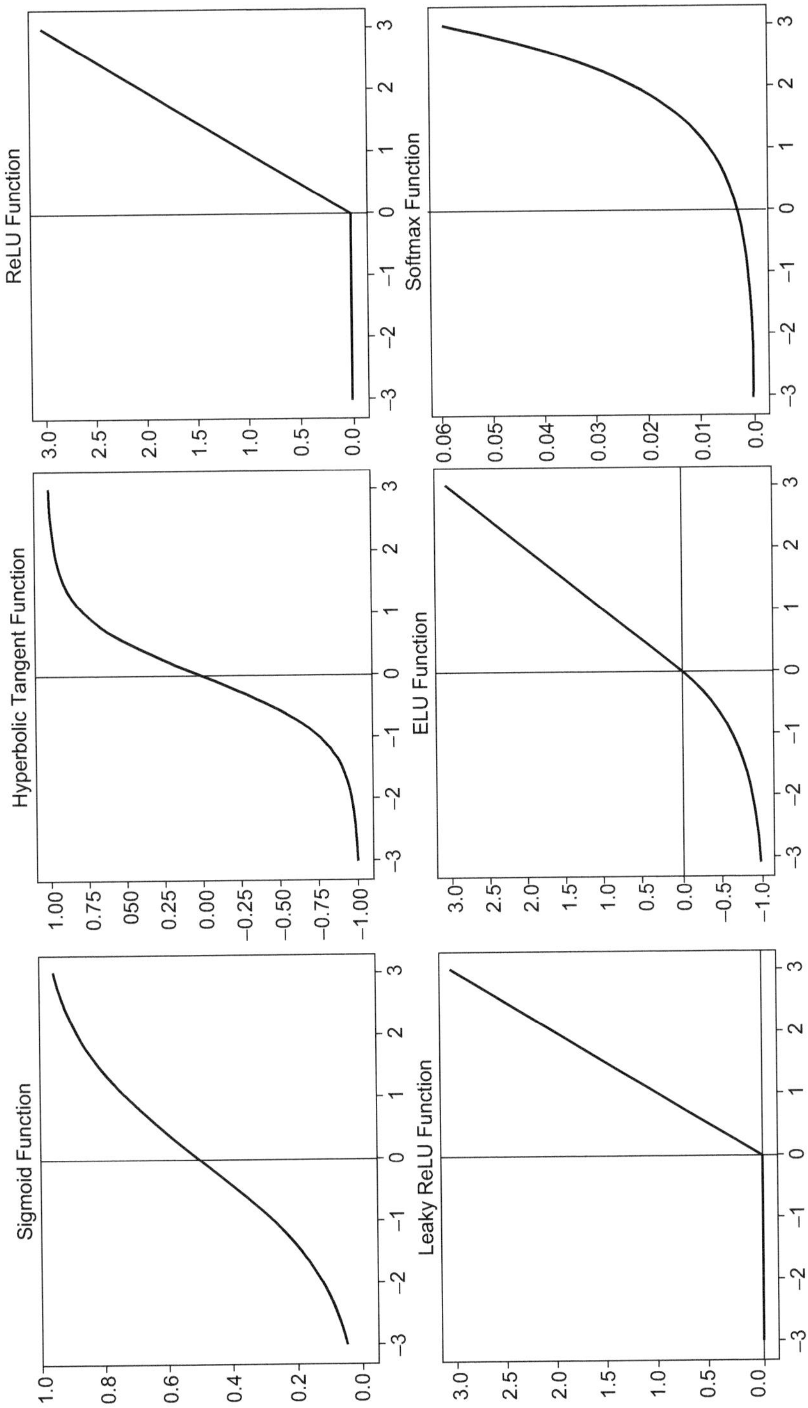


**Figure 9.2: Visualization of common neural network activation functions.**

- **Output Layer**: This produces the final output, representing the network's prediction or decision.

The process of calculating the output of a neural network is known as forward propagation. It involves the following steps:

**Combining Inputs with Weights**: Each neuron receives inputs, multiplies them by their weights, and adds a bias term. Mathematically, this is represented as:

$$z = \sum_{i=1}^{n} (w_i \cdot x_i) + b$$

where, $w_i$ is the weight, $x_i$ is the input, $b$ is the bias, and $n$ is the number of inputs to the neuron.

**Applying Activation Function:** The combined value $z$ is then passed through an activation function $\phi$:

$$a = \phi(z)$$

where, $a$ is the output of the neuron.

## Mechanism of Backpropagation

As indicated in Chapter 7, backpropagation works by calculating the gradient (rate of change) of the loss function (error measure) with respect to each weight in the neural network. This is done in a backward manner, starting from the output layer, and moving towards the input layer.

### *Key Steps in Backpropagation*

- **Forward Pass:** This computes the output of the neural network (forward propagation) and the resulting error or loss.
- **Compute Gradients:** They calculate the gradient of the loss function with respect to each weight. This is done by applying the chain rule of calculus.
- **Update Weights:** Adjust the weights and biases in the network in a direction that reduces the error. This is typically done using gradient descent.

## Gradient Descent in Neural Networks

Gradient descent is an essential optimization algorithm used in training neural networks. It is a technique for minimizing the cost (or loss) function, a measure of how far a model's predictions are from the actual values. The efficiency and effectiveness of neural network training largely depends on how well gradient descent is implemented and tuned.

### *The Concept of Gradient Descent*

The idea behind gradient descent is to iteratively adjust the parameters (weights and biases) of the neural network to minimize the cost function. The gradient (or derivative) of the cost function with respect to each parameter indicates the direction in which the parameter should be adjusted to decrease the cost.

#### *The Cost Function*

A common cost function used in neural networks, especially for regression tasks, is the Mean Squared Error (MSE), defined as:

$$MSE = \frac{1}{N}\sum_{i=1}^{N}(y_i - \hat{y}_i)^2.$$

> **The cost or loss function could be Mean Squared Error, Cross-Entropy Loss, or any other function suitable for measuring the performance of the network for a specific task.**

#### *Gradient Descent Formula*

This adjusts the weights and biases in the direction that reduces the error. The adjustment is proportional to the gradient and a learning rate $a$. The core formula for updating each parameter in gradient descent is:

$$\theta_{new} = \theta_{old} - a \cdot \nabla_{\theta} J(\theta)$$

where,

- $\theta$ represents the parameters (weights/biases) of the neural network.
- $a$ is the learning rate, a hyperparameter that controls the size of the steps taken towards the minimum.
- $J(\theta)$ is the cost function.
- $\nabla_{\theta} J(\theta)$ is the gradient of the cost function with respect to its parameters.

The learning rate is a crucial hyperparameter in gradient descent. If it is too small, the algorithm will converge slowly, taking a long time to reach the minimum. If it is too large, the algorithm might overshoot the minimum, leading to divergence.

#### *Types of Gradient Descent*

- **Batch Gradient Descent**: Computes the gradient of the cost function using the entire dataset. This method is precise but can be very slow and computationally expensive with large datasets.
- **Stochastic Gradient Descent (SGD):** Computes the gradient for each sample in the dataset. This method is much faster and can help escape local minima, but the frequent updates lead to a high variance in the cost function.

**Mini-batch Gradient Descent**: A compromise between batch and stochastic gradient descent. It computes the gradient on small batches of data. This method balances the efficiency of SGD with the precision of batch gradient descent.

### *Gradient Descent in Practice*

In practice, gradient descent is implemented through an iterative process:

- Initialize the network parameters with initial values.
- Calculate the gradient of the cost function with respect to each parameter.
- Update the parameters using the gradient descent formula.
- Repeat steps 2 and 3 until the cost function converges to a minimum.

Gradient descent enables models to learn from data and improve their accuracy. Its implementation and tuning are critical for the successful training of neural networks, particularly in how it balances the speed of convergence with the risk of overshooting or getting stuck in local minima.

## Network Architectures in Neural Networks

Neural networks obtain their computational power from a parallel and interconnected structure rather than from single neurons operating independently. This leads to various neural network architectures, each suited for different types of tasks and data processing.

### Single Layer of Neurons

Consider the architecture of a single layer of neurons, as illustrated in Figure 9.3. In such a network, there are '$p + 1$' inputs feeding into '$s$' neurons. Each neuron in this layer can utilize a distinct transfer or activation function, allowing for flexibility in how inputs are processed. The network's weights are organized in

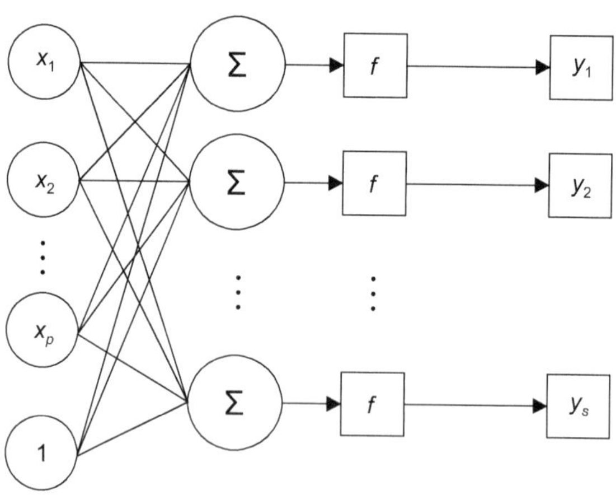


**Figure 9.3: The structure of a layer of neurons.**

a matrix, with each weight corresponding to the connection strength between an input and a neuron. In this matrix, the row index indicates the neuron's position, while the column index corresponds to the specific input. The input value 1 is used to accommodate for the bias of the network.

### Multiple Layers of Neurons

A more complex structure is found in networks with multiple layers of neurons. These networks typically consist of an input layer, several hidden layers, and an output layer. The input layer serves as the entry point for external inputs, and the output layer produces the network's final output. Hidden layers, each with their own weights, biases, and transfer functions, perform the bulk of processing. The weights of a layer are denoted as $w_{ij}$, where $i$ indicates the layer number and $j$ represents the connected neuron.

Using multiple layers of nonlinear units significantly enhances the network's capabilities compared to a single-layer architecture. For example, a multilayer network can approximate complex functions using a combination of sigmoid and linear functions across its layers. Such multilayered networks are suitable for tasks like pattern classification and function approximation, including modelling and prediction.

One example of a multilayer neural network is the multilayer perceptron (Figure 9.4), known for its versatility and effectiveness in various applications.

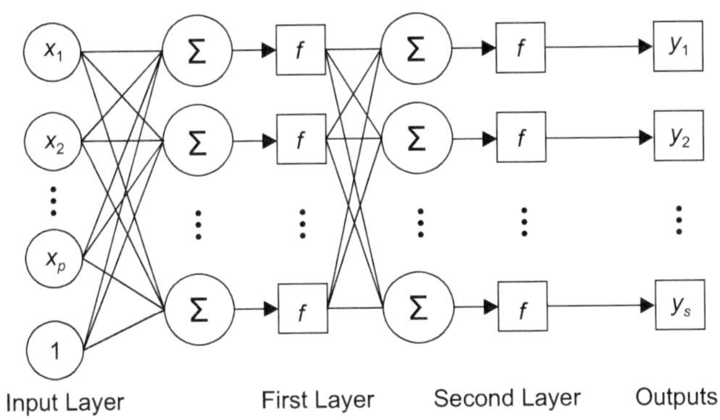


**Figure 9.4: The structure of a multiple layer of neurons.**

## *Training Neural Networks in Pytorch*

When designing a neural network for classification tasks, several key hyperparameters need to be considered. These parameters may vary slightly depending on whether the task is binary classification or multiclass classification.

### Hyperparameters

- Input Layer Shape (in_features):
  - Binary Classification: The number of input features corresponds to the number of variables in your dataset. For instance, in a lameness disease prediction model this could be the number of animal characteristics like age, breed, weight, and activity levels.
  - Multiclass Classification: The input layer shape is identical to that in binary classification, as it is determined by the number of features in your dataset, not the number of classes.
- Hidden Layer(s):
  - Both Binary and Multiclass: The number of hidden layers is determined based on the complexity of the problem. At least one hidden layer is typical, but more complex problems may require multiple layers. There is no strict upper limit on the number of layers, but adding too many can lead to overfitting.
- Neurons per Hidden Layer:
  - Both Binary and Multiclass: The number of neurons in each hidden layer is problem-specific and generally ranges from 10 to 512. This parameter should be tuned based on the complexity of the task and the amount of available data.
- Output Layer Shape (out_features):
  - Binary Classification: The output layer typically has a single neuron, representing the probability of belonging to one of the two classes.
  - Multiclass Classification: The number of neurons in the output layer equals the number of classes.
- Hidden Layer Activation Function:
  - Both Binary and Multiclass: ReLU (Rectified Linear Unit) is a common choice for the activation function in hidden layers, though other functions like tanh or Leaky ReLU can also be used.
- Output Activation Function:
  - Binary Classification: The sigmoid function is used to output a probability between 0 and 1, indicating the likelihood of the instance belonging to one class.
  - Multiclass Classification: The softmax function is used to output a probability distribution across all classes.
- Loss Function:
  - Binary Classification: Binary crossentropy loss is suitable as it compares the predicted probabilities with the actual binary labels.

- Multiclass Classification: Cross entropy loss is used, which is a generalization of binary crossentropy for multiple classes.

- Optimizer:
  - Both Binary and Multiclass: Common choices include Stochastic Gradient Descent (SGD) and Adam. The choice depends on the specific problem and dataset.

Understanding these hyperparameters is necessary for designing an effective neural network for classification tasks. While some parameters like the input layer shape and the number of hidden layers and neurons per layer are largely dependent on the specific problem and the dataset, others like the choice of activation functions and loss functions vary based on the type of classification task (binary vs. multi-class). The optimizer is generally chosen based on experimentation and what works best for the specific problem and dataset.

### *Binary Classification Model Example*

For a binary classification task, we use PyTorch's **torch.nn** (for a comprehensive list we suggest you look at the official PyTorch website at https://pytorch.org/docs/stable/nn.html) . We can create a simple feedforward neural network with three layers (input, hidden, and output) as follows. This example will be followed by the appropriate loss function and optimizer for binary classification, and then we will explain the modifications needed for multi-class classification.

```
# Binary classification Model

import torch
import torch.nn as nn
import torch.optim as optim

# To define the model, we use torch.nn.Sequential()
model = nn.Sequential(
    nn.Linear(in_features=10, out_features=20), # First layer with
                                                  10 inputs and
                                                  20 outputs
    nn.ReLU(),               # ReLU activation function
    nn.Sequential(20,10),    # Second layer with 20 inputs and 10
                               outputs
    nn.ReLU(),               # ReLU activation function
    nn.Linear(10,1)          # Output layer with 10 inputs and 1 output
                               (binary)
    nn.Sigmoid()             # Sigmoid activation function for binary
                               classification
)
```

```
# Loss function for binary classification
criterion = nn.BCELoss()

# Optimizer
optimizer =optim.Adam(model.parameters(), lr=0.001)
```

Setting up a simple neural network for a classification task using PyTorch:

- Importing Necessary Libraries
  - import torch: This is the main PyTorch package, which provides tensor computation (like NumPy) with strong GPU acceleration.
  - import torch.nn: A subpackage that contains modules and classes to help create and train neural networks. It includes definitions for various layers, activation functions, and more.
  - import torch.optim: This subpackage provides implementations of various optimization algorithms, which are used to update the weights and biases during training (have a look at https://pytorch.org/docs/stable/optim.html).
- Defining the Neural Network Model
  - nn.Sequential: This is a sequential container that chains together the specified modules and layers in the order they are passed. It allows for the creation of models in a step-by-step fashion.
  - nn.Linear(in_features = 10, out_features = 20): This is a linear or fully connected layer. It applies a linear transformation to the incoming data. Here, in_features = 10 means this layer expects each input sample to have 10 features (or neurons), and out_features = 20 means this layer will output 20 features.
  - nn.ReLU(): This is the Rectified Linear Unit (ReLU) activation function.
  - nn.Linear(20, 10): This is another linear layer which takes the 20 features output by the previous layer and transforms them into 10 features.
  - nn.Sigmoid(): The Sigmoid activation function is used in the final layer for binary classification. It squashes the output between 0 and 1, making it suitable for probability prediction.
- Setting up the loss function
  - nn.BCELoss(): This is the binary cross-entropy loss function. It is suitable for models where the last layer is a sigmoid. If you are interested in other loss functions have a look at the official Pytorch website at https://pytorch.org/docs/stable/nn.html#loss-functions.

- Configuring the Optimizer
  - optim.Adam(...): Adam optimization is an alternative to classical stochastic gradient descent that iteratively updates network weights with training data. It is often more efficient than standard SGD.
  - model.parameters(): This retrieves all the weights and biases in the model, which the optimizer will update during training.
  - lr = 0.001: This sets the learning rate for the optimizer.

*Multiclass Classification Model Example:*

```
# Multiclass classification model

model = nn.Sequential(
    nn.Linear(in_features=10, out_features=20),
    nn.ReLU(),
    nn.Linear(20, 10),
    nn.ReLU(),
    nn.Linear(10, 3) # Assume 3 classes for multiclass classification
    # No activation function here
)

# Loss function for multiclass classification
criterion = nn.CrossEntropyLoss()

# Optimizer remains the same
optimizer = optim.Adam(model.parameters(), lr=0.001)
```

For multiclass classification, the primary differences would be in the output layer and the loss function:

1. The output layer should have as many neurons as there are classes. For example, if there are 3 classes, you would have nn.Linear(10, 3).
2. Instead of using nn.Sigmoid(), you would not apply an activation function in the output layer because nn.CrossEntropyLoss() applies the softmax function internally.
3. Use nn.CrossEntropyLoss() as the loss function.

### Neural Networks for Classification with Non-image Data

Neural networks, particularly those implemented using frameworks like PyTorch, are widely recognized for their effectiveness in image processing and computer vision tasks. However, the versatility of neural networks extends far beyond image data. They can be equally powerful in analyzing and making predictions from various types of data, including time-series data, text, and in our case, accelerometer data.

### *Example 1: Binary Classification*

In our first example, we address a binary classification problem using accelerometer data which involves categorizing data into two distinct groups. Our dataset contain accelerometer readings labeled as *'active'* or *'inactive'*, representing two different states or behaviors (*binary_data.csv*).

### *Example 2: Multiclass Classification*

The second example is concerned with multiclass classification. We utilize the *sheep_data.csv* as a reference, where accelerometer readings are used to classify different behaviors of sheep.

*Data Preparation and Loading*

For both examples, the first step involves *loading* and *preprocessing* the accelerometer data. Data from CSV files is first read into *Pandas DataFrames*. The process includes splitting the data into *features* (accelerometer readings) and *labels* (activity classifications), followed by data encoding for categorical labels and data normalization or standardization as required.

We then convert the preprocessed data into PyTorch *tensors*, which are the fundamental data structures used in PyTorch for building neural networks. Tensors are similar to NumPy arrays but with additional capabilities that are conducive to GPU acceleration.

*Utilizing PyTorch's DataLoader*

To efficiently manage this data during model training, we use PyTorch's *DataLoader*. It allows us to batch the data, shuffle it for each epoch (which helps in generalizing the model), and leverage parallel processing.

## Practical Implementation Using PyTorch

### *Binary Classification Problem*

To follow along, you can find the Jupyter notebook of this example in our GitHub Repository in folder Chapter 9 under the name Chapter_9_Binary_Classification_Neural_Networks.ipynb.

Let's walk through the code and understand each part:

- **Setting up the Environment**

```
import torch
print(torch.__version__)
```

  - Importing PyTorch: The import torch statement is used to import the PyTorch library. The printed version (torch._version_) helps ensure that you have the correct version installed.

```
if torch.cuda.is_available():
    device = "cuda"
elif torch.backends.mps.is_available():
    device = "mps"
else:
    device = "cpu"
```

- Device Selection: This code checks if CUDA (NVIDIA GPU) or MPS (Apple Silicon GPU) is available for PyTorch to use. If so, computations will be directed to the GPU for faster processing. If neither is available, it defaults to using the CPU.

- **Data Loading and Preprocessing**

```
import pandas as pd
from sklearn.model_selection import train_test_split
from sklearn.preprocessing import StandardScaler, LabelEncoder

df = pd.read_csv('binary_data.csv')

X = df.iloc[:, :-1].values
y = df.iloc[:, -1].values

#'y' is our categorical labels array
label_encoder = LabelEncoder()
encoded_labels = label_encoder.fit_transform(y)
encoded_labels

# Split the data
X_train, X_test, y_train, y_test = train_test_split(X, encoded_
    labels, test_size=0.2, random_state=42)

# Fit the scaler on the training data and transform both training
and testing data
scaler = StandardScaler()
X_train = scaler.fit_transform(X_train)
X_test = scaler.transform(X_test)
```

- We load our dataset using Pandas and split it into features (X) and labels (y).
- We use LabelEncoder to encode categorical labels into numerical format, which is necessary for neural network training.
- We split the dataset into training and testing sets.
- StandardScaler is used to standardize the features.

- **Converting Data to PyTorch Tensors**

```
# Turning training the data into tensors
X_train = torch.from_numpy(X_train).type(torch.float)
y_train = torch.from_numpy(y_train).type(torch.long)
X_test = torch.from_numpy(X_test).type(torch.float)
y_test = torch.from_numpy(y_test).type(torch.long)
```

  - We have to convert the NumPy arrays into PyTorch tensors and ensure they have the correct data types (float for features and long for labels).

- **Defining the Neural Network Model**

```
import torch.nn as nn

# Define a  model class that inherits from nn.Module
class BinaryClassifier(nn.Module):
    def __init__(self):
        super().__init__()

        # First linear layer with 82 inputs and 64 outputs
        self.fc1 = nn.Linear(82, 64)

        # Additional linear layers to increase model complexity
        self.fc2 = nn.Linear(64, 32)  # Second layer
        self.fc3 = nn.Linear(32, 64)  # Third layer

        # Final output layer for binary classification
        # Outputs a single value from 64 input features
        self.fc4 = nn.Linear(64, 1)

        # Activation function to introduce non-linearity
        self.relu = nn.ReLU()

    # Define the forward pass through the network
    def forward(self, x):
        # Apply layers with ReLU activation functions in between for
           non-linearity
        x = self.relu(self.fc1(x))
        x = self.relu(self.fc2(x))
        x = self.relu(self.fc3(x))

        # No activation function in the final layer
        x = self.fc4(x)
        return x

# 'device' is defined (e.g., as 'cuda' or 'cpu')
# Create an instance of the model and send it to the specified
device
```

```
binary_model = BinaryClassifier().to(device)
binary_model
```

- We define a neural network model class BinaryClassifier inherited from nn.Module.
- The model consists of linear layers and ReLU activations, ending in a single output neuron for binary classification.
- We move the model to the chosen computation device (GPU or CPU).

- **Setting Loss Function and Optimizer**

```
# loss and optimizer function

criterion = torch.nn.BCEWithLogitsLoss()
optimizer = torch.optim.Adam(params=binary_model.parameters(),
    lr=0.01)
```

- BCEWithLogitsLoss is used as the loss function, suitable for binary classification tasks.
- The Adam optimizer is chosen for optimizing the model parameters.

- **Training the Binary Classification Model**

```
torch.manual_seed(42)
epochs = 5000

# Put all data on target device
X_train, y_train = X_train.to(device), y_train.to(device).float()
X_test, y_test = X_test.to(device), y_test.to(device).float()

for epoch in range(epochs):
    # Training Phase
    binary_model.train()
    y_logits = binary_model(X_train).squeeze()
    y_pred = torch.round(torch.sigmoid(y_logits))

    # Calculate loss
    loss_value = criterion(y_logits, y_train)

    # Calculate accuracy
    correct_train = (y_pred == y_train).sum().item()
    train_accuracy = 100 * correct_train / len(y_train)

    # Zero gradients, backward pass, optimize
    optimizer.zero_grad()
```

```
        loss_value.backward()
        optimizer.step()

        # Testing Phase
        binary_model.eval()
        with torch.no_grad():
            test_logits = binary_model(X_test).squeeze()
            test_pred = torch.round(torch.sigmoid(test_logits))
            test_loss = criterion(test_logits, y_test)

            correct_test = (test_pred == y_test).sum().item()
            test_accuracy = 100 * correct_test / len(y_test)
        if epoch % 100 == 0:
            print(f"Epoch: {epoch} | Train Loss: {loss_value.item()},
            Train Accuracy:
            {train_accuracy:.3f}% | Test Loss: {test_loss.item()}, Test
                  Accuracy: {test_accuracy:.3f}%")

binary_model.train()
```

- torch.manual_seed(42) sets the seed for generating random numbers. This is done to ensure reproducibility of the results.
- epochs = 5000 defines the number of times the entire dataset will pass through the neural network.
- The feature tensors (X_train, X_test) and label tensors (y_train, y_test) are moved to the specified device (CPU or GPU).
- .float() converts the tensors to floating-point numbers, which is the required data type for computation in neural networks.
- The loop iterates over the dataset epochs times. Each iteration is an epoch.
- binary_model.train() sets the model to training mode.
- The model predicts outputs (y_logits) for the training data. .squeeze() removes any extra dimensions.
- y_pred is obtained by applying the sigmoid function (to convert logits to probabilities) and rounding off to get binary predictions.
- The loss is calculated using the criterion, which compares the model's predictions with the true labels.
- Training accuracy is calculated by comparing predicted labels with true labels and computing the percentage of correct predictions.
- Gradients are zeroed with optimizer.zero_grad() to prevent accumulation from previous iterations.
- loss_value.backward() computes the gradient of the loss with respect to the model parameters.

- optimizer.step() updates the model parameters based on the calculated gradients.
- binary_model.eval() sets the model to evaluation mode.
- torch.no_grad() ensures that gradients are not calculated during testing, which saves memory and computations.
- Similar to the training phase, the model makes predictions on the test set, and test loss and accuracy are calculated.
- Printing the training and testing loss and accuracy every 100 epochs to monitor the model's performance.
- binary_model.train(): This resets the model to training mode for any further training steps.

- **Final Evaluation Step**

```
import numpy as np

# Evaluating the model
binary_model.eval()

with torch.no_grad():
    y_preds = torch.round(torch.sigmoid(binary_model(X_test))).
    squeeze()

# Ensure y_preds is of the same data type and device as y_test_encoded
    y_preds = y_preds.type_as(y_test)

# Calculate the number of correct predictions
correct_predictions = (y_preds == y_test).sum()

# Calculate the accuracy percentage
accuracy_percentage = 100 * correct_predictions.item() / len(y_test)

print(f"Accuracy: {accuracy_percentage:.2f}%")

# Output
Accuracy: 99.98%
```

Purpose of the Separate Evaluation Step:

- Consolidated Model Evaluation: While our training loop includes test accuracy calculation at each epoch, performing a separate evaluation after training provides a consolidated, single measure of how well the model performs on the test set. This is particularly useful for reporting and analyzing the final performance of the model.

- Model State and Gradient Tracking: By explicitly setting binary_model.eval() and using torch.no_grad(), we ensure the model is in evaluation mode and not tracking gradients. This is standard practice when evaluating a model as it can lead to reduced memory usage and improved computational efficiency.
- Data Type and Device Consistency: The line y_preds = y_preds.type_as(y_test) ensures that the predictions and test labels are of the same data type and on the same device. This is important for accurately comparing these tensors and calculating metrics like accuracy.
- Final Accuracy Calculation: This step calculates the overall accuracy of the model on the entire test set, which is a common metric for evaluating classification models. It gives us a straightforward percentage indicating how often the model predicted correctly.

### *Multiclass Classification Problem*

Now that we have a solid understanding of training a binary classifier, we will proceed to address a multiclass classification problem. In the upcoming section, we will explore how to adapt our approach to handle multiple classes.

As shown in the accompanying Jupyter notebook (Chapter_9_Multiclass_Classification_Neural Networks_with_PyTorch.ipynb) the loading and preprocessing of the multiclass dataset for our neural network model follow the same principles as those applied to the binary dataset. Therefore, we will not repeat those steps here. Instead, we will focus on the subsequent stages of developing our multiclass classification model using PyTorch.

We begin by defining the structure of our neural network. The *MulticlassClassifier* is a PyTorch nn.Module subclass, designed to handle multiclass classification problems. The model comprises several linear layers, followed by ReLU activation functions. The final layer's number of nodes corresponds to the number of classes in our dataset. Here is the model definition:

```
import torch.nn as nn

class MulticlassClassifier(nn.Module):
    def __init__(self, input_size, num_classes):
        super(MulticlassClassifier, self).__init__()
        # Define the layers of the neural network
        self.fc1 = nn.Linear(input_size, 64)  # First linear layer
        self.fc2 = nn.Linear(64, 128)         # Second linear layer
        self.fc3 = nn.Linear(128, 64)         # Third linear layer
        self.fc4 = nn.Linear(64, num_classes) # Output layer

        # Activation function
```

```
        self.relu = nn.ReLU()

    def forward(self, x):
        # Forward pass through the network
        x = self.relu(self.fc1(x))  # Activation function after
                                     first layer
        x = self.relu(self.fc2(x))  # Activation function after
                                     second layer
        x = self.relu(self.fc3(x))  # Activation function after
                                     third layer
        x = self.fc4(x)              # No activation function in the
                                     output layer
        return x
```

Setting up a MulticlassClassifier neural network for a multiclass classification problem with specific configurations for the number of input features, number of classes, loss function, and optimizer.

```
num_features = 82
num_classes = 5

multiclass_model = MulticlassClassifier(num_features, num_classes).
    to(device)

# loss and optimizer function
criterion = nn.CrossEntropyLoss()
optimizer = torch.optim.Adam(multiclass_model.parameters(), lr=0.001)
```

- num_features = 82: This specifies that the input data has 82 features.
- num_classes = 5: This indicates that there are 5 distinct classes that the model needs to classify.
- MulticlassClassifier(num_features, num_classes): This creates an instance of the MulticlassClassifier with the specified number of input features and classes.
- criterion = nn.CrossEntropyLoss(): This line sets up the loss function for the model. CrossEntropyLoss is a standard loss function for multiclass classification tasks in PyTorch.
- optimizer = torch.optim.Adam(multiclass_model.parameters(), lr = 0.001): This line defines the optimizer for the training process. The learning rate (lr) is set to 0.001, a common starting point that can be adjusted based on the specific requirements of your training process.

### Training and Testing the Model

```
import torch

# Assuming multiclass_model, criterion (CrossEntropyLoss), optimizer
    are defined
# Also, assuming X_train, y_train, X_test, y_test are your datasets

torch.manual_seed(42)

# Number of epochs
# You can adjust this number
epochs = 1000

X_train, y_train = X_train.to(device), y_train.to(device)
X_test, y_test = X_test.to(device), y_test.to(device)

for epoch in range(epochs):
    # Train the model
    multiclass_model.train()

    # Forward pass
    y_log = multiclass_model(X_train)
    # Convert logits to prediction labels for accuracy calculation
    y_pred = torch.argmax(y_log, dim=1)

    # loss and Accuracy
    loss = criterion(y_log, y_train)
    correct_train = (y_pred == y_train).sum().item()
    train_accuracy = 100 * correct_train / len(y_train)

    # Zero gradients
    optimizer.zero_grad()

    # Perform backpropagation
    loss.backward()

    # Optimizer step
    optimizer.step()

    #Testing the model using .eval()
    multiclass_model.eval()
    with torch.no_grad():
        # Forward pass on test data
        test_log = multiclass_model(X_test)
        test_pred = torch.argmax(test_log, dim=1)

        # Test loss and test accuracy
        test_loss = criterion(test_log, y_test)
        correct_test = (test_pred == y_test).sum().item()
        test_accuracy = 100 * correct_test / len(y_test)
```

```
    # Print training status
    if epoch % 10 == 0:
        print(f"At Epoch: {epoch} | Training Loss: {loss:.5f},
          Training Accuracy:
          {train_accuracy:.2f}% | Test Loss: {test_loss:.5f}, Test
          Accuracy:
          {test_accuracy:.2f}%")
```

Since we have already worked with a binary classifier and this process is similar, we will focus on highlighting the differences that apply to multiclass classification:

- For multiclass classification, the model's output (y_log) consists of logits for each class. To get the predicted class labels, *torch.argmax* is used, which selects the class with the highest logit value. This is different from the binary case, where the output is typically a single probability score per instance.
- The criterion here is CrossEntropyLoss, but in the binary case it was BCEWithLogitsLoss(). CrossEntropyLoss is appropriate for multiclass problems as it calculates the loss by applying a softmax to the logits to convert them into probabilities.
- Similar to the training phase, during the test phase, the model outputs logits for each class, and torch.argmax is used to determine the predicted class.

The final step in evaluating our MulticlassClassifier involves generating a detailed classification report, which provides insights beyond mere accuracy.

```
from sklearn.metrics import classification_report

# Retrieve class names from label encoder
class_names = label_encoder.classes_

# Set model to evaluation mode
multiclass_model.eval()

# No gradient computation in evaluation to save memory and
computations
with torch.no_grad():
    outputs = multiclass_model(X_test)
    _, predicted = torch.max(outputs.data, 1)

# Convert the tensors back to numpy arrays for sklearn compatibility
y_test_np = y_test.cpu().numpy()
predicted_np = predicted.cpu().numpy()
```

```
# Generate the classification report
report = classification_report(y_test_np, predicted_np, target_
names=class_names)

print(report)
# Output
    precision    recall   f1-score   support
      grazing       1.00       1.00       1.00      3186
      resting       0.97       0.97       0.97      5568
   scratching       0.92       0.90       0.91       116
     standing       0.95       0.94       0.95      2800
      walking       0.99       1.00       0.99      1256

     accuracy                             0.97     12926
    macro avg       0.97       0.96       0.96     12926
 weighted avg       0.97       0.97       0.97     12926
```

- class_names = label_encoder.classes_: This retrieves the original class names that were encoded by LabelEncoder. These names provide a human-readable format for the classes in the report.
- multiclass_model.eval(): This sets the model to evaluation mode, which is important for certain layers like dropout layers to behave correctly during inference.
- torch.no_grad(): This ensures that no gradients are computed, reducing memory usage and speeding up computations.
- outputs = multiclass_model(X_test): This generates predictions for the test set. The model outputs logits for each class.
- _, predicted = torch.max(outputs.data, 1): This determines the predicted class for each example in the test set by selecting the class with the highest logit value.
- The tensors (y_test and predicted) are converted to numpy arrays to work with classification_report from scikit-learn.
- classification_report(y_test_np, predicted_np, target_names = class_names): This produces a classification report, which includes metrics like precision, recall, and F1-score for each class based on the predictions.
- print(report): This prints the classification report to the console.

**How to Get Evaluation Metrics using PyTorch:**

- Accuracy: torchmetrics.Accuracy().
- Precision: torchmetrics.Precision().
- Recall: torchmetrics.Recall().

- F1-score: torchmetrics.F1Score().
- Confusion matrix: torchmetrics.ConfusionMatrix().

This completes our introduction to neural networks and their application to both binary and multiclass classification problems using PyTorch. Through these sections, we have seen how to structure neural network models for different types of classification tasks, how to train, and finally evaluate these models.

## Convolutional Neural Networks

This section provides an exploration of CNNs, a class of deep neural networks highly effective in analyzing visual imagery, time-series data, and other structured data. We will particularly focus on applying CNNs to accelerometer data.

### *Introduction to CNNs*

CNNs (Bengio, 2009; LeCun et al., 2015) are a class of deep neural networks, primarily used in analyzing visual imagery, and other data-intensive applications like time-series analysis. CNNs are known for their ability to learn spatial hierarchies of features automatically and adaptively from input data.

**Convolutional Operation:** The core of a CNN is the convolutional operation. This involves sliding a filter or kernel over the input data and computing the dot product of the filter with the input data at each position. A convolutional layer consists of several such filters, and each filter generates a feature map. Mathematically, this operation for a single filter can be expressed as:

$$F(I)_{x,y} = \sum_{i=-k}^{k} \sum_{j=-k}^{k} I_{x+i,y+j} \cdot K_{i,j}$$

where, $F(I)_{x,y}$ is the feature map, $I$ is the input data, $K$ is the kernel, and $k$ is the kernel size.

**Role of Convolution**: The convolutional layer acts as a feature extractor that learns the spatial hierarchies in the data. Lower layers might learn basic features like edges, while deeper layers can learn more complex features like shapes or specific objects in the case of image data.

*Key Components*

- **Convolutional Layers:** These layers perform the convolution operation, extracting features from the input data.
- **Pooling Layers**: Following convolutional layers, pooling layers (like max pooling) are used to reduce the spatial dimensions (width and height) of the input volume for the next convolutional layer. It helps in reducing the computational load, memory usage, and also help in making the detection of features invariant to scale and orientation changes.

- **Fully Connected Layers**: At the end of a CNN architecture, fully connected layers are used where neurons are connected to all activations in the previous layer. These layers are typically used to flatten the output and produce the final output predictions. In a classification task, the last fully connected layer uses a softmax function to output probabilities of classes.

*The Role of CNNs in Pattern Recognition and Feature Extraction*

CNNs are exceptional at picking up patterns in the input data which are spatially and temporally correlated. They are able to learn both low-level and high-level features, making them efficient for image and video recognition, image classification, image analysis, and time-series forecasting. The learned features are robust to common variations in the input, making CNNs versatile for various applications.

*Historical Background*

The foundational principles of CNNs were established in the 1980s with the development of the Neocognitron by Kunihiko Fukushima (Fukushima, 1980). However, CNNs came into view in the 1990s following the introduction of LeNet-5 by Yann LeCun, which was used for digit recognition tasks.

*Key Milestones and Influential Models*

The evolution of CNNs is marked by several key models that have significantly influenced the field of deep learning. Below is a summary of these models:

- LeNet (1998):
  - **Developer**: Yann LeCun et al.
  - **Features**: One of the first successful applications of CNNs. It had a relatively simple architecture with 5 layers alternating between convolutional and pooling layers. Used tanh/sigmoid activation functions.
  - **Applications**: Recognizing handwritten and machine-printed characters.

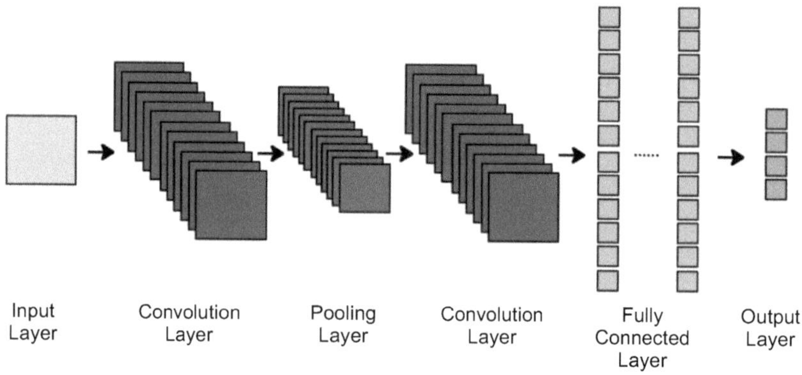


**Figure 9.5: Schematic of a simple CNN architecture.**

- AlexNet (2012):
  - **Developer**: Alex Krizhevsky, Ilya Sutskever, and Geoffrey Hinton.
  - **Features**: A deeper and wider architecture compared to LeNet. It introduced the use of the ReLU activation function, implemented dropout layers, and utilized GPUs for training.
  - **Applications**: Large-scale image recognition tasks.
- ZFNet (2013):
  - **Developer**: Matthew Zeiler and Rob Fergus.
  - **Features**: This is built on the architecture of AlexNet but with different filter sizes and numbers of filters. Additionally, visualization techniques are introduced for understanding the network.
  - **Applications**: ImageNet classification.
- VGGNet (2014):
  - **Developer**: Karen Simonyan and Andrew Zisserman.
  - **Features**: It is characterized by deeper networks with smaller filters (3×3), and standardized depth across all convolutional layers. It comes in multiple configurations like VGG16 and VGG19.
  - **Applications**: Large-scale image recognition.
- ResNet (2015):
  - **Developer**: Kaiming He et al.
  - **Features**: It introduces "skip connections" or "shortcuts" to enable the training of deeper networks. It is available in multiple configurations like ResNet-50, ResNet-101, and ResNet-152.
  - **Applications**: Large-scale image recognition; notably won the 1st place in the ILSVRC 2015.
- GoogleLeNet (2014):
  - **Developer**: Christian Szegedy et al. at Google.
  - **Features**: Introduced the Inception module, allowing for more efficient computation and enabling deeper networks. Evolved through multiple versions including Inception v1, v2, v3, and v4.
  - **Applications**: Large-scale image recognition; won 1st place in the ILSVRC 2014.
- MobileNets (2017):
  - **Developer**: Andrew Howard et al.
  - **Features**: Designed for mobile and embedded vision applications. Uses depthwise separable convolutions to reduce the model size and complexity.

– **Applications**: Mobile and embedded vision applications, real-time object detection.

Each of these models has contributed to the development and understanding of CNNs, pushing the boundaries of what is possible in the field of computer vision and beyond. They serve as foundations and inspirations for contemporary deep learning research and applications.

## Architectural Components of CNNs

### *Convolutional Layers*

Convolutional layers are the building blocks of CNNs. As mentioned before, the primary operation in a convolutional layer is the convolution operation, which mathematically combines two functions to produce a third function.

**Stride and Padding:** Two important concepts in the convolution operation are stride and padding. The stride controls how the filter convolves around the input volume, while padding involves adding zeros around the input feature map to control the spatial size of the output volumes.

*Filters/Kernels and Feature Map Generation*

**Filters/Kernels**: A filter (or kernel) in a convolutional layer is a small matrix used to extract features such as edges, textures, or more complex patterns in deeper layers. By convolving each filter across the input volume's width and height, and calculating the dot product of the filter's entries with the input, a two-dimensional activation map for that filter is generated.

**Feature Maps:** As a result of the convolution operation, a feature map is generated. This map is essentially the output of one filter applied to the previous layer. A CNN layer consists of several such filters (each learning different features), and thus produces a stack of feature maps. These feature maps then act as input to the subsequent layers.

**Learning Feature Representations**: The network learns to activate certain filters more than others depending on the input data.

To provide a deeper understanding of how convolutional layers work, let's illustrate with a concrete example. We will explore how an input matrix is transformed by the convolution operation, focusing on the concepts of padding and stride.

*Example of Convolution Operation*

Let's consider a simple example where we have a 5×5 grayscale image (represented as a matrix of pixel values) and a 3×3 filter (or kernel).

- Input Image (5×5 matrix)

$$\begin{pmatrix} 255 & 255 & 255 & 0 & 0 \\ 0 & 255 & 255 & 255 & 0 \\ 0 & 0 & 255 & 255 & 255 \\ 0 & 0 & 255 & 255 & 0 \\ 0 & 255 & 255 & 0 & 0 \end{pmatrix}$$

- Filter (3×3 matrix)

$$\begin{pmatrix} 1 & 0 & -1 \\ 1 & 0 & -1 \\ 1 & 0 & -1 \end{pmatrix}$$

This filter is a simple edge detection filter, often used in image processing for highlighting vertical edges. Let's break down how it works:

- Positive Weights (1s on the Left Side): The positive values on the left side of the filter are designed to react strongly to light areas of the image that are aligned with these weights.
- Negative Weights (–1s on the Right Side): The negative values on the right side of the filter react to dark areas of the image.
- Zero Weights (0s in the Center): The zeros in the middle column of the filter do not influence the result.

When this filter is convolved over an image, it computes the difference in intensity between the left and right sides of the filter. This makes it effective at detecting vertical edges, where there is a significant intensity change from left to right. In areas of the image with a strong vertical edge, this filter produces high values in the feature map, highlighting those edges.

In summary, this filter is a basic example of an edge detection filter in image processing, specifically tuned to highlight vertical edges due to its pattern of positive, zero, and negative weights.

Figure 9.6 illustrates the convolution process in a CNN using a simple 5×5 grayscale image, a 3×3 edge-detection filter, and the resulting feature map. The leftmost image represents the input matrix (or image), the middle image shows the applied 3×3 vertical edge-detection filter, and the rightmost image displays the resulting feature map after convolution, highlighting detected edges.

**Transformation with Stride and Padding**

Let's assume a stride of 1 and no padding (valid padding). Here is how the first convolution operation would look:

- Position the filter over the top-left corner of the image.
- Perform element-wise multiplication and sum the results
- Move the filter one pixel to the right (stride = 1) and repeat the process.

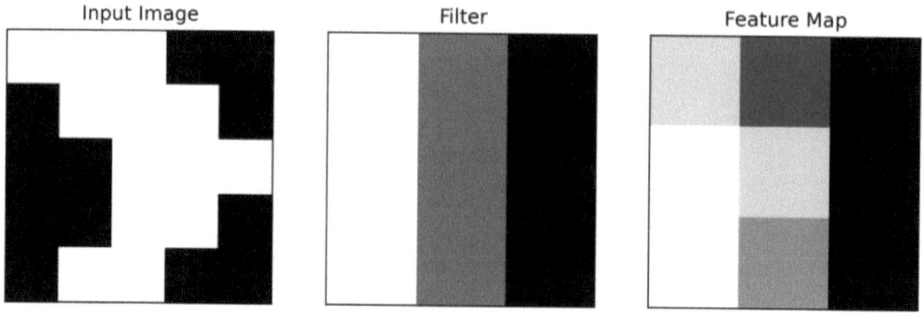


**Figure 9.6: Visualization of convolution operation in a CNN.**

After completing the operation across the entire image, you get a feature map. If the stride is 1 and no padding is used, the output feature map will have a smaller dimension than the input image. In our case, the output feature map will be a 3×3 matrix, as the 3×3 filter can only move three steps horizontally and vertically across the original 5×5 input.

*Result with Padding*

If we apply padding, for instance, a padding of 1 (adding a border of zeros around the input image), the input matrix becomes a 7×7 matrix, with the original image in the middle. The convolution operation with the same filter and a stride of 1 would then produce a 5×5 feature map, preserving the original input size.

By manipulating the stride and padding parameters, you can control the size and resolution of the feature maps produced by the convolutional layers. This plays a significant role in the architecture of the CNN and the level of feature extraction it can perform.

### *Activation Functions in CNNs*

In CNNs activation functions play a crucial role, just as they do in other types of neural networks. As we mentioned before, these functions introduce non-linearity into the model, allowing it to learn more complex patterns in the data.

### *Pooling Layers*

Pooling layers are another critical component in CNNs positioned typically after convolutional layers. Their primary role is to reduce the spatial dimensions, specifically, the width and height of the input feature maps. This dimensionality reduction is crucial for several reasons:

- Reducing Computational Complexity: By downsizing the input's spatial dimensions, pooling layers decrease the number of parameters and computations required in the network. This efficiency is vital in handling large and complex inputs, like high-resolution images, and contributes to faster training times.

- Decreasing Memory Usage: Smaller feature maps mean less memory is needed to store intermediate representations of the data as it passes through the network.
- Mitigating Overfitting: Lowering the number of parameters in the model helps in reducing the chance of overfitting, where a model becomes too closely fitted to the training data and fails to generalize well to new data.
- Enhancing Feature Extraction: Pooling layers contribute to the network's ability to be invariant to small translations and distortions. For example, max pooling ensures that the presence of specific features is recognized regardless of minor shifts or rotations in the input image. This invariance is crucial for the model to recognize objects and patterns robustly, irrespective of their specific positioning or orientation in the input.

*Types of Pooling*

- Max Pooling: This is a commonly used type of pooling in CNNs. Max pooling operates by sliding a window (often 2×2) over the input and taking the maximum value from each section of the feature map covered by the window. By doing so, it effectively captures the most prominent feature in each portion of the feature map, ensuring that these features are retained even as the dimensionality of the data is reduced.
- Average Pooling: In contrast to max pooling, average pooling calculates the average of all values within its window on the feature map. It produces a smoothed, averaged representation of the features. This type of pooling is less common but can be useful in scenarios where retaining the background information is as important as the prominent features.

### Fully Connected Layers in CNNs

Fully Connected (FC) layers are typically positioned towards the end of the network. In these layers, neurons are interconnected with every neuron in the preceding layer, hence the term *fully connected*. FC layers integrate and interpret the features extracted by the convolutional and pooling layers.

**Integration of Features:** The FC layers take the high-dimensional output from the convolutional and pooling layers, which typically consists of localized feature maps, and transform it into a one-dimensional vector. This transformation, known as *flattening*, is crucial for synthesizing the features across the entire input space, allowing the network to understand the global context rather than just localized features.

**Decision Making:** FC layers serve as the decision-making component of the CNN. They synthesize the learned features and map them to the desired output. This output could be class labels in a classification task, continuous values in regression, or even more complex outputs in advanced applications. For instance, in an image classification task, these layers might take the features extracted from previous layers (such as edges, textures, and so on) and use them to identify the presence of an object like a sheep or a pig in the image.

**Hierarchy of Feature Learning:** The sequence of convolutional layers followed by pooling layers in a CNN establishes a hierarchical pattern learning mechanism. The initial layers tend to capture basic patterns like edges, and as we go deeper into the network, subsequent layers build upon these to recognize more intricate and complex patterns. The FC layers, being at the end of this hierarchy, play a pivotal role in interpreting these patterns in the context of the specific task at hand.

### *Output Layer*

The output layer in a CNN is the final layer, responsible for producing the final result or prediction of the network. Its design and function are directly linked to the specific task that the CNN is intended to perform.

**Considerations in Designing the Output Layer**

Number of Neurons: The number of neurons in the output layer corresponds to the number of classes in a classification task or the dimensionality of the output in a regression task.

Activation Function: The choice of activation function in the output layer is crucial and depends on the specific type of problem. For classification, softmax is common, while for regression tasks, linear or other suitable activation functions are used.

Loss Function Association: The design of the output layer is closely linked to the choice of the loss function during the training process. For instance, a softmax output layer is often paired with a cross-entropy loss function in classification tasks.

## Regularization Techniques in Deep Learning

Regularization techniques in deep learning are essential for preventing overfitting. The following are techniques that modify the learning process to produce models that improve performance of CNN.

### *The Concept and Application of Dropout Layers*

Dropout layers, introduced by Nitish Srivastava, Geoffrey Hinton and their colleagues (Srivastava et al., 2014), have become a cornerstone in the design and training of deep neural networks. They provide a simple yet effective mechanism for preventing overfitting.

**Concept of Dropout**

- Random Deactivation: In a dropout layer, a random set of neurons is *dropped out* or deactivated during each training iteration. This means that these neurons do not participate in forward propagation and backpropagation during that particular training pass. Essentially, they are temporarily removed from the network.

- Probability of Dropout: Each neuron has a fixed probability $p$ of being dropped out during training. This probability is a hyperparameter and is typically set between 0.2 and 0.5. A dropout rate of 0.5, for example, means each neuron has a 50% chance of being excluded in each training iteration.
- Impact on Neuron Dependency: By randomly excluding different subsets of neurons, dropout reduces the network's reliance on specific neurons. This encourages the model to learn more robust and redundant representations of the data, as it cannot depend on the presence of particular neurons to correct mistakes.

**Application of Dropout Layers**

- Where to Apply: Dropout is most commonly applied to fully connected layers of neural networks, as these layers are more prone to overfitting due to their high number of parameters. However, it can also be used in convolutional layers, even though with a lower dropout rate.
- Training vs. Inference: During training, neurons are dropped out randomly according to the specified probability. However, during inference (or testing), dropout is disabled, and the entire network is used. To compensate for the larger number of active neurons at test time, the weights learned during training are scaled down. This scaling is usually handled automatically by deep learning frameworks.
- Hyperparameter Tuning: The dropout rate ($p$) is a crucial hyperparameter that needs to be tuned according to the specific dataset and model architecture. It is often determined through cross-validation.

**Benefits and Considerations**

- Reduces Overfitting: Dropout decreases the risk of overfitting by encouraging the learning of more generalizable features.
- Implicit Ensembling: The random exclusion of neurons simulates the effect of training different multiple networks, which is similar to an ensemble method.
- Simplicity and Effectiveness: Dropout is easy to implement and has been proven effective across various applications and network architectures.

### *L1 and L2 Regularization (Weight Regularization)*

L1 and L2 Regularization techniques are encountered multiple times across the book. Here, as we focus on CNNs it is crucial to revisit these techniques, emphasizing their significance in preventing overfitting.

**L1 Regularization (Lasso)**

Mechanism: L1 regularization adds a penalty equal to the absolute value of the magnitude of the weights to the loss function. Mathematically, the L1 loss for a neural network is given by:

$$L_{L1} = L_{original} + \lambda \sum_{i} |w_i|$$

where, $L_{original}$ is the original loss function (such as cross-entropy loss in classification tasks), represents each weight in the network, and λ is the regularization strength, a hyperparameter that controls the degree of regularization.

Effect: L1 regularization encourages the model weights to be sparse, meaning many of the weights will be zero or close to zero.

**L2 Regularization (Ridge)**

Mechanism: L2 regularization adds a penalty equal to the square of the magnitude of the weights. The L2 loss for a neural network is defined as:

$$L_{L2} = L_{original} + \lambda \sum_{i} w_i^2.$$

Effect: Unlike L1, L2 regularization spreads the error among all the weights and tends to drive the weights to small, non-zero values. This technique is effective in handling the overfitting by penalizing the weights with larger magnitudes.

**Application in CNNs**

Incorporation in Loss Function: In CNNs, these regularization terms are added to the loss function during training. As the network learns from the data, the regularization term discourages the model from fitting too closely to the training data by penalizing large weights.

Impact on Learning: Both L1 and L2 regularization influence the training process by minimizing the loss to fit the data and keep the model weights as small as possible. This duality helps the model in generalizing better to unseen data.

**Choice Between L1 and L2**

- L1 regularization is often chosen when we want to impose sparsity on the weights (i.e., making irrelevant weights equal to zero).
- L2 regularization is usually preferred in most scenarios as it tends to perform better in minimizing overfitting without increasing sparsity.

Hyperparameter Tuning: The regularization strength λ is a hyperparameter that requires tuning. Depending on its value, the regularization can have a more or less significant impact on the weights of the network.

**L1 vs. L2 Regularization**

- L1 Regularization: Leads to sparsity and is useful when we have a high number of features, some of which may be irrelevant for prediction.
- L2 Regularization: Tends to give better prediction accuracy as it does not enforce sparse weights.

### *Early Stopping*

Early stopping is a regularization technique used during the training of a deep learning model to prevent overfitting. It involves monitoring the model's performance on a validation set and stopping the training process once the model's performance ceases to improve.

**Concept and Mechanism**

- Monitoring Validation Loss: Early stopping focuses on tracking the loss on a validation dataset, which is separate from the training data. The idea is to identify the point during training when the model's performance on the validation set starts to degrade, indicating overfitting on the training data.
- Stopping Criterion: Training is stopped when the validation loss has not improved for a predetermined number of epochs. This number is often referred to as *patience*. The model weights associated with the lowest validation loss are typically saved and used for future predictions.

**Implementation**

- Training Loop Adjustment: In the training loop, after each epoch, the model is evaluated on the validation set. The loss on this validation set is then compared to the best loss observed in previous epochs.
- Patience Hyperparameter: If the validation loss fails to improve for a consecutive number of epochs (defined by the patience parameter), the training process is halted.
- Model Checkpointing: Often, early stopping is used in conjunction with model checkpointing, where the model weights are saved whenever an improvement in validation loss is observed. This ensures that the best performing model is retained, even if the model's performance degrades in subsequent epochs.

**Benefits**

- Prevents Overfitting: By stopping training early at the right moment, it is ensured that the model does not overfit to the training data, enhancing its ability to generalize to new, unseen data.
- Saves Time and Resources: It reduces unnecessary training time and computational resources by stopping the training process once further training ceases to yield better results on the validation set.

**Considerations**

- Validation Set Selection: The effectiveness of early stopping heavily relies on having a representative validation set. If the validation data does not reflect the distribution of the unseen test set or real-world data, early stopping might halt training too early or too late.
- Patience Setting: Setting the right patience is crucial. Too little patience may stop training prematurely, while too much patience might delay the stopping, leading to overfitting. This hyperparameter often requires tuning based on the specific dataset and model architecture.

### *Data Augmentation*

Data augmentation is a widely used regularization technique in deep learning, particularly effective in tasks like image and speech recognition. It involves artificially expanding the training dataset by applying various transformations to the existing data, thereby creating additional, altered versions of the data points.

**Concept and Purpose**

- Expanding Dataset: Data augmentation increases the size and diversity of the training dataset without the need to collect new data. This is achieved by applying a series of transformations that alter the data in realistic ways.
- Preventing Overfitting: By introducing a variety of modified versions of the original data, data augmentation helps the model learn to generalize better.

**Implementation and Considerations**

- Automated Augmentation: Many deep learning frameworks provide tools to automate the process of data augmentation, applying transformations randomly during training.
- Careful Design: The choice and extent of augmentations should be relevant to the problem domain. For instance, flipping images horizontally might not be appropriate for digit recognition, as it would change the meaning of numbers like 6 and 9.
- Impact on Training: While data augmentation can significantly improve model robustness and generalization, it also increases the computational workload during training, as it effectively enlarges the training dataset.

**Benefits**

- Enhanced Generalization: Models trained on augmented data are less likely to overfit and usually perform better on unseen data.
- Improved Robustness: Augmentation can make models more robust to variations and distortions in input data, which is crucial for real-world applications.

## CNNs in Practice

In this section, we will look into how to prepare accelerometer data for analysis using CNNs with PyTorch (Refer to Chapter_9_CNN.ipynb in the GitHub repository).

To process the accelerometer data from a CSV file for use in a CNN with PyTorch, we will follow these steps:

- Load the dataset from the CSV file.
- Normalize the feature data.
- Encode the categorical labels.
- Reshape the data for CNN input. CNNs expect input data in a certain shape. For PyTorch, the data should be shaped as [batch_size, channels, length], where batch_size is the number of samples in a batch, channels corresponds to the number of sensor axes (usually 3 for accelerometer data), and length is the number of time steps in each segment.
- Convert the data to PyTorch tensors. PyTorch works with tensors, so the preprocessed data needs to be converted into tensor format. This can be done using PyTorch's tensor utilities.
- Create a dataset and a DataLoader for training. In PyTorch, datasets and dataloaders are used to efficiently load data during the training and testing process. The TensorDataset and DataLoader classes are particularly useful for handling batches of data and simplifying the training loop.

Let's start with the code:

**Step 1: Load the Dataset**

```
# 0. Importing the libraries

import pandas as pd
from sklearn.preprocessing import StandardScaler, LabelEncoder
from sklearn.model_selection import train_test_split
import torch
from torch.utils.data import TensorDataset, DataLoader

# 1. Load the dataset
df = pd.read_csv('sheep_data.csv')
df.shape

# Output
(64626, 83)
```

### Step 2: Normalize the Features

```
# 2. Normalise the features
# Separate features and labels
X = df.iloc[:, :-1]  # All columns except the last are features
y = df.iloc[:, -1]   # The last column is the label

# Normalize the features
scaler = StandardScaler()
X_normalized = scaler.fit_transform(X)
```

### Step 3: Encode the Labels

```
# 3. Encode the categorical labels

encoder = LabelEncoder()
y_encoded = encoder.fit_transform(y)
label_names = encoder.classes_
label_names, y_encoded

# Output
(array(['grazing', 'resting', 'scratching', 'standing', 'walking'],
       dtype=object),
 array([0, 0, 0, ..., 4, 4, 4]))
```

### Step 4: Reshape the Data for CNN Input

```
# 4. Reshape data for CNN input (Batch, Channels, Length)

X_reshaped = X_normalized.reshape(-1,1,82)
X_reshaped.shape

# Output
(64626, 1, 82)
```

### Step 5: Convert Data to Tensors

```
# 5. Convert to PyTorch tensors

X_tensor = torch.tensor(X_reshaped, dtype=torch.float32)
y_tensor = torch.tensor(y_encoded, dtype=torch.long)
X_tensor.shape, y_tensor.shape

# Output
(torch.Size([64626, 1, 82]), torch.Size([64626]))
```

**Step 6: Create Dataset and DataLoader**

```
# 6. Create Dataset and DataLoader

dataset = TensorDataset(X_tensor, y_tensor)

# Split the dataset into training and temporary dataset
train_dataset, temp_dataset = train_test_split(dataset, test_
    size=0.3, random_state=42)

# Split the temporary dataset into validation and test datasets
val_dataset, test_dataset = train_test_split(temp_dataset, test_
    size=0.5, random_state=42)

# Create DataLoaders for training, validation, and test sets
train_loader = DataLoader(train_dataset, batch_size=64,
    shuffle=True)
val_loader = DataLoader(val_dataset, batch_size=64)
test_loader = DataLoader(test_dataset, batch_size=64)
```

**Step 7: Define the CNN Model**

After preparing our dataset, the next critical step is defining the structure of CNN. Here, we introduce the AccelerometerCNN class, a custom model extending PyTorch's nn.Module. This model is tailored to handle the specific nature of our accelerometer data.

```
# 7. Define the CNN model

import torch.nn as nn
import torch.nn.functional as F
import torch.optim as optim

class AccelerometerCNN(nn.Module):
    def __init__(self):
        super(AccelerometerCNN, self).__init__()
        # Assuming input shape [batch_size, 1, 82]
        self.conv1 = nn.Conv1d(1, 32, kernel_size=3, padding=1)
        # Output: [batch_size, 32, 82]
        self.pool = nn.MaxPool1d(2) # Output: [batch_size, 32, 41]
        self.conv2 = nn.Conv1d(32, 64, kernel_size=3, padding=1)
        # Output: [batch_size, 64, 41]
        # Apply pooling again: [batch_size, 64, 20]
        self.dropout = nn.Dropout(p=0.5)
        self.fc1 = nn.Linear(64 * 20, 128) # Fully connected layers
        self.fc2 = nn.Linear(128, 5) # Output layer, 5 classes
```

```
    def forward(self, x):
        x = self.pool(F.relu(self.conv1(x)))
        x = self.pool(F.relu(self.conv2(x)))
        x = x.view(-1, 64 * 20) # Flatten for fully connected layer
        x = self.dropout(F.relu(self.fc1(x)))
        x = self.fc2(x)
        return x
```

Key Components of the AccelerometerCNN Class:

- Initialization Method (_init_):
  - We begin by initializing our AccelerometerCNN class, inherited from nn.Module, which provides a base for building neural networks in PyTorch.
  - The network structure comprises convolutional layers (nn.Conv1d), pooling layers (nn.MaxPool1d), a dropout layer (nn.Dropout), and fully connected layers (nn.Linear).
- Convolutional Layers:
  - self.conv1: The first convolutional layer takes a single-channel input (as our data has one feature per time step) and applies 32 filters. The kernel_size is set to 3, and padding is 1. This layer aims to extract features from the input data.
  - self.conv2: A second convolutional layer increases the depth to 64 filters, further enhancing the network's ability to learn complex features.
- Pooling Layer:
  - self.pool: A max pooling layer with a window of size 2. Pooling reduces the spatial dimensions (time steps in our case) by half, decreasing the computation required for subsequent layers and helping to extract dominant features.
- Dropout Layer:
  - self.dropout: A dropout layer with a dropout probability of 0.5. This layer randomly zeroes some of the elements of the input tensor with probability p during training, preventing overfitting.
- Fully Connected Layers:
  - self.fc1: The first fully connected layer reduces the dimensionality from the flattened convolutional layer to 128 neurons.
  - self.fc2: The final output layer maps these 128 features to our 5 classes, corresponding to different activities or outcomes in the accelerometer data.

*Forward Pass Method* (*forward*):

- The forward method defines how the input data x passes through the model.
- It sequentially applies the first convolutional layer, pooling, second convolutional layer, and another round of pooling.
- The output is then flattened and passed through the dropout and fully connected layers.
- The final output from self.fc2 is what the network uses to make its predictions.

This model exemplifies a typical CNN structure suitable for time-series data like accelerometer readings, emphasizing feature extraction through convolutional and pooling layers and classification through fully connected layers. The inclusion of dropout aids in regularization, helping the model generalize better to unseen data.

**Step 8: Setup Early Stopping**

```
# 8. Setup Early Stopping

class EarlyStopping:
    def __init__(self, patience=10, min_delta=0):
        self.patience = patience
        self.min_delta = min_delta
        self.counter = 0
        self.best_loss = None
        self.early_stop = False

    def __call__(self, val_loss):
        if self.best_loss is None:
            self.best_loss = val_loss
        elif val_loss > self.best_loss - self.min_delta:
            self.counter += 1
            if self.counter >= self.patience:
                self.early_stop = True
        else:
            self.best_loss = val_loss
            self.counter = 0
```

Understanding the EarlyStopping Class:

- Class Definition:
  - EarlyStopping is defined as a Python class. It is designed to be called at the end of each epoch during training to determine whether the training process should continue.
- Initialization (_init_method):

- patience: This parameter defines the number of epochs to wait for an improvement in the validation loss before stopping the training. A higher patience gives the model more room to improve, while a lower patience makes the stopping criterion more stringent.
- min_delta: This parameter sets the minimum change in the validation loss to qualify as an improvement. This helps to ignore very small changes in loss and focus on significant improvements.
- counter: Used to keep track of how many epochs have passed without an improvement in validation loss.
- best_loss: Stores the best (lowest) validation loss observed so far during training.
- early_stop: A boolean flag that signals whether training should be stopped.

- The_call_Method:
  - This method is invoked after each epoch with the current validation loss.
  - If the current validation loss has not improved by at least min_delta compared to best_loss, the counter is incremented. If it has improved, best_loss is updated, and the counter is reset.
  - If the counter reaches the patience threshold, early_stop is set to True, indicating that training should be halted.

**Step 9: Initialize the Model, Loss Function, Optimizer, and Move to GPU**

```
# 9. Initialize the Model, Loss Function, Optimizer, and Move to GPU

# Initialize model
model = AccelerometerCNN()

# Move model to GPU if available
device = torch.device("cuda" if torch.cuda.is_available() else "cpu")
model.to(device)

# Loss function with L2 Regularization (Weight Decay)
criterion = nn.CrossEntropyLoss()
optimizer = optim.Adam(model.parameters(), lr=0.001, weight_
    decay=1e-5) # L2 regularization

# Early stopping
early_stopping = EarlyStopping(patience=20, min_delta=0.01)
```

In this step, we set up the foundational elements required for training our CNN model. This involves initializing the model, setting up the loss function and optimizer, and ensuring the model utilizes the GPU for faster computation, if available.

- Model Creation: An instance of the AccelerometerCNN class is created. This instance, model, encapsulates our CNN architecture and is the object we will train and use for predictions.
- Device Selection: This step checks if a GPU with CUDA support is available. If it is, device is set to use the GPU; otherwise, it falls back to the CPU.
- Model to Device: The model.to(device) command moves our model to the chosen device (GPU or CPU), ensuring all computations of the model are performed on that device.
- Loss Function: nn.CrossEntropyLoss is used as the loss function, appropriate for multi-class classification tasks. It combines a softmax layer and a cross-entropy loss in one single class, simplifying the model architecture and the training loop.
- Optimizer: The Adam optimizer is chosen for its efficiency in handling sparse gradients and adaptive learning rates. The learning rate (lr) is set to 0.001, a common choice for many tasks.
- L2 Regularization: weight_decay in the Adam optimizer is used for L2 regularization. It adds a penalty on the size of the weights, helping to prevent overfitting by discouraging overly complex models.
- Early Stopping Setup: An instance of the EarlyStopping class is created with a patience of 20 epochs and a minimum delta (min_delta) of 0.01. This means the training will stop if the validation loss does not improve by at least 0.01 for 20 consecutive epochs.

**Step 10: Model Training and Evaluation on Validation Set**

This step covers the training of the AccelerometerCNN model and its evaluation using the validation set. The goal is to optimize the model parameters based on the training data and to gauge its performance on unseen data (validation set) to prevent overfitting.

```
# 10. Model training and evaluation on validation set

from sklearn.metrics import classification_report, accuracy_score

max_epochs = 1000

for epoch in range(max_epochs):
    model.train()
    train_loss = 0.0
    correct = 0
```

```
    total = 0

    for data, target in train_loader:
        data, target = data.to(device), target.to(device)

        optimizer.zero_grad()
        output = model(data)
        loss = criterion(output, target)
        train_loss += loss.item()
        _, predicted = torch.max(output.data, 1)
        total += target.size(0)
        correct += (predicted == target).sum().item()
        loss.backward()
        optimizer.step()

    train_loss /= len(train_loader)
    train_accuracy = 100 * correct / total

    model.eval()
    val_loss = 0.0
    correct = 0
    total = 0
    with torch.no_grad():
        for data, target in val_loader:
            data, target = data.to(device), target.to(device)
            output = model(data)
            loss = criterion(output, target)
            val_loss += loss.item()
            _, predicted = torch.max(output.data, 1)
            total += target.size(0)
            correct += (predicted == target).sum().item()

    val_loss /= len(val_loader)
    val_accuracy = 100 * correct / total

    print(f'Epoch {epoch+1}, Train Loss: {train_loss:.4f}, Train
          Accuracy:
          {train_accuracy:.2f}%,'f'Val Loss: {val_loss:.4f}, Val
          Accuracy:
          {val_accuracy:.2f}%')

        early_stopping(val_loss)
    if early_stopping.early_stop:
        print("Early stopping triggered")
        break
```

Training and Validation Loop Explained:

- Setting Maximum Epochs: We define max_epochs as 1000, indicating the maximum number of training iterations over the entire dataset unless early stopping criteria are met.

- Training Loop (Epoch-wise):
  - model.train(): Puts the model in training mode, enabling certain layers like dropout layers to function correctly during training.
  - Training Process: For each batch of data in train_loader, the model performs a forward pass, calculates the loss, performs a backward pass, and updates the weights.
  - Tracking Training Loss and Accuracy: The training loss and accuracy are calculated for each epoch. The accuracy is determined by comparing the model's predictions to the actual labels.
- Validation Loop:
  - model.eval(): Puts the model in evaluation mode, disabling dropout layers for consistent performance.
  - Validation Process: Similar to the training loop but without backpropagation and weight updates. It's crucial for monitoring the model's performance on unseen data.
  - No Gradient Calculation: torch.no_grad() is used to ensure gradients are not calculated during the validation pass, reducing memory usage and speeding up computations.
- Print Metrics: At the end of each epoch, the training loss, training accuracy, validation loss, and validation accuracy are printed. This provides insights into the model's learning progress.
- Early Stopping Check: After each epoch, we check if the validation loss has stopped improving using the early_stopping call. If the model doesn't improve for a specified number of epochs (patience), training is stopped. This helps in preventing overfitting.
- Breaking the Loop: If early stopping is triggered, the loop breaks, concluding the training process.

**Step 11: Evaluate the Model on the Test Set**

After training and validating the CNN model, the final step is to evaluate its performance on the test dataset, which consists of data that the model has never seen during the training process.

```
# 11. Evaluate the model

model.eval()
y_true = []
y_pred = []

with torch.no_grad():
```

```
    for data, target in test_loader:
        data, target = data.to(device), target.to(device)
        output = model(data)
        _, predicted = torch.max(output.data, 1)
        y_true.extend(target.cpu().numpy())
        y_pred.extend(predicted.cpu().numpy())

# Classification report with names of the labels
target_names = encoder.classes_
classification_report = classification_report(y_true, y_pred,
                        target_names=target_names)
print("Classification Report:\n", classification_report)

# Calculate accuracy
accuracy = accuracy_score(y_true, y_pred)
print("Accuracy:", accuracy)

# Output
Classification Report:

              precision    recall  f1-score   support
     grazing       0.99      1.00      1.00      2416
     resting       0.98      0.97      0.98      4166
  scratching       0.96      0.89      0.92        88
    standing       0.95      0.96      0.96      2058
     walking       1.00      1.00      1.00       966
    accuracy                           0.98      9694
   macro avg       0.98      0.96      0.97      9694
weighted avg       0.98      0.98      0.98      9694

Accuracy: 0.9789560552919332
```

Process of Model Evaluation:

- Set Model to Evaluation Mode:
  - model.eval(): This ensures that layers like dropout and batch normalization work in evaluation mode rather than training mode.
- Initialize Lists for True and Predicted Labels:
  - y_true and y_pred lists are initialized to store the actual and predicted labels, respectively.
- Evaluate the Model on Test Data:
  - The model iterates over the test_loader without gradient calculation (torch.no_grad()), as we don't need to update the model's weights during evaluation.
  - For each batch in the test_loader, the model predicts the output, and the predicted classes are determined by finding the index of the maximum output score (using torch.max).

  - The true and predicted labels are appended to y_true and y_pred lists, respectively, after moving them to the CPU (cpu().numpy()).
- Generate Classification Report:
  - classification_report from Scikit-learn provides a detailed analysis of the model's performance, including metrics like precision, recall, and F1-score for each class, as well as overall accuracy.
  - target_names are the names of the classes obtained from the LabelEncoder used during data preprocessing.
- Calculate Overall Accuracy:
  - accuracy_score computes the overall accuracy of the model on the test dataset.
- Output and Interpretation:
  - The classification report and accuracy score are printed, providing a detailed overview of the model's performance.
  - In the provided example, the model achieves an accuracy of approximately 97.9% (which is not bad). The precision, recall, and F1-score for each class (like grazing and resting,) indicate how well the model performs for each specific category.

This final evaluation step is a definitive measure of the model's ability to handle new data and is crucial for assessing its practical utility. The detailed metrics provided in the classification report offer insights into the strengths and weaknesses of the model, guiding potential improvements or adjustments for future iterations or similar projects.

## Recurrent Neural Networks

### *Introduction to RNNs*

In the field of AAR, the ability to accurately model and interpret sequential data, such as that derived from sensor-equipped wearables, is vital. RNNs, with their natural feedback loops, stand at the forefront of this technological advancement. These networks are adept at maintaining a form of memory through hidden states, allowing them to utilize past information to influence future input processing. This characteristic is crucial when dealing with time series data or any form of sequential information, as it ensures the temporal dependencies within the data are captured and considered throughout the analysis.

### *Understanding Sequential Data*

Sequential data is inherently ordered. This order is not simply a collection of standalone points but a narrative where each point is influenced by its predecessors.

In the context of RNNs, this narrative structure allows the network to learn from the entire sequence of data, rather than viewing each data point in isolation.

At each timestep of processing, a RNN combines the current input vector with the previous timestep's hidden state to compute a new hidden state. This process is repeated across the sequence, enabling the network to effectively remember and incorporate past information to inform current and future decisions (Figure 9.7). It is this ability to capture temporal dependencies that makes RNNs particularly suited to tasks involving time series data, such as monitoring animal activities through sensors.

However, traditional RNNs are not without their challenges. They can be difficult to train and are prone to issues like vanishing or exploding gradients, convergence issues, complicating their application in tasks requiring modeling of long-term dependencies. To address these limitations, advancements in RNN architecture have been introduced, including LSTM and Gated Recurrent Units, which will be explored in subsequent sections.

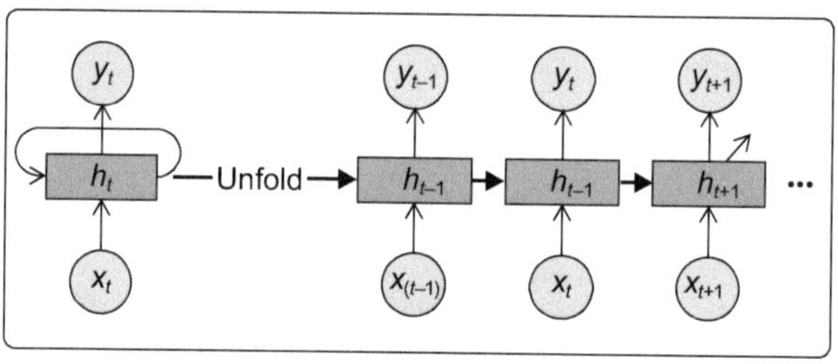


**Figure 9.7: Unfolding of a Recurrent Neural Network (RNN) over time steps, illustrating the flow of information from one unit to the next with the incorporation of the previous hidden state into the current input.**

## *Recurrent Neural Networks: A Primer*

RNNs are designed to recognize patterns in sequences such as text data, genomes, time-series data, or, appropriate to our discussion, sensor data from farm animals. Unlike traditional neural networks, which assume all inputs (and outputs) are independent of each other, RNNs possess the unique feature of memory. They are capable of remembering previous inputs in order to influence the output or the current state. This makes them ideal for applications where the context or the sequence of data points is important.

### The Basic Structure of a RNN

A RNN processes sequences by iterating through the elements and maintaining a state that contains information relative to what it has seen so far. In essence, at each step of a sequence, the RNN performs two tasks: it updates its state and produces an output.

RNNs are designed with the capability to remember past information, integrating it with current inputs to produce outputs. This memory component allows RNNs to maintain a form of internal state that captures information about the sequence it has processed so far. In contrast to feedforward neural networks that handle inputs independently, RNNs utilize feedback loops, enabling them to perform both prediction and classification tasks by leveraging static and temporal information within the input sequence.

Mathematically, this behavior can be described as follows:

**State Update**: The new state $h_t$ is a function $f$ of the current input $x_i$ and the previous state $h_{t-1}$:

$$h_t = f(h_{t-1}, x_t).$$

**Output Generation**: The output at each step $y_t$ can be calculated using the current state $h_t$:

$$y_t = g(h_t)$$

where,

- $h_t$ is the state at time $t$.
- $x_t$ is the input at time $t$.
- $y_t$ is the output at time $t$.
- $f$ and $g$ are the functions learned during training.

### Limitations and Improvements

While RNNs are powerful, they are not without limitations. They can struggle with long sequences due to the vanishing gradient problem, where the influence of inputs diminishes as the distance between them increases. To address this, more complex variants like LSTM networks and GRUs have been developed. These architectures introduce mechanisms to better control the flow of information, making them more effective at capturing long-range dependencies.

The phenomena of vanishing and exploding gradients are significant challenges encountered during the training of deep neural networks, particularly those that involve sequences such as RNNs.

**Vanishing Gradients**: This problem occurs when the gradient of the loss function shrinks exponentially as it propagates backward through the layers during training. As a result, the weights in the earlier layers of the network receive very small updates or none at all. This leads to a situation where the learning stalls because the gradient is too small to cause significant changes in the weights – it virtually *vanishes*. This is especially problematic for RNNs when dealing with long sequences, as the gradients from later stages have to be propagated through many time steps to affect the earlier stages, and they may become very small by the time they arrive.

**Exploding Gradients:** Conversely, the exploding gradients issue arises from the accumulation of large error gradients, leading to markedly large updates to the neural network's weights during the train process. This can result in an unstable model, with the model weights diverging and possibly resulting in NaN values due to numerical instability. In RNNs, this is also more likely to occur with long sequences, where gradients can compound over many time steps and grow exponentially.

Both of these issues make it difficult for the network to learn, as they lead to either very slow learning or to divergence of the model weights. Several techniques have been devised to mitigate these problems:

- Gradient clipping: This involves scaling down gradients when they exceed a certain threshold to prevent them from exploding.
- Weight Initialization: Careful initialization of weights can prevent gradients from vanishing or exploding at the start of training.
- Using LSTM/GRU cells: These variants of RNNs include gating mechanisms that help to control the flow of gradients and are less susceptible to the vanishing gradients problem.
- Batch Normalization: Though less common in RNNs, this technique can help maintain stable gradients throughout the network.

Understanding and mitigating vanishing and exploding gradients is crucial for training deep neural networks effectively, especially when dealing with sequential data that requires capturing long-range dependencies.

## Long Short-Term Memory (LSTM) Networks

LSTMs, are RNNs capable of learning long-term dependencies. They were introduced by Hochreiter & Schmidhuber in 1997 (Hochreiter & Schmidhuber, 1997) to specifically address the vanishing gradient problem that affects standard RNNs. LSTMs are designed to avoid the long-term dependency problem, allowing them to remember information for prolonged periods.

### Architecture

LSTMs are centered around the concept of a cell state, which flows directly through the network's entire chain, experiencing minimal linear interactions. This design allows information to flow along it unchanged if necessary. LSTMs possess the capability to modify the cell state by either adding or removing information, a process thoroughly controlled by mechanisms known as gates.

Figure 9.8 illustrates the architecture of an LSTM unit. It shows the input $x_t$ and the previous hidden state $h_{t-1}$ feeding into three gates: the forget gate $f_t$, the input gate $i_t$, and the output gate $o_t$, represented by boxes with the sigmoid activation function $\sigma$. Additionally, the tanh function creates a new candidate cell state $\tilde{C}_t$.

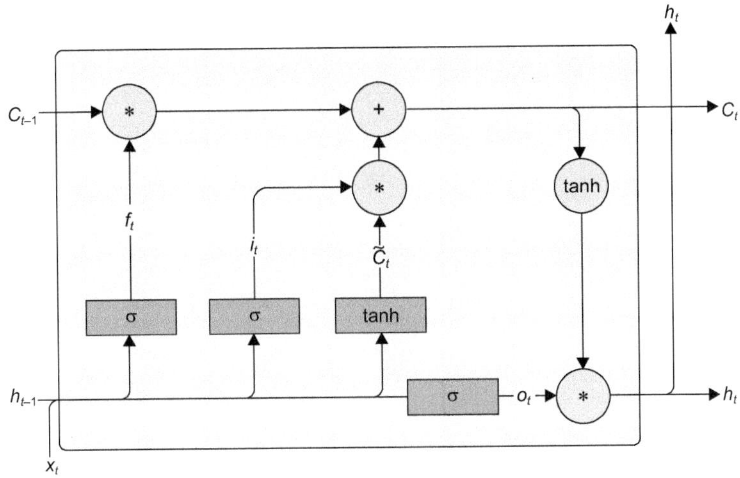


**Figure 9.8: Schematic representation of a Long Short-Term Memory (LSTM) unit, illustrating the flow of information through its various gates and the state updates.**

These gates regulate the update and output of the new cell state $C_t$ and the new hidden state $h_t$, utilizing operations like element-wise multiplication and addition, leading to the final cell and hidden state outputs.

### Gates

Gates serve as selective channels for information flow, consisting of a sigmoid neural network layer and a pointwise multiplication operation to either permit or block information. The sigmoid layer outputs numbers between zero and one, describing how much of each component should be let through. The LSTM has three of these gates to protect and control the cell state:

- **Forget Gate ($f_t$):**

This decides what information should be discarded from the cell state. It looks at the previous hidden state $h_{t-1}$ and the current input $x_t$ and generates a value ranging from 0 to 1 for each element in the cell state $C_{t-1}$, where 1 signifies fully retaining the element, and 0 indicates entirely discarding it.

$$f_t = \sigma(W_f \cdot [h_{t-1}, x_t] + b_f)$$

- **Input Gate ($i_t$):**

This decides what new information will be stored in the cell state. It involves two parts: a sigmoid layer that decides which values to update, and a tanh layer that creates a vector of new candidate values $\tilde{C}_t$ that could be added to the state.

$$f_t = \sigma(W_f \cdot [h_{t-1}, x_t] + b_f)$$
$$\tilde{C}_t = tanh(W_C \cdot [h_{t-1}, x_t] + b_C)$$

- **Output Gate ($o_t$):**

This decides what the next hidden state should be. The hidden state contains information about previous inputs. The output gate looks at the previous hidden state and the current input and decides what the next hidden state should be.

$$o_t = \sigma(W_o \cdot [h_{t-1}, x_t] + b_o)$$
$$h_t = o_t * tanh(C_t)$$

The cell state is updated as follows:

$$C_t = f_t * C_{t-1} + i_t * \tilde{C}_t$$

where,

- $x_t$ is the input at time step $t$.
- $h_t$ is the hidden state at time step $t$.
- $C_t$ is the cell state at time step $t$.
- $W$ and $b$ are the weights and biases for each gate.
- $\sigma$ is the sigmoid function.
- tanh is the hyperbolic tangent function.
- $*$ denotes pointwise multiplication.

The LSTM's ability to maintain a cell state over time, along with the controlled modifications through the gates, effectively allows it to learn when to *remember* and *forget* certain pieces of information, making it suitable for a variety of complex sequential tasks.

## Gated Recurrent Units (GRUs)

GRUs are another type of RNN similar to LSTMs but with a simpler structure. They were introduced by (Cho et al., 2014). GRUs are designed to help solve the problem of long-term dependencies in sequential data. GRUs combine the forget and input gates into a single *update gate* and merge the cell state and hidden state, resulting in a model that is easier to work with and faster to compute than LSTMs.

### The GRU Architecture

The GRU modifies the standard recurrent unit by incorporating gating units. These gates effectively control the flow of information. They determine which information is to be sent to the output and what should continue to the next step of the sequence, allowing the model to keep or discard information across many time steps. A simple GRU architecture is presented in Figure 9.9.

A GRU has two gates:

- **Update Gate**: This gate determines the extent of information from past time steps that should be carried forward to future stages. It helps the model

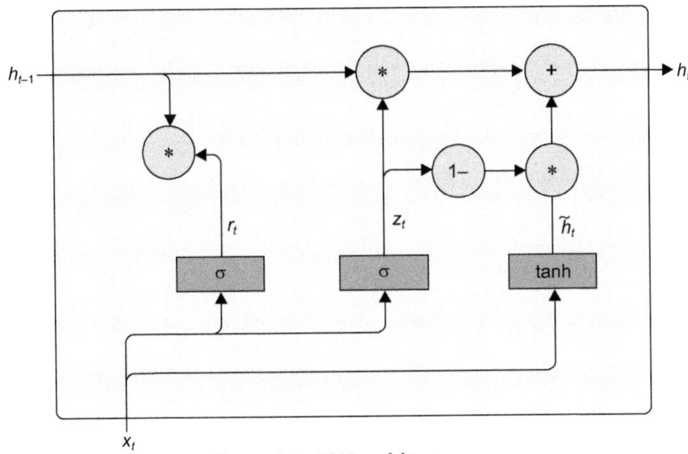


**Figure 9.9: GRU architecture.**

determine the level of influence the previous state should have on the current state. Mathematically, this can be represented as:

$$z_t = \sigma(W_z \cdot [h_{t-1}, x_t] + b_z).$$

Here, $z_t$ is the update gate vector, σ is the sigmoid function, $W_z$ is the weight matrix for the update gate, $h_{t-1}$ is the previous hidden state, $x_t$ is the input at time $t$, and $b_z$ is the bias for the update gate.

- **Reset Gate**: This gate controls the amount of past information to be discarded, deciding on the extent of previous data that should be forgotten. The reset gate can be expressed as:

$$r_t = \sigma(W_r \cdot [h_{t-1}, x_t] + b_r)$$

where, $r_t$ is the reset gate vector, $W_t$ and $b_r$ and are the weight matrix and bias for the reset gate, respectively.

The current memory content utilizes the reset gate to store the relevant information from the past. It is then combined with the update gate to form the final output of the GRU.

The new hidden state $h_t$ is a combination of the old hidden state $h_{t-1}$ and the candidate hidden state $\tilde{h}_t$ and is computed as:

$$h_t = (1 - z_t) * h_{t-1} + z_t * \tilde{h}_t.$$

The candidate hidden state $\tilde{h}_t$ is calculated using the current input and the reset gate:

$$\tilde{h}_t = tanh(W \cdot [r_t * h_{t-1}, x_t] + b).$$

In this equation, $W$ and $b$ are the weights and biases applied to the candidate hidden state, and denotes an element-wise multiplication.

**Advantages of GRUs**

- Simplicity: GRUs have fewer parameters than LSTMs because they lack an output gate.
- Efficiency: GRUs are generally faster to compute and train due to their simpler structure.
- Flexibility: They have shown competitive performance with LSTMs on a variety of tasks.

## LSTM Networks in Practice Using PyTorch

### *Running LSTM for Activity Recognition Using PyTorch*

In this section, we will walk you through the process of running a Long Short-Term Memory (LSTM) model using PyTorch for activity recognition. We will be using a dataset containing sensor data from farm animals, sheep and goats, to demonstrate the implementation. The dataset and the LSTM example can be found in the Chapter_9 folder, with the Jupyter notebook titled Chapter_9_LSTM.ipynb.

To get started, we have defined a Python function called read_csv_files_from_folder(folder_path). This function is responsible for reading and joining data from multiple CSV files located in a specified folder. It returns a single Pandas DataFrame containing all the data.

Then, we call this function to read data from the "data" folder and assign it to the variable df.

Next, we select specific columns from the DataFrame to work with. These columns include 'label', 'animal_ID', 'timestamp_ms', and the accelerometer data 'ax', 'ay', and 'az'. Data integrity is essential, so we remove rows with missing values (NaN) using the dropna() method. This ensures that our DataFrame is free from missing data.

LSTM models require ordered data, especially when dealing with sequential data. Therefore, we sort the DataFrame based on 'animal_ID' and 'timestamp_ms' in ascending order. This step is important for maintaining the chronological order of our data.

In our dataset, we had a variety of labels that represented different activities performed by animals. Some of these labels were quite common, such as "standing", "grazing", and "walking", while others were less common, including "scratch_biting", "fighting", and "shaking". These less common labels had relatively fewer instances compared to the major activities.

To simplify our classification problem and potentially improve the model's performance, we decided to group together certain labels under a single category called "other". Specifically, the labels "scratch_biting", "fighting", and "shaking" were selected to be assigned as "other". The rationale behind this decision is that these activities might not have as much data available for the model to learn effectively and grouping them together could help balance the dataset and improve the model's ability to recognize major activities.

To implement this change, we performed the following steps:

- Defined a list called labels_to_be_assigned_as_other containing the labels we wanted to group as "other".
- Updated the labels in the DataFrame by applying a lambda function to the "label" column. The lambda function checked if a label was in the labels_to_be_assigned_as_other list. If it was, the label was replaced with "other"; otherwise, it remained unchanged.

By doing this, we effectively transformed our dataset to have fewer distinct labels, with the less common activities now grouped under the more general "other" category. This preprocessing step can help streamline our classification task and potentially lead to better model performance when distinguishing between major activities and the combined "other" category.

Then, we create several constants that will be used throughout our code, including the number of time steps (n_time_steps), the number of features (n_features), the number of classes (n_classes), the number of epochs (n_epochs), batch size (batch_size), learning rate (learning_rate), and L2 loss (l2_loss).

The following code is used for the above steps:

```
import pandas as pd
import os
import glob

def read_csv_files_from_folder(folder_path):
    """
    Reads all CSV files from a specified folder and merges them into
    a single DataFrame.

    Parameters:
    folder_path (str): The path to the folder containing CSV files.
    Returns:
    pd.DataFrame: A Pandas DataFrame containing the merged data or
    None if no data is found.
    """
    try:
        # Create an empty list to store DataFrames
        dfs = []

        # Use glob to get a list of all CSV files in the folder
        csv_files = glob.glob(os.path.join(folder_path, '*.csv'))
```

```
        if not csv_files:
            print("No CSV files found in the specified folder.")
            return None

        # Read and append each CSV file to the list
        for csv_file in csv_files:
            df = pd.read_csv(csv_file)
            if not df.empty:
                dfs.append(df)

        if not dfs:
            print("No valid data found in the CSV files.")
            return None

        # Concatenate all DataFrames vertically
        merged_df = pd.concat(dfs, ignore_index=True)

        return merged_df
    except FileNotFoundError:
        print("The specified folder or CSV files were not found.")
        return None
    except Exception as e:
        print(f"An error occurred: {str(e)}")
        return None

# Reading the dataset using our custom function
df = read_csv_files_from_folder("data")

# Selecting specific columns from the dataset
selected_columns = ['label', 'animal_ID', 'timestamp_ms', 'ax',
    'ay', 'az']
new_df = df[selected_columns]

new_df.head()

# Output:
   label   animal_ID timestamp_ms    ax           ay          az
0  walking      G1       1         1.57538     4.34787    -9.27514
1  walking      G1       6         1.47962     4.30477    -9.31105
2  walking      G1       11        1.36469     4.24492    -9.42118
3  walking      G1       16        1.21386     4.22816    -9.59835
4  walking      G1       21        1.07021     4.29520    -9.67257

# Import libraries
%matplotlib inline
import numpy as np
import matplotlib.pyplot as plt
```

```
import seaborn as sns
import warnings
warnings.filterwarnings('ignore')

# Remove null values if any
df = new_df.dropna()
df.shape

# Output:
(13778153, 6)

# Arranging data in ascending order using animal_ID and timestamp_ms
df = df.sort_values(by = ['animal_ID', 'timestamp_ms'], ignore_
    index=True)

# Defining a list of labels to be assigned as "other"
labels_to_be_assigned_as_other = ["scratch_biting", "fighting",
    "shaking"]

# Update the labels in the DataFrame
df['label'] = df['label'].apply(lambda x: 'other' if x in
                labels_to_be_assigned_as_other else x)

# Now, the labels in the DataFrame have been updated, and the
specified labels are assigned as "other"

df['label'].value_counts()

# Output:
standing          6031069
grazing           3735835
walking           2109794
lying              984396
trotting           408051
running            324775
other              184233
Name: label, dtype: int64
```

The core of our data preparation involves segmenting it into smaller windows that will be fed into the LSTM model. We iterate through the DataFrame, creating segments of 1000 time steps each, based on the ‘ax’, ‘ay’, and ‘az’ features. The label for each segment is determined by finding the mode of the ‘label’ column within that segment. Finally, we reshaped the segments and encoded the categorical labels.

```
# Importing stats from scipy
from scipy import stats

# Defining constants and hyperparameters
random_seed = 42
n_time_steps = 1000 # the data is at 200Hz, and we want 5 seconds
n_features = 3 # Number of features (x, y, z)
step = 200 # Step size for segmenting
n_classes = 7
n_epochs = 50
l2_loss = 0.0015

segments = []
labels = []

# Iterate over the data with a moving window of 1000 time steps

for i in range(0,  df.shape[0]- n_time_steps, step):
    x = df['ax'].values[i: i + n_time_steps]
    y = df['ay'].values[i: i + n_time_steps]
    z = df['az'].values[i: i + n_time_steps]

    # Calculate the mode of the 'label' column within the segment
    label = stats.mode(df['label'][i: i + n_time_steps])[0][0]

    # Create a segment that includes data from all three axes
    segment = [x, y, z]

    segments.append(segment)
    labels.append(label)

# Reshape the segments into a 3D array
reshaped_segments = np.asarray(segments, dtype=np.float32).
reshape(-1, n_time_steps, n_features)

from sklearn.preprocessing import LabelEncoder

# Initialize the LabelEncoder
label_encoder = LabelEncoder()

# Fit the LabelEncoder on your list of labels
label_encoder.fit(labels)
# Transform the original labels into encoded labels
encoded_labels = label_encoder.transform(labels)

reshaped_segments.shape, encoded_labels.shape

# Output
((68886, 1000, 3), (68886,))
```

Following these preprocessing steps, our data is now in the form of reshaped segments, with the shape of (68886, 1000, 3). This means we have approximately 69,000 thousand segments, each containing 1000 time steps and 3 features ('x', 'y', 'z'). Additionally, we have label encoded labels ready for classification.

In the following code we split and prepare our data for use in PyTorch:

```
import torch
import torch.nn as nn
from torch.utils.data import TensorDataset, DataLoader
from sklearn.model_selection import train_test_split
from sklearn.preprocessing import StandardScaler

# Define constants and hyperparameters
input_size = n_features
hidden_size = 128
num_layers = 1  # Number of layers
num_classes = n_classes
dropout_prob = 0.5
learning_rate = 0.001
batch_size = 256

# Split the data into training, validation, and test sets
X_train, X_temp, y_train, y_temp = train_test_split(reshaped_segments,
                                  encoded_labels, test_size=0.3,
                                  random_state=random_seed)
X_val, X_test, y_val, y_test = train_test_split(X_temp, y_temp,
test_size=0.5, random_state=random_seed)

# Normalize the data using StandardScaler
scaler = StandardScaler()
X_train = scaler.fit_transform(X_train.reshape(-1, n_features)).
    reshape(-1, n_time_steps, n_features)
X_val = scaler.transform(X_val.reshape(-1, n_features)).reshape(-1,
    n_time_steps, n_features)
X_test = scaler.transform(X_test.reshape(-1, n_features)).
    reshape(-1, n_time_steps, n_features)

# Convert NumPy arrays to PyTorch tensors
X_train_tensor = torch.tensor(X_train, dtype=torch.float32)
y_train_tensor = torch.tensor(y_train, dtype=torch.int64)

X_val_tensor = torch.tensor(X_val, dtype=torch.float32)
y_val_tensor = torch.tensor(y_val, dtype=torch.int64)

X_test_tensor = torch.tensor(X_test, dtype=torch.float32)
y_test_tensor = torch.tensor(y_test, dtype=torch.int64)
```

```
# Create TensorDatasets
train_dataset = TensorDataset(X_train_tensor, y_train_tensor)
val_dataset = TensorDataset(X_val_tensor, y_val_tensor)
test_dataset = TensorDataset(X_test_tensor, y_test_tensor)

# Create DataLoader for batching
train_loader = DataLoader(train_dataset, batch_size=batch_size,
    shuffle=True)
val_loader = DataLoader(val_dataset, batch_size=batch_size)
test_loader = DataLoader(test_dataset, batch_size=batch_size)
```

The above python example involves splitting the data into training, validation, and test sets, normalizing it, and preparing it for use in PyTorch.

In our code:

- train_test_split(reshaped_segments, encoded_labels, test_size = 0.3, random_state = random_seed):
  - We use the train_test_split function to split our preprocessed data into training, validation, and test sets. reshaped_segments contain the sensor data and encoded_labels contains the corresponding encoded labels.
  - test_size = 0.3: We have specified that 30% of the data should be allocated for testing, leaving 70% for training and validation.
  - random_state = random_seed: Setting the random seed ensures reproducibility, meaning that every time you run the code with the same seed, you'll get the same data split.
- Data Normalization: We created an instance of StandardScaler() to standardize our data.
- Converting to PyTorch Tensors: We convert our normalized data from NumPy arrays into PyTorch tensors using torch.tensor(). We specify the data type for the tensors, such as torch.float32 for the sensor data and torch.int64 for the labels. The data type matters for the operations performed by the model.
- Creating Data Loaders: Machine learning models are often trained in batches, not on entire datasets. Data loaders are used to efficiently load and batch the data during training.
  - We create TensorDataset objects, which combine the input data tensors and their corresponding label tensors.
  - We use DataLoader objects to batch the data from the datasets. These data loaders automatically divide our data into batches of the specified size, making it suitable for training deep learning models.

By performing these steps, we have organized our data into training, validation, and test sets, normalized the input data, and converted it into a format suitable for PyTorch. This prepares the data for efficient model training and evaluation.

## LSTM Model Class

```
class LSTMModel(nn.Module):
   def __init__(self, input_size, hidden_size, num_layers, num_
   classes, seq_length, dropout_prob=0.5):
         super(LSTMModel, self).__init__()
         self.num_classes = num_classes
         self.num_layers = num_layers
         self.input_size = input_size
         self.hidden_size = hidden_size
         self.seq_length = seq_length

         self.lstm = nn.LSTM(input_size=input_size, hidden_
         size=hidden_size, num_layers=num_layers, batch_first=True)
         self.fc_1 = nn.Linear(hidden_size, 128)
         self.relu = nn.ReLU()
         self.fc = nn.Linear(128, num_classes)

    def forward(self, x):
         h_0 = torch.zeros(self.num_layers, x.size(0),
              self.hidden_size).to(x.device)
         c_0 = torch.zeros(self.num_layers, x.size(0),
              self.hidden_size).to(x.device)

        # Reshape the input data to match the expected format
              (batch_size,
        seq_length, input_size)
        x = x.view(-1, self.seq_length, self.input_size)

        # Propagate input through LSTM
        output, (hn, cn) = self.lstm(x, (h_0, c_0))
        hn = hn.view(-1, self.hidden_size)
        out = self.relu(hn)
        out = self.fc_1(out)
        out = self.relu(out)
        out = self.fc(out)
        return out

 # Define the model and parameters
 input_size = n_features
 hidden_size = 128
 num_layers = 1
 num_classes = n_classes
 seq_length = 1000

 model = LSTMModel(input_size, hidden_size, num_layers, num_classes,
 seq_length)
```

The LSTM (Long Short-Term Memory) model, defined here, is a type of recurrent neural network (RNN) commonly used for sequence prediction tasks. It consists of an *input layer*, an *LSTM layer*, and two *fully connected* (dense) layers. The input layer accepts data in the format of (*batch_size, sequence_length, input_size*), where batch_size represents the number of samples in each batch, *sequence_length* denotes the length of the input sequence, and *input_size* represents the number of features in each time step.

The LSTM layer processes the input sequence, capturing temporal dependencies, and producing hidden states. The hidden states are then fed into a fully connected layer with a ReLU activation function, followed by another fully connected layer, which outputs the final predictions. The model is parameterized by *input_size*, *hidden_size*, *num_layers*, and *num_classes*, where hidden_size determines the number of units in the LSTM layer, *num_layers* specify the number of LSTM layers stacked on top of each other, and *num_classes* represent the number of output classes for classification tasks. In this specific implementation, the model is designed to operate on sequences of length 1000 (5 second windows).

### Model Training Loop

```
# Check if a GPU is available
device = torch.device('cuda' if torch.cuda.is_available() else 'cpu')

# Move the model to the GPU
model.to(device)

# Define the loss and optimizer
criterion = nn.CrossEntropyLoss()
optimizer = torch.optim.Adam(model.parameters(), lr=learning_rate,
      weight_decay=l2_loss)

# Initialize lists to store training and validation loss, accuracy
train_losses = []
val_losses = []
train_accuracies = []
val_accuracies = []

# Training loop
for epoch in range(n_epochs):
    model.train()
    total_loss = 0.0
    correct_train = 0
    total_train = 0

    for inputs, labels in train_loader:
        inputs, labels = inputs.to(device), labels.to(device) #
        Move data to GPU
```

```
        optimizer.zero_grad()
        outputs = model(inputs)
        loss = criterion(outputs, labels)
        loss.backward()
        optimizer.step()

        total_loss += loss.item()
        _, predicted = torch.max(outputs.data, 1)
        total_train += labels.size(0)
        correct_train += (predicted == labels).sum().item()

    average_loss = total_loss / len(train_loader)
    train_accuracy = correct_train / total_train

    # Validation
    model.eval()
    total_loss = 0.0
    correct_val = 0
    total_val = 0

    with torch.no_grad():
      for inputs, labels in val_loader:
            inputs, labels = inputs.to(device), labels.to(device)
            # Move data to
               GPU
            outputs = model(inputs)
            loss = criterion(outputs, labels)

            total_loss += loss.item()
            _, predicted = torch.max(outputs.data, 1)
            total_val += labels.size(0)
            correct_val += (predicted == labels).sum().item()

    val_accuracy = correct_val / total_val
    average_val_loss = total_loss / len(val_loader)

    print(f'Epoch [{epoch + 1}/{n_epochs}], '
            f'Training Loss: {average_loss:.4f}, Training Accuracy:
            {train_accuracy:.4f}, '
            f'Validation Loss: {average_val_loss:.4f}, Validation
               Accuracy:
            {val_accuracy:.4f}')

    # Store the loss and accuracy for plotting
    train_losses.append(average_loss)
    val_losses.append(average_val_loss)
    train_accuracies.append(train_accuracy)
    val_accuracies.append(val_accuracy)

print('Training finished!')
```

In the above code, we prepare our model for training on a GPU if available, by checking for its presence and moving the model to the GPU using PyTorch. We define our loss function, which is the Cross Entropy Loss, and our optimizer, which is the Adam optimizer. Additionally, we initialize lists to keep track of training and validation losses, as well as accuracies, to monitor the model's performance during training. The code then enters a training loop, where the model iterates over the specified number of epochs. Within each epoch, the model is set to training mode, and we calculate the loss and accuracy on the training dataset. We also evaluate the model's performance on the validation dataset to prevent overfitting. Finally, the training and validation losses and accuracies are printed for each epoch, and these values are stored for later visualization. Once training ends, a message indicating the end of training is displayed.

### Plotting the Losses and Accuracies

```
import matplotlib.pyplot as plt

# Plotting losses
plt.figure(figsize=(10, 5))
plt.plot(train_losses, label='Train Loss')
plt.plot(val_losses, label='Validation Loss')
plt.title('Training and Validation Losses')
plt.xlabel('Epoch')
plt.ylabel('Loss')
plt.legend()
plt.grid(True)

# Plotting accuracies
plt.figure(figsize=(10, 5))
plt.plot(train_accuracies, label='Train Accuracy')
plt.plot(val_accuracies, label='Validation Accuracy')
plt.title('Training and Validation Accuracies')
plt.xlabel('Epoch')
plt.ylabel('Accuracy')
plt.legend()
plt.grid(True)

# Adjust layout to prevent overlapping
plt.tight_layout()

# Save the combined plot as an image file
plt.savefig('accuracy_loss.png', dpi=300)

# Show the combined plot
plt.show()
```

In our code, we plotted the results to visually track the training and validation accuracies and losses across 50 epochs. By plotting these metrics, we can easily observe how our machine learning model performed over time, providing valuable insights into its learning dynamics.

The first plot of Figure 9.10 illustrates the path of training and validation losses over epochs. Initially, both exhibit a downward trend, indicating the model's ability to minimize errors on both training and validation data. The second plot reveals the progression of training and validation accuracies over epochs.

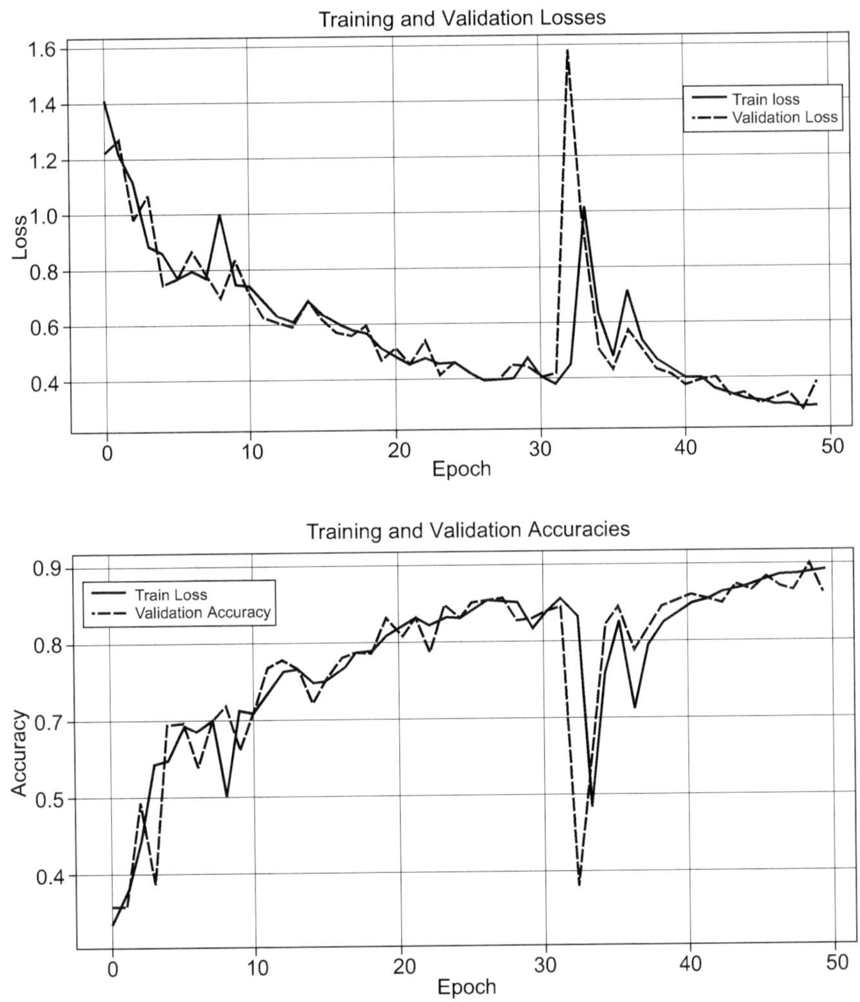


**Figure 9.10: Training and validation losses and accuracies plotted over 50 epochs.**

### Model Evaluation on the Test Set

```
model.eval()
total_test = 0
correct_test = 0

with torch.no_grad():
    for inputs, labels in test_loader:
    inputs, labels = inputs.to(device), labels.to(device)
        # Move data to GPU
        outputs = model(inputs)
        _, predicted = torch.max(outputs.data, 1)
        total_test += labels.size(0)
        correct_test += (predicted == labels).sum().item()

test_accuracy = correct_test / total_test
print(f'Test Accuracy: {test_accuracy:.4f}')

# Output
Test Accuracy: 0.8619
```

Finally, we evaluate the trained model on a separate test dataset to assess its performance on unseen data. The model is put into evaluation mode using *model.eval()*, ensuring that certain layers (like dropout) behave differently during evaluation compared to training. We then initialize variables to keep track of the total number of test samples (*total_test*) and the number of correctly predicted samples (*correct_test*). Within a *torch.no_grad()* block to disable gradient calculation, we iterate over batches of test data. For each batch, we move the data to the GPU (if available), pass it through the model to obtain predictions, and compute the number of correct predictions. After processing all test data, we calculate the test accuracy by dividing the number of correctly predicted samples by the total number of test samples. Finally, the test accuracy is printed to assess the model's performance on unseen data, indicating that the model correctly predicted approximately 86.19% of the test samples.

As we wrap up our discussion on using LSTM models for analyzing accelerometer data to recognize farm animal activities, it is key to note that we have covered foundational aspects to get you started with LSTMs in PyTorch. From data preprocessing, model architecture setup, training, to evaluation, we have provided a step-by-step guide to introduce you to the process.

## Deep Learning Applications in Farm Animal Activity Recognition

Recent advancements in wearable technology for livestock have led to innovative studies comparing neural network models for behavior monitoring. For example, Arablouei et al. used MLPs over traditional machine learning algorithms, such as decision trees and support-vector machines, in classifying cattle behaviors from accelerometer data, achieving an accuracy of 93.4% (Arablouei et al., 2021). This groundwork laid the foundation for further exploration that refined behavior recognition models. Specifically, they used a deep learning approach that combines signal filtering with MLP for classification, attaining 95.68% accuracy, and a multimodal approach integrating accelerometer and global navigation satellite system (GNSS) data, which showed results with an 88.47% accuracy for collar-based datasets (Arablouei et al., 2023; Arablouei et al., 2023).

Hosseininoorbin et al. used a deep sequential neural network that utilizes time-frequency domain data for cattle behavior classification, based on a dataset from tri-axial accelerometers, magnetometers, and gyroscopes on beef steers' collars, achieving a F1-score of 94.9% for three behavior classes and 89.3% for nine behavior classes (Hosseininoorbin et al., 2021). Kamminga et al. trained classifiers using shared training data and feature-representation, demonstrating that Multitask Learning significantly improves classifier performance, and compared seven classifiers on resource usage and activity recognition in real-world data from goats and sheep, with a Deep Neural Network achieving 94% accuracy for both species (Kamminga et al., 2017).

The exploration of CNNs in animal activity recognition utilizing wearable sensors has also advanced the field. Eerdekens and colleagues, through a series of studies from 2020 to 2022, have effectively employed CNNs alongside accelerometer data for monitoring behaviors in horses and dogs, achieving prediction accuracies above 97% (Eerdekens et al., 2020, 2021, 2022). These studies highlight the robustness of CNNs, especially when augmented with statistical features, underscoring their performance over traditional methods relying on raw sensor data.

Similarly, Kleanthous et al. and Mao et al. have contributed to the domain by developing CNN models tailored for sheep and horses, respectively. Kleanthous et al. utilized Deep Transfer Learning using CNNs combined with a feature engineering approach to extract handcrafted features, which led to an accuracy of 98.55% in identifying sheep activities (Kleanthous et al., 2022). Mao et al. crafted a new CNN architecture utilizing accelerometer and gyroscope data to identify horse activities (Mao et al., 2022).

In the realm of canine activity recognition, Kasnesis et al. and Hussain et al. explored the efficacy of CNNs using data from accelerometers and gyroscopes attached to different body parts. Kasnesis et al. developed a deep-learning system for Search and Rescue (SaR) of dogs, incorporating wearable devices and cloud infrastructure to monitor and analyze the dog's activity, audio signals, and location, using CNNs for activity and sound recognition. The system, validated in two SaR scenarios, achieved a F1-score of over 99% in detecting victims and providing real-time alerts to rescuers (Kasnesis et al., 2022). Hussain et al. utilized a one-dimensional convolution CNN with raw sensor data features, obtaining a 96.85% accuracy in detecting dog behaviors (Hussain et al., 2022). A recent study has expanded the application of CNNs to monitor hens activities. Shahbazi et al. claimed that they reached nearly 100% accuracy in classifying hens activity levels using body-worn inertial measurement unit sensors, focusing on broader activity categories (Shahbazi et al., 2023).

Current studies also have underscored the efficacy of RNNs, particularly LSTM and GRU variants, in the domain of AAR using wearable sensors. These studies have leveraged the sophisticated capabilities of RNNs to recognize and classify complex patterns of behavior based on sensor data, achieving significant accuracy in identifying various animal activities. For instance, research by Peng et al. (Peng et al., 2019) has been pioneering in applying LSTMs to cattle behavior recognition, utilizing nine-axial-motion data from collar-attached inertial measurement unit sensors. Their work has demonstrated the potential of LSTMs in distinguishing between multiple cattle behaviors with high accuracy, thereby offering insights into cattle health and welfare. In one study, the LSTM model achieved 88.7% accuracy in identifying activities such as feeding, licking salt, and headbutting.

Further studies have expanded on this foundation, exploring the use of deep residual bidirectional LSTMs for early identification of diseases in cattle, with one notable study reporting a classification accuracy of 94.9% (Wu et al., 2022). Such high levels of accuracy highlight the potential of RNNs in early disease detection and prevention, significantly impacting animal welfare and farm management practices.

Comparative analyses have also shown that RNNs can outperform conventional CNN models in classifying cattle behaviors. These RNN models have been praised for their efficiency, requiring fewer computational resources while still achieving comparable or superior accuracy. A two-layer bidirectional GRU model, for example, achieved accuracy rates of 89.5% and 80% on collar- and ear-attached sensor datasets, respectively (Wang et al., 2023).

Moreover, the application of RNNs extends beyond cattle to include other species, such as dogs, where LSTM-based methods have been employed to detect activities using motion data from accelerometers (Chambers et al., 2021). These methods have successfully classified various dog activities with a high degree of accuracy, further illustrating the versatility and effectiveness of RNNs in animal activity recognition.

This overview of deep learning applications in the field of farm animals demonstrates the significant advances made in leveraging neural networks, including MLPs, CNNs, and RNNs, for accurate and efficient behavior classification across various species. These advances set a solid foundation for the practical implementation of deep learning models in animal welfare and management.

## Summary

In Chapter 9, we delve into the critical role that various neural network architectures play in the domain of deep learning, with a specific focus on their application to wearable sensor data for the purpose of animal activity recognition.

We start by introducing the foundational Multilayer Perceptron Neural Networks, explaining their basic structure and operational principles as the groundwork for more complex neural network architectures. The discussion then extends to Convolutional Neural Networks, showing how they can be adeptly applied to analyze wearable sensor data to identify spatial patterns. Further, we explore Recurrent Neural Networks and Long Short-Term Memory networks, emphasizing their ability to handle time-series data, which is often generated by wearable sensors. Recognizing the importance of efficiency in processing sequences, we also introduce Gated Recurrent Units, presenting them as a streamlined alternative to LSTMs.

A significant portion of this chapter is dedicated to demonstrating practical applications using PyTorch. We showed how neural networks can be employed for both binary and multiclass classification tasks, using NNs, CNNs, and LSTMs, thus providing readers with hands-on examples. These examples are particularly focused on classifying animal activities based on accelerometer data, aiming to equip readers with the knowledge and skills to apply deep learning techniques effectively in the context of wearable sensor data analysis.

Through this comprehensive analysis and practical demonstration, our goal is to bridge the gap between theoretical understanding to practical application of deep learning in wearable sensor data analysis, enhancing our collective ability to interpret and understand animal behavior through technology.

## Final Remarks

As we conclude our exploration into machine learning and deep learning for sensor data analysis in farm animal activity recognition, we would like to express our appreciation to you for engaging with this book. Throughout this book, we have covered a broad spectrum of topics, from the basics of animal behavior and machine learning to the intricacies of data collection, preprocessing, feature selection, and various learning techniques, concluding with an insightful discussion on deep learning. Practical applications in Python have been a foundation of each chapter, aiming to provide a hands-on understanding of the concepts discussed.

We recognize the dynamic nature of the field and the possibility of updates or corrections to the code examples provided. Your feedback is invaluable to us, and we welcome any corrections or suggestions you may have. Feel free to contact us with your feedback or questions at https://github.com/nkcAna/WSDpython.

Given the introductory nature of this book, we encourage you to delve deeper into the latest literature for the most current advancements in the field. The fast-paced evolution of machine learning and deep learning technologies means there is always something new to learn. Practicing with Python and building upon the foundational knowledge acquired here will further enhance your skills and understanding.

This book is intended to serve as a stepping stone into the vast and exciting world of machine learning and deep learning in animal behavior analysis. We hope it has sparked your curiosity and equipped you with the tools to continue your learning journey. Thank you for joining us on this journey.

# References

Abramson, N., Braverman, D., & Sebestyen, G. (1963). Pattern recognition and machine learning. *IEEE Transactions on Information Theory*, *9*(4), 257–261. https://doi.org/10.1109/TIT.1963.1057854

Arablouei, R., Currie, L., Kusy, B., Ingham, A., Greenwood, P. L., & Bishop-Hurley, G. (2021). In-situ classification of cattle behavior using accelerometry data. *Computers and Electronics in Agriculture*, *183*. https://doi.org/10.1016/j.compag.2021.106045

Arablouei, R., Wang, L., Currie, L., Yates, J., Alvarenga, F. A. P., & Bishop-Hurley, G. J. (2023). Animal behavior classification via deep learning on embedded systems. *Computers and Electronics in Agriculture*, *207*. https://doi.org/10.1016/j.compag.2023.107707

Arablouei, R., Wang, Z., Bishop-Hurley, G. J., & Liu, J. (2023). Multimodal sensor data fusion for in-situ classification of animal behavior using accelerometry and GNSS data. *Smart Agricultural Technology*, *4*. https://doi.org/10.1016/j.atech.2022.100163

Arcidiacono, C., Porto, S. M. C. C., Mancino, M., & Cascone, G. (2017). Development of a threshold-based classifier for real-time recognition of cow feeding and standing behavioural activities from accelerometer data. *Computers and Electronics in Agriculture*, *134*, 124–134. https://doi.org/10.1016/j.compag.2017.01.021

Barwick, J., Lamb, D., Dobos, R., Schneider, D., Welch, M., & Trotter, M. (2018). Predicting lameness in sheep activity using tri-axial acceleration signals. *Animals*, *8*(1), 1–16. https://doi.org/10.3390/ani8010012

Belkina, A. C., Ciccolella, C. O., Anno, R., Halpert, R., Spidlen, J., & Snyder-Cappione, J. E. (n.d.). *Automated optimized parameters for T-distributed stochastic neighbor embedding improve visualization and analysis of large datasets*. https://doi.org/10.1038/s41467-019-13055-y

Bengio, Y. (2009). Learning deep architectures for AI. *Foundations and Trends in Machine Learning*, *2*(1), 1–27. https://doi.org/10.1561/2200000006

Benos, L., Tagarakis, A. C., Dolias, G., Berruto, R., Kateris, D., & Bochtis, D. (2021). Machine learning in agriculture: A comprehensive updated review. In *Sensors* (Vol. 21, Issue 11, p. 3758). Multidisciplinary Digital Publishing Institute. https://doi.org/10.3390/s21113758

Bishop, C. M. (1995). *Neural networks for pattern recognition*. Oxford university press.

Bishop, C. M. (2006). *Pattern recognition and machine learning* (1st ed., Vol. 1). Springer New York, NY.

Borgelt, C. (2005). An implementation of the FP-growth algorithm. *Proceedings of the ACM SIGKDD International Conference on Knowledge Discovery and Data Mining*. https://doi.org/10.1145/1133905.1133907

Breiman, L. (2001). Random forests. *Machine Learning*, *45*(1), 5–32. https://doi.org/10.1023/A:1010933404324/METRICS

Broom, D. M. (2010). Animal welfare: An aspect of care, sustainability, and food quality required by the public. *Journal of Veterinary Medical Education*, *37*(1), 83–88. https://doi.org/10.3138/JVME.37.1.83

Chambers, R. D., Yoder, N. C., Carson, A. B., Junge, C., Allen, D. E., Prescott, L. M., Bradley, S., Wymore, G., Lloyd, K., & Lyle, S. (2021). Deep learning classification of canine behavior using a single collar-mounted accelerometer: Real-world validation. *Animals, 11*(6). https://doi.org/10.3390/ani11061549

Chandola, V., Banerjee, A., & Kumar, V. (2009). Anomaly detection: A survey. In *ACM Computing Surveys* (Vol. 41, Issue 3). https://doi.org/10.1145/1541880.1541882

Chawla, N. V., Bowyer, K. W., Hall, L. O., & Kegelmeyer, W. P. (2002). SMOTE: Synthetic Minority Over-sampling Technique. *Journal of Artificial Intelligence Research, 16*, 321–357. https://doi.org/10.1613/JAIR.953

Cho, K., van Merriënboer, B., Bahdanau, D., & Bengio, Y. (2014). On the properties of neural machine translation: Encoder–decoder approaches. *Proceedings of SSST 2014 - 8th Workshop on Syntax, Semantics and Structure in Statistical Translation*. https://doi.org/10.3115/v1/w14-4012

Cichocki, A., & Unbehauen, R. (1993). *Neural Networks for Optimization and Signal Processing*. Wiley and Sons Ltd.

Cunningham, P., & Delany, S. J. (2021). k-Nearest Neighbour Classifiers - A Tutorial. *ACM Computing Surveys (CSUR), 54*(6). https://doi.org/10.1145/3459665

David B. Parker. (1985). Learning-logic: Casting the cortex of the human brain in silicon. *Center for Computational Research in Economics and Management Science, MIT*.

Dietterich, T. G. (2000). Ensemble methods in machine learning. *Lecture Notes in Computer Science (Including Subseries Lecture Notes in Artificial Intelligence and Lecture Notes in Bioinformatics), 1857 LNCS*. https://doi.org/10.1007/3-540-45014-9_1

Donders, A. R. T., van der Heijden, G. J. M. G., Stijnen, T., & Moons, K. G. M. (2006). Review: A gentle introduction to imputation of missing values. *Journal of Clinical Epidemiology, 59*(10). https://doi.org/10.1016/j.jclinepi.2006.01.014

Eerdekens, A., Callaert, A., Deruyck, M., Martens, L., & Joseph, W. (2022). Dog's Behaviour Classification Based on Wearable Sensor Accelerometer Data. *5th Conference on Cloud and Internet of Things, CIoT 2022*. https://doi.org/10.1109/CIoT53061.2022.9766553

Eerdekens, A., Deruyck, M., Fontaine, J., Martens, L., De Poorter, E., Plets, D., & Joseph, W. (2021). A framework for energy-efficient equine activity recognition with leg accelerometers. *Computers and Electronics in Agriculture, 183*. https://doi.org/10.1016/j.compag.2021.106020

Eerdekens, A., Deruyck, M., Fontaine, J., Martens, L., Poorter, E. De, Plets, D., & Joseph, W. (2020). Resampling and Data Augmentation for Equines' Behaviour Classification Based on Wearable Sensor Accelerometer Data Using a Convolutional Neural Network. *2020 International Conference on Omni-Layer Intelligent Systems, COINS 2020*. https://doi.org/10.1109/COINS49042.2020.9191639

Ester, M., Kriegel, H.-P., Sander, J., & Xu, X. (1996). A Density-Based Algorithm for Discovering Clusters in Large Spatial Databases with Noise. *Proceedings of the Second International Conference on Knowledge Discovery and Data Mining*, 226–231.

Fogarty, E. S., Swain, D. L., Cronin, G. M., & Trotter, M. (2019). A systematic review of the potential uses of on-animal sensors to monitor the welfare of sheep evaluated using the five domains model as a framework. *Animal Welfare*, *28*(4), 407–420. https://doi.org/10.7120/09627286.28.4.407

Friedman, J., Hastie, T., & Tibshirani, R. (2010). Regularization paths for generalized linear models via coordinate descent. *Journal of Statistical Software*, *33*(1). https://doi.org/10.18637/jss.v033.i01

Fukushima, K. (1980). Neocognitron: A self-organizing neural network model for a mechanism of pattern recognition unaffected by shift in position. *Biological Cybernetics*, *36*(4). https://doi.org/10.1007/BF00344251

García, R., Aguilar, J., Toro, M., Pinto, A., & Rodríguez, P. (2020). A systematic literature review on the use of machine learning in precision livestock farming. *Computers and Electronics in Agriculture*, *179*, 105826. https://doi.org/10.1016/J.COMPAG.2020.105826

Gaye, B., Zhang, D., & Wulamu, A. (2021). Improvement of Support Vector Machine Algorithm in Big Data Background. *Mathematical Problems in Engineering*, *2021*. https://doi.org/10.1155/2021/5594899

Géron, A., & Russell, Rudolph. (2019). Machine learning step-by-step guide to implement machine learning algorithms with Python. *O'Reilly Media, Inc*.

Goldberg, X. (2009). Introduction to semi-supervised learning. *Synthesis Lectures on Artificial Intelligence and Machine Learning*, *6*. https://doi.org/10.2200/S00196ED1V01Y200906AIM006

González, L. A., Bishop-Hurley, G. J., Handcock, R. N., & Crossman, C. (2015). Behavioral classification of data from collars containing motion sensors in grazing cattle. *Computers and Electronics in Agriculture*, *110*, 91–102. https://doi.org/10.1016/j.compag.2014.10.018

Goodfellow, I. J., Pouget-Abadie, J., Mirza, M., Xu, B., Warde-Farley, D., Ozair, S., Courville, A., & Bengio, Y. (2014). Generative adversarial nets. *Advances in Neural Information Processing Systems*, *3*(January), 2672–2680. https://doi.org/10.3156/jsoft.29.5_177_2

Gutierrez-Galan, D., Dominguez-Morales, J. P., Cerezuela-Escudero, E., Rios-Navarro, A., Tapiador-Morales, R., Rivas-Perez, M., Dominguez-Morales, M., Jimenez-Fernandez, A., & Linares-Barranco, A. (2018). Embedded neural network for real-time animal behavior classification. *Neurocomputing*, *272*, 17–26. https://doi.org/10.1016/j.neucom.2017.03.090

Hahsler, M., Grün, B., & Hornik, K. (2005). Arules - A computational environment for mining association rules and frequent item sets. *Journal of Statistical Software*, *14*. https://doi.org/10.18637/jss.v014.i15

Hartigan, J. A., & Wong, M. A. (1979). Algorithm AS 136: A K-Means Clustering Algorithm. *Applied Statistics*, *28*(1), 100. https://doi.org/10.2307/2346830

Hearst, M. A., Dumais, S. T., Osuna, E., Platt, J., & Scholkopf, B. (1998). Support vector machines. *IEEE Intelligent Systems and Their Applications*, *13*(4), 18–28. https://doi.org/10.1109/5254.708428

Hertz, J., Krogh, A., Palmer, R. G., & Horner, H. (1991). Introduction to the Theory of Neural Computation . *Physics Today*, *44*(12). https://doi.org/10.1063/1.2810360

Hilbe, J. M. (2009). *Logistic regression models*. CRC Press. https://www.routledge.com/Logistic-Regression-Models/Hilbe/p/book/9781138106710

Hochreiter, S., & Schmidhuber, J. (1997). Long Short-Term Memory. *Neural Computation, 9*(8). https://doi.org/10.1162/neco.1997.9.8.1735

Hosmer, D. W., Lemeshow, S., & Sturdivant, R. X. (1989). Applied Logistic Regression, 3rd Edition. *Wiley Series in Probability and Statistics*, 528. https://www.wiley.com/en-us/Applied+Logistic+Regression%2C+3rd+Edition-p-9780470582473

Hosseininoorbin, S., Layeghy, S., Kusy, B., Jurdak, R., Bishop-Hurley, G. J., Greenwood, P. L., & Portmann, M. (2021). Deep learning-based cattle behaviour classification using joint time-frequency data representation. *Computers and Electronics in Agriculture, 187*. https://doi.org/10.1016/j.compag.2021.106241

Hussain, A., Ali, S., Abdullah, & Kim, H. C. (2022). Activity Detection for the Wellbeing of Dogs Using Wearable Sensors Based on Deep Learning. *IEEE Access, 10*. https://doi.org/10.1109/ACCESS.2022.3174813

Iliyasu, R., & Etikan, I. (2021). Comparison of quota sampling and stratified random sampling. *Biometrics & Biostatistics International Journal, 10*(1). https://doi.org/10.15406/bbij.2021.10.00326

Johnson, A. A., Ott, M. Q., & Dogucu, M. (2022). *Bayes Rules! : An Introduction to Applied Bayesian Modeling*. https://doi.org/10.1201/9780429288340

Jurafsky, D., & Martin, J. (2014). Speech and Language Processing. In *Speech and Language Processing*. (Vol. 3).

Jurafsky, D., & Martin, J. H. (2009). Book Review Speech and Language Processing ( second edition ). *Computational Linguistics*.

Kaelbling, L. P., Littman, M. L., & Moore, A. W. (1996). Reinforcement Learning: A Survey. *Journal of Artificial Intelligence Research, 4*, 237–285. https://doi.org/10.1613/JAIR.301

Kaler, J., Mitsch, J., Vázquez-Diosdado, J. A., Bollard, N., Dottorini, T., & Ellis, K. A. (2020). Automated detection of lameness in sheep using machine learning approaches: novel insights into behavioural differences among lame and non-lame sheep. *Royal Society Open Science, 7*(1), 190824. https://doi.org/10.1098/rsos.190824

Kamminga, J. W., Bisby, H. C., Le, D. V., Meratnia, N., & Havinga, P. J. M. (2017). Generic Online Animal Activity Recognition on Collar Tags. *Proceedings of the 2017 ACM International Joint Conference on Pervasive and Ubiquitous Computing and Proceedings of the 2017 ACM International Symposium on Wearable Computers on - UbiComp '17, October*, 597–606. https://doi.org/10.1145/3123024.3124407

Kasnesis, P., Doulgerakis, V., Uzunidis, D., Kogias, D. G., Funcia, S. I., González, M. B., Giannousis, C., & Patrikakis, C. Z. (2022). Deep Learning Empowered Wearable-Based Behavior Recognition for Search and Rescue Dogs. *Sensors, 22*(3). https://doi.org/10.3390/s22030993

Kaur, J., & Madan, N. (2015). Association Rule Mining: A Survey. *International Journal of Hybrid Information Technology, 8*(7). https://doi.org/10.14257/ijhit.2015.8.7.22

Kleanthous, N., Hussain, A., Khan, W., Sneddon, J., & Liatsis, P. (2022). Deep transfer learning in sheep activity recognition using accelerometer data. *Expert Systems with Applications, 207*, 117925. https://doi.org/10.1016/j.eswa.2022.117925

Kleanthous, N., Hussain, A., Mason, A., & Sneddon, J. (2019). Data Science Approaches for the Analysis of Animal Behaviours. In *Lecture Notes in Computer Science (including subseries Lecture Notes in Artificial Intelligence and Lecture Notes in Bioinformatics): Vol. 11645 LNAI* (pp. 411–422). https://doi.org/10.1007/978-3-030-26766-7_38

Kleanthous, N., Hussain, A., Mason, A., Sneddon, J., Shaw, A., Fergus, P., Chalmers, C., & Al-Jumeily, D. (2018). Machine Learning Techniques for Classification of Livestock Behavior. In *Lecture Notes in Computer Science (including subseries Lecture Notes in Artificial Intelligence and Lecture Notes in Bioinformatics): Vol. 11304 LNCS* (pp. 304–315). https://doi.org/10.1007/978-3-030-04212-7_26

Kohonen, T. (1989). *Self-Organization and Associative Memory* (3rd ed., Vol. 8). Springer Berlin Heidelberg. https://doi.org/10.1007/978-3-642-88163-3

LeCun, Y., Bengio, Y., & Hinton, G. (2015). Deep learning. *Nature*, *521*(7553), 436–444. https://doi.org/10.1038/nature14539

Lehner, P. N. (1996). *Handbook of ethological methods*. Cambridge University Press.

Li, P., Stuart, E. A., & Allison, D. B. (2015). Multiple imputation: A flexible tool for handling missing data. In *JAMA - Journal of the American Medical Association* (Vol. 314, Issue 18). https://doi.org/10.1001/jama.2015.15281

Liakos, K. G., Busato, P., Moshou, D., Pearson, S., & Bochtis, D. (2018). Machine learning in agriculture: A review. In *Sensors (Switzerland)* (Vol. 18, Issue 8, p. 2674). Multidisciplinary Digital Publishing Institute. https://doi.org/10.3390/s18082674

Liu, Q., Zhai, J. W., Zhang, Z. Z., Zhong, S., Zhou, Q., Zhang, P., & Xu, J. (2018). A Survey on Deep Reinforcement Learning. In *Jisuanji Xuebao/Chinese Journal of Computers* (Vol. 41, Issue 1). https://doi.org/10.11897/SP.J.1016.2018.00001

Lundberg, S. M., & Lee, S. I. (2017). A unified approach to interpreting model predictions. *Advances in Neural Information Processing Systems*, *2017-December*.

machinelearningmastery. (2019). Overfitting and Underfitting With Machine Learning Algorithms. *Machinelearningmastery*.

Mao, A., Huang, E., Gan, H., & Liu, K. (2022). FedAAR: A Novel Federated Learning Framework for Animal Activity Recognition with Wearable Sensors. *Animals*, *12*(16). https://doi.org/10.3390/ani12162142

Marjani, M., Nasaruddin, F., Gani, A., Karim, A., Hashem, I. A. T., Siddiqa, A., & Yaqoob, I. (2017). Big IoT Data Analytics: Architecture, Opportunities, and Open Research Challenges. *IEEE Access*, *5*, 5247–5261. https://doi.org/10.1109/ACCESS.2017.2689040

Neethirajan, S. (2020). The role of sensors, big data and machine learning in modern animal farming. In *Sensing and Bio-Sensing Research* (Vol. 29, p. 100367). Elsevier. https://doi.org/10.1016/j.sbsr.2020.100367

Ostertagová, E. (2012). Modelling using Polynomial Regression. *Procedia Engineering*, *48*, 500–506. https://doi.org/10.1016/J.PROENG.2012.09.545

Pan, S. J., & Yang, Q. (2010). A survey on transfer learning. In *IEEE Transactions on Knowledge and Data Engineering* (Vol. 22, Issue 10, pp. 1345–1359). https://doi.org/10.1109/TKDE.2009.191

Paschalidis, I. C., & Chen, Y. (2010). Statistical anomaly detection with sensor networks. *ACM Transactions on Sensor Networks*, *7*(2). https://doi.org/10.1145/1824766.1824773

Peng, Y., Kondo, N., Fujiura, T., Suzuki, T., Wulandari, Yoshioka, H., & Itoyama, E. (2019). Classification of multiple cattle behavior patterns using a recurrent neural network with long short-term memory and inertial measurement units. *Computers and Electronics in Agriculture*, *157*. https://doi.org/10.1016/j.compag.2018.12.023

Pu, G., Wang, L., Shen, J., & Dong, F. (2021). A hybrid unsupervised clustering-based anomaly detection method. *Tsinghua Science and Technology*, *26*(2). https://doi.org/10.26599/TST.2019.9010051

Quinlan, J. R. (1986). Induction of decision trees. *Machine Learning*, *1*(1), 81–106. https://doi.org/10.1007/bf00116251

Rast, W., Kimmig, S. E., Giese, L., & Berger, A. (2020). Machine learning goes wild: Using data from captive individuals to infer wildlife behaviours. *PLoS ONE*, *15*(5). https://doi.org/10.1371/journal.pone.0227317

Reddy, G. T., Reddy, M. P. K., Lakshmanna, K., Kaluri, R., Rajput, D. S., Srivastava, G., & Baker, T. (2020). Analysis of Dimensionality Reduction Techniques on Big Data. *IEEE Access*, *8*, 54776–54788. https://doi.org/10.1109/ACCESS.2020.2980942

Ribeiro, M. T., Singh, S., & Guestrin, C. (2016). "Why Should I Trust You?": Explaining the Predictions of Any Classifier. *NAACL-HLT 2016 - 2016 Conference of the North American Chapter of the Association for Computational Linguistics: Human Language Technologies, Proceedings of the Demonstrations Session*, 97–101. https://doi.org/10.18653/v1/n16-3020

Rumelhart, D. E., Hinton, G. E., & Williams, R. J. (1986). Learning representations by back-propagating errors. *Nature*, *323*(6088). https://doi.org/10.1038/323533a0

Safavian, S. R., & Landgrebe, D. (1991). A Survey of Decision Tree Classifier Methodology. *IEEE Transactions on Systems, Man and Cybernetics*, *21*(3), 660–674. https://doi.org/10.1109/21.97458

Samariya, D., & Thakkar, A. (2023). A Comprehensive Survey of Anomaly Detection Algorithms. In *Annals of Data Science* (Vol. 10, Issue 3). https://doi.org/10.1007/s40745-021-00362-9

Santosh Kumar, M. B., & Balakrishnan, K. (2019). Development of a model recommender system for agriculture using apriori algorithm. *Advances in Intelligent Systems and Computing*, *768*. https://doi.org/10.1007/978-981-13-0617-4_15

Sayed, A. H. (2023). Q-Learning. In *Inference and Learning from Data*. https://doi.org/10.1017/9781009218245.022

Schafer, J. L., & Olsen, M. K. (1998). Multiple imputation for multivariate missing-data problems: A data analyst's perspective. In *Multivariate Behavioral Research* (Vol. 33, Issue 4). https://doi.org/10.1207/s15327906mbr3304_5

Schlecht, E., Hülsebusch, C., Mahler, F., & Becker, K. (2004). The use of differentially corrected global positioning system to monitor activities of cattle at pasture. *Applied Animal Behaviour Science*, *85*(3), 185–202. https://doi.org/10.1016/j.applanim.2003.11.003

Schwager, M., Anderson, D. M., Butler, Z., & Rus, D. (2007). Robust classification of animal tracking data. *Computers and Electronics in Agriculture*, *56*(1), 46–59. https://doi.org/10.1016/j.compag.2007.01.002

Seber, G. A. F. (George A. F., & Lee, A. J. (2003). *Linear regression analysis*. 557. https://www.wiley.com/en-ie/Linear+Regression+Analysis%2C+2nd+Edition-p-9780471415404

Shahbazi, M., Mohammadi, K., Derakhshani, S. M., & Groot Koerkamp, P. W. G. (2023). Deep Learning for Laying Hen Activity Recognition Using Wearable Sensors. *Agriculture (Switzerland)*, *13*(3). https://doi.org/10.3390/agriculture13030738

Song, Y. Y., & Lu, Y. (2015). Decision tree methods: applications for classification and prediction. *Shanghai Archives of Psychiatry*, *27*(2), 130. https://doi.org/10.11919/J.ISSN.1002-0829.215044

Srivastava, N., Hinton, G., Krizhevsky, A., Sutskever, I., & Salakhutdinov, R. (2014). Dropout: A simple way to prevent neural networks from overfitting. *Journal of Machine Learning Research*, *15*.

Sutton, C. D. (2005). Classification and Regression Trees, Bagging, and Boosting. *Handbook of Statistics*, *24*, 303–329. https://doi.org/10.1016/S0169-7161(04)24011-1

Sutton, R. S., McAllester, D., Singh, S., & Mansour, Y. (2000). Policy gradient methods for reinforcement learning with function approximation. *Advances in Neural Information Processing Systems*.

Telikani, A., Gandomi, A. H., & Shahbahrami, A. (2020). A survey of evolutionary computation for association rule mining. *Information Sciences*, *524*. https://doi.org/10.1016/j.ins.2020.02.073

*The Nobel Prize in Physiology or Medicine 1973 - NobelPrize.org*. (n.d.). Retrieved August 29, 2023, from https://www.nobelprize.org/prizes/medicine/1973/summary/

Tran, D. N., Nguyen, T. N., Khanh, P. C. P., & Trana, D. T. (2021). An IoT-based Design Using Accelerometers in Animal Behavior Recognition Systems. *IEEE Sensors Journal*, 1–1. https://doi.org/10.1109/JSEN.2021.3051194

Ungar, E. D., Henkin, Z., Gutman, M., Dolev, A., Genizi, A., & Ganskopp, D. (2005). Inference of Animal Activity From GPS Collar Data on Free-Ranging Cattle. *Rangeland Ecology & Management*, *58*(3), 256–266. https://doi.org/10.2111/1551-5028(2005)58[256:IOAAFG]2.0.CO;2

Valletta, J. J., Torney, C., Kings, M., Thornton, A., & Madden, J. (2017). Applications of machine learning in animal behaviour studies. *Animal Behaviour*, *124*, 203–220. https://doi.org/10.1016/j.anbehav.2016.12.005

van Engelen, J. E., & Hoos, H. H. (2020). A survey on semi-supervised learning. *Machine Learning*, *109*(2). https://doi.org/10.1007/s10994-019-05855-6

Varga, B., Kulcsár, B., & Chehreghani, M. H. (2023). Deep Q-learning: A robust control approach. *International Journal of Robust and Nonlinear Control*, *33*(1). https://doi.org/10.1002/rnc.6457

Walker, R. T., & Hill, H. M. (2020). Behavioral Ecology. *Encyclopedia of Personality and Individual Differences*, 406–408. https://doi.org/10.1007/978-3-319-24612-3_1610

Wang, L., Arablouei, R., Alvarenga, F. A. P., & Bishop-Hurley, G. J. (2023). Classifying animal behavior from accelerometry data via recurrent neural networks. *Computers and Electronics in Agriculture, 206*. https://doi.org/10.1016/j.compag.2023.107647

Wang, Y., Yao, H., & Zhao, S. (2015). *Auto-encoder based dimensionality reduction*. https://doi.org/10.1016/j.neucom.2015.08.104

Werbos, P. (1974). *Beyond Regression: New Tools for Prediction and Analysis in the Behavioral Science. Thesis (Ph. D.). Appl. Math. Harvard University.*

Wu, Y., Liu, M., Peng, Z., Liu, M., Wang, M., & Peng, Y. (2022). Recognising Cattle Behaviour with Deep Residual Bidirectional LSTM Model Using a Wearable Movement Monitoring Collar. *Agriculture (Switzerland), 12*(8). https://doi.org/10.3390/agriculture12081237

Zou, H., & Hastie, T. (2005). Regularization and variable selection via the elastic net. *Journal of the Royal Statistical Society. Series B: Statistical Methodology, 67*(2). https://doi.org/10.1111/j.1467-9868.2005.00503.x

# Index